The Urban Book Series

Igor Vojnovic, Department of Geography, Michigan State University, East Lansing, MI, USA

Claudia Yamu, Department of Spatial Planning and Environment, University of Groningen, Groningen, Groningen, The Netherlands

Qunshan Zhao, School of Social and Political Sciences, University of Glasgow, Glasgow, UK

The Urban Book Series is a resource for urban studies and geography research worldwide. It provides a unique and innovative resource for the latest developments in the field, nurturing a comprehensive and encompassing publication venue for urban studies, urban geography, planning and regional development.

The series publishes peer-reviewed volumes related to urbanization, sustainability, urban environments, sustainable urbanism, governance, globalization, urban and sustainable development, spatial and area studies, urban management, transport systems, urban infrastructure, urban dynamics, green cities and urban landscapes. It also invites research which documents urbanization processes and urban dynamics on a national, regional and local level, welcoming case studies, as well as comparative and applied research.

The series will appeal to urbanists, geographers, planners, engineers, architects, policy makers, and to all of those interested in a wide-ranging overview of contemporary urban studies and innovations in the field. It accepts monographs, edited volumes and textbooks.

Indexed by Scopus.

More information about this series at https://link.springer.com/bookseries/14773

Camilla Perrone
Editor

Critical Planning and Design

Roots, Pathways, and Frames

Springer

Editor
Camilla Perrone
Department of Architecture
University of Florence
Firenze, Italy

ISSN 2365-757X ISSN 2365-7588 (electronic)
The Urban Book Series
ISBN 978-3-030-93106-3 ISBN 978-3-030-93107-0 (eBook)
https://doi.org/10.1007/978-3-030-93107-0

This Springer imprint is published by the registered company Springer Nature Switzerland AG
The registered company address is: Gewerbestrasse 11, 6330 Cham, Switzerland

This book is dedicated to Giancarlo Paba, Professor of Urban and Regional Planning at the University of Florence and passionate about roots and critical theories of planning.

Constantly and restlessly fostering radical change in society and cities throughout his lifetime Giancarlo Paba was, first of all, an insurgent, energetic, and mindful intellectual, tormented by the search for happiness (of others, the city, the wider area). This book and the research that will follow is inspired by Giancarlo Paba's idea of a "usable past" (the roots of planning), that motivates contemporary actions and, like roots of trees, encourages and generates new articulations of a dissenting thought.

Preface

If there is something that this book makes clear, is that the planning debate is multifaceted and has its in many traditions of thought that it is important not to forget, to rediscover in some cases, to reinterpret in others as inspirational of the contemporary discourses on planning.

The book does not have a historical-philological intent, nor does it intend to build an overarching perspective on critical planning and design. It is neither a manual nor an exhaustive register of contributions on the contemporary approach to critical planning, if there is one.

It does not wear the flavour of the postmodern debate of the beginning of the millennium but draws inspiration from the sensitivity of postmodern thought towards all forms of diversity often obliterated by a dominant or rational-comprehensive culture. From postmodern thought and, in particular, from the words of Bhikhu Parekh in Rethinking Multiculturalism (Harvard University Press, 2000), the book borrows the idea that "cultures are not superstructures built upon identical and unchanging foundations, or manifestation of a common human essence, but unique human creations that reconstitute and give different meaning and orientation to those properties that all human beings share in common" (122).

With this perspective, the book suggests a re-education of the "gaze" towards a multiple and diverse planning past—be it represented by giants' thought that needs unpacking of certain consolidated simplifications or by minor authors obscured by the dominant culture of a specific moment of planning history—which might become usable in the contemporary planning and design thought.

This gaze owes critical and radical approaches to urban studies and planning, although it keeps this scenario in the background to give space to a journey through foundational books (and the authors' thinking) that shows how theories, interpretations, and perspectives often from other disciplines have crossed time and are today influential ideas and inspirations of planning thought, which is radical, critical, alternative to a mainstream thinking.

The book is not a systematic but exploratory attempt to initiate a conversation based on an informed discourse on critical planning and design. It paves the way for reflecting about what is or can be a critical planning and design approach inspired by critical urban theory and developed along influential and radical planning thinking aimed at pursuing a socially and environmentally just society. In this line, contemporary authors re-read fundamental and influential books that count for each of them—written over the last two centuries—that contributed more or less implicitly to outlining a field of critical planning in dialogue with a usable past from which to draw generative and dissenting ideas.

This journey through some roots, pathways, and frames of critical planning and design is neither linear nor exhaustive and includes many non-planner scholars whose contribution proved to be crucial for the development of a critical perspective on planning and design. It is a trajectory with some stops inspired by the thought of contemporary authors (the authors of this book) who have built and advanced their scientific-critical baggage on influential thinkers' shoulders making this influential contribution their own "usable past". This past is useful insofar as it allows scholars to reflect on the past conditions that shaped planning experience and theories, and to grasp the potential for change that lays within the society and needs an enriched critical approach.

We can say that this book is a book about the books of other scholars that takes the reader on a journey (one of many possible) through a multifaced universe of critical urban studies.

The book borrows and keeps critical urban theory in the background to critically engage the readers in reflections on the dialectics of social, historical, and radical change—and its agents that could realize the possibilities suppressed under current socio-political and urban conditions. In this sense, the book engages a plural and multifaceted perspective to contribute to the reflection on contemporary usable planning perspectives that aims at changing the status quo in favour of a more just and less unequal socio-spatial and environmental transformation, and urbanization processes.

The book represents the first milestone of a cultural manifesto of the research Laboratory of Critical Planning and Design founded in 2015 at the University of Florence by Camilla Perrone under the scientific supervision of Giancarlo Paba. The Laboratory comprises a group of researchers from the Department of Architecture at the University of Florence, in collaboration with a network of international scholars with which it aims to develop a *critical view* on urban and regional planning, in particular, along the following research paths:

- *Roots*: critical exploration of the *roots* of planning; *genealogy* of the "restless" and radical concepts of planning; emphasis on *dissenting imaginations*, and eccentric and alternative planning traditions.
- *Theories*: critical discussion on planning theories; *difference-sensitive* planning approach; *multi-agent* planning and interactive design; the concept of *co-evolution* applied to the relations between nature, territory, and settlements; planning approaches based on *self-organization* and *insurgent urbanism*.
- *Practices*: interactive design experiences; *radical participatory planning* practices; involvement of the inhabitants in the rebuilding of *shared urban and regional codes*.

The book collects the results of a promoted lectures series held in Florence, 2018, on critical thinkers of planning (or related disciplines on which planning draws on in contemporary developments) and the fundamental books that disseminated their ideas.

As well as the book, the laboratory intertwines a planning approach such as "societal guidance" and "social transformation" with a critical distance through which it focuses on an approach in dialogue (even managing the conflict) with both sides of the ridge. The discourse on space is faced both in its social dimension—as a sphere of negotiation of power and urban restructuring under capitalism—and its territorial dimension as a socially produced space—as defined by the geographer Claude Raffestin, who linked Henri Lefebvre's theory of space to the conception of Foucault's power—and a neo-ecosystem that brings together the milieus of nature, the environment built and human beings.

Borrowing inspirational contributions of Roger Keil, Neil Brenner, and Christian Schmid, the urbanization question has been at the centre stage of the lab's interest in recent years with a specific focus on fixity and motion dialectics in urban space under capitalism. In response to the challenges of contemporary extended urbanization processes critically debated by these scholars and to the impact of the COVID-19 pandemic as a component of the climate change and the rebellion of the earth to the Anthropocene, the Critical Planning and Design Laboratory is developing a model of multi-agent critical planning, according to a form of agency extended to "significant others" and "other terrestrials".

This is done with a focus on radical approaches to planning and on some disciplinary roots of interactive planning and design. A second milestone book will follow this argument.

Firenze, Italy Camilla Perrone

Acknowledgements

I sincerely thank the wonderful and knowledgeable team of authors, who have made an enormous effort to dress their approaches with a critical perspective and contribute valuably to the roots, pathways, and frames of critical planning and design as approached in this book. They were very patient and generous in contributing to an editorial project with long and complicated gestation.

My gratitude goes to the INURA network (International Network for Urban Research and Action), which has contributed since 1999 to nourish my critical and proactive gaze on urbanization processes and the social and spatial effects of these. The countless urban immersions made with Inura's friends and colleagues have settled over time a necessary demand for spatial justice, social empowerment, and widespread urban and territorial happiness. Indeed, the book would not have had the same value if Roger Keil, Ute Lehrer, Michael Edwards, Christian Schmid, Stefan Kipfer, and other inurians had not contributed.

A special thank goes to Giancarlo Paba (to whom I dedicated the first chapter), who introduced me to critical thinking in planning and without whom this book would never have been born.

Finally, sincere and dutiful thanks go to Springer Publisher, who accompanied me in the book's composition with patience, professionalism, and kindness, always looking for a way to overcome the many obstacles that occurred during the production of this book.

Book Description

Critical urban planning and design is a field of research and action inspired by critical urban theory and developed along with ideas, theories, and designs that prove to be radical, alternative, dialectical to the mainstream history of planning.

The book recounts and recombines within a subjective trajectory the indefatigable work on the roots of planning in search for untold planning histories and dissenting imaginations as a generative source to deconstruct the modernist epistemologies of planning and to examine alternative planning pathways.

The book is not a systematic but exploratory attempt to initiate a conversation based on an informed discourse on critical planning and design. It paves the way for reflecting about what is or can be a critical planning and design approach inspired by critical urban theory and developed along influential and radical planning thinking aimed at pursuing a socially and environmentally just society. In this line, contemporary authors re-read fundamental and influential books that count for each of them—written over the last two centuries—that contributed more or less implicitly to outlining a field of critical planning in dialogue with a usable past from which to draw generative and dissenting ideas.

The journey through roots, pathways, and frames of critical planning and design traced in the book shows the influential contribution of thinkers and their books in digging in and arguing on the antagonistic relationship to existing urban formation and the action on it, from different perspectives. Sometimes they are placed in the critical urban theory; sometimes, they are antagonistic to it. In other cases, they search for a theoretical domain to contribute with hints on planning and design approaches for just and sustainable forms of urbanization.

This journey is non-linear and includes many non-planner scholars whose contribution proved crucial for developing a critical perspective on planning and design. Likewise, it's not exhaustive and has no ambition in this sense. Contrarily, it shows one of the possible journeys through an immense alternative planning and design domain, which is very much needed for contemporary mainstream planning.

A past made of ideas, theories, and design, and the relationship with it that each scholar engages play a crucial role in this book. The past in question is a dynamic and continuously reframed past according to contemporary situated and contested issues.

Thus, theories from this past become usable in discovering situated urbanization pathways and patterns and possible alternative and radical planning approaches.

Including thoughtful and critical arguments on influential thinkers of the past two centuries, international research, and contributions from prominent urban scholars, this volume will enable students, scholars, and researchers of planning, design, political science, geographical, environmental, and urban studies to better understand how interrelated, economic, political, and cultural processes form and transform urban environments and reflect about which planning posture might be.

Contents

Editor and Contributors

About the Editor

Camilla Perrone, PhD in Urban, Regional and Environmental Design (2002), is an Associate Professor of Urban and Regional Planning at the University of Florence (UniFi) Italy. Her research interests include critical planning, interactive design and urban governance, urban diversity, multiscalar urbanization, urban political ecology and self-organising city. She is a member of the International Network for Urban Research and Action (INURA). Her publications: "DiverCity" (FrancoAngeli 2010); "Space Matters" (with Giancarlo Paba), in A. Balducci et. al., Post-Metropolitan Territories (Routledge 2017); "Città in movimento" (Guerini, forthcoming). Perrone is also the editor (with Giancarlo Paba) of "Critical Planning & design" (Springer, 2020). She is the Founding Director of the Research Laboratory of Critical Planning and Design at the University of Florence.

Contributors

Peter Ache Nijmegen School of Management, Radboud University, Nijmegen, Netherlands

Giovanni Attili Urbanism Division, Department of Civil, Building and Environmental Engineering, Sapienza University of Rome, Rome, Italy

Alessandro Balducci Department of Architecture and Urban Studies, Politecnico di Milano, Milan, Italy

Lidia Decandia Department of Architecture, Design and Urbanism, Sassary University (Alghero Campus), Sassary, Italy

Michael Edwards The Bartlett School of Planning, Faculty of the Built Environment UCL, London, UK

Roger Keil Faculty of Environmental and Urban Change, York University, Toronto, Canada

Stefan Kipfer Faculty of Environmental and Urban Change, York University, Toronto, Canada

Ute Lehrer Faculty of Environmental and Urban Change, York University, Toronto, Canada

Francesco Lo Piccolo Department of Architecture, University of Palermo, Palermo, Italy

Alberto Magnaghi Department of Urban and Regional Planning, University of Florence, Florence, Italy

Luigi Mazza Department of Architecture and Urban Studies, Politecnico di Milano, Milan, Italy

Francesco Domenico Moccia Department of Architecture, Federico II University, Naples, Italy

Stefano Moroni Urban Planning and Design Division, Department of Architecture and Urban Studies, Politecnico di Milano, Milan, Italy

Gabriele Pasqui Department of Architecture and Urban Studies, Politecnico di Milano, Milan, Italy

Camilla Perrone Urban and Regional Planning Divison, Department of Architecture, University of Florence, Florence, Italy

Paolo Perulli Economic Sociology, University of Piedmont Orientale, Vercelli, Italy;
Urban Sociology, Architecture Academy of Mendrisio, Mendrisio, Switzerland

Paolo Pileri Department of Architecture and Urban Studies, Politecnico di Milano, Milan, Italy

Barbara Pizzo Urban Planning Studies Division, Department of History, Design and Restoration of Architecture - DSDRA, Sapienza University of Rome, Rome, Italy

Cristiana Rossignolo Interuniversity Department of Regional and Urban Studies and Planning, Polytechnic of Turin, Turin, Italy

Christian Schmid Department of Architecture, ETH Zürich, Zürich, Switzerland

Chapter 1
Critical Planning and Design: Walking Through Roots and Dissenting Imaginations

Camilla Perrone

Abstract The field of Critical planning and design introduced in this chapter and in the book contribute to the intellectual universe of critical urban studies. However, it borrow and hook strands of literature that are unrelated or new to the critical urban theory's core. In particular, this contribution portrays critical planning and design, suggesting some pathways for intertwining critical urban theory with an idea of engaged "planning and design" approach that implies discovering an alternative history of planning counter-posed to a self-referential, manly masculine, and Eurocentric modernist account of planning. To this end, it suggests to explores and unpack theories while digging into the influential thinkers' contributions to build one's "usable past" as an inspiration for contemporary reflexive thinking, radical acting, and critical planning for a socially just and sustainable form of urbanization. In particular, this contribution suggests a posture to dig into the boundless planning literature in an alternative way. This is focused on searching (as rediscovering, re-reading, reframing) for the roots of disciplines from an anti-dogmatic perspective biased toward the future. This posture is twofold end consists of radicalizing problems and solutions and going to the roots of concepts and stories of planning while critically unpacking mainstream paradigms. The search for roots of the planning discipline implies engaging in a relationship with the past. Finally, the contribution hint that walking through roots, pathways, and frames of critical planning and design, is allowed by two attitudes: a plural understanding of critical and the ability to discover alternative planning and design histories that turn to be radical and insurgent.

Keywords Critical planning · Critical design · Critical urban theory · Alternative planning · Planning history · Insurgent planning practices · Usable past

C. Perrone (✉)
Urban and Regional Planning, Department of Architecture, University of Florence, Florence, Italy
e-mail: camilla.perrone@unifi.it

C. Perrone (ed.), *Critical Planning and Design*, The Urban Book Series,
https://doi.org/10.1007/978-3-030-93107-0_1

1.1 Walking Through Roots, Pathways, and Frames

The field of planning and urban studies is immense and boundless today, and it has widened dramatically in recent decades. It has become more complicated, as researchers have pursued the complexity and variety of subjects that can be studied within the field. Interactions with other fields of study, which now include all the physical, natural, and human sciences, have multiplied.

The literature consulted by planners and scholars has also grown exponentially. Our books and essays require a multiplicity of scientific and cultural frames, and they mobilize research and information, dissect a growing plethora of case studies, and multiply bibliographic references. The digital libraries of the great universities provide world literature instantaneously; the syllabuses have become increasingly articulated and efficient, which improves learning; the readings provide the essential contributions deemed necessary for each field of study.

Critical planning and design suggest a posture to dig into the boundless planning literature in an alternative way. This is focused on the search (as rediscovering, re-reading, reframing) for the roots of disciplines from an anti-dogmatic perspective biased toward the future. This is a twofold effort. On the one hand, it asks for the radicalization of problems and solutions. It is going to the roots of concepts and stories of planning, while critically unpacking mainstream paradigms and dogmatic approaches. On the other hand, this search for roots of the planning discipline implies engaging in a relationship with the past. This is not like working as historians, who philologically recount history. Rather, it implies building one's "usable past" (Blake 1990; Mumford 1950; Paba 2003) as an inspiration for contemporary reflexive thinking, radical acting, and critical planning for a socially just and sustainable form of urbanization. This exercise helps in becoming aware of one's scientific posture (built-in humility and prudence; Paba 2003), consolidates a conceptual and practical background, and enables taking steps forward each time, toward unpredictable and captivating prospects oriented toward specific historical conditions and contexts. As the roots of trees develop into branches, this usable past possibly generates new articulations of thought which are able to stay inside the contextually specific time/space history and embed the dialectics of social and historical change. The arrangement of these two attitudes—going to the roots of theories and building one's usable past—inspires the ethical urgency of a continuous verification of ideas through social interaction and the constantly reinventing of critical theory in relation to the unevenly evolving political-economic geographies of historically specific formation of capitalistic urbanization and the conflicts it embeds (Brenner and Keil 2005).

Critical urban theory as defined by Brenner (2012) nourishes this approach and represents the theoretical background (a theoretical root) with respect to which most of the authors of the book refer their contributions or develop a dialectic approach to it: "critical theory differs fundamentally from what might be termed "mainstream" urban theory—for example, the approaches inherited from the Chicago School of urban sociology, or those deployed within technocratic or neoliberal forms of policy science. [...] critical urban theory involves the critique of ideology (including social-scientific ideologies) and the critique of power, inequality, injustice, and exploitation, at once within and among cities" (11).

However, although inspired by the critical urban theory, the book includes perspectives that are not fully attributable to the field of critical urban theory as delineated by the writings of leftist or radical urban scholars during the post-1968 period such as Henri Lefebvre, David Harvey, Manuel Castells, Peter Marcuse, and the others who have been influenced by them (Brenner 2012; Katznelson 1993; Merrifield 2002).

Indeed, walking through roots, pathways, and frames of planning pertains to the critical attitude of going back to the origins of planning history, which are nowadays of clear actuality, such as the awareness of the differentiated nature of plans and projects' targets, the importance of self-government and self-management practices; an idea of planning sensitive to all the components of land and environment: natural and cultural, organic and inorganic, human and non-human.

The final effort is to pave the way to an engaged and critical approach to planning which is intertwined, inspired, and influenced by critical urban theory.

In doing this (latching back to planning roots), an alternative history of planning is being built. In recent years, an impressive number of works have built, in contrast to the official history of planning, a counter-history of the city's planning and transformation, which is increasingly richer branched and documented. It is a secondary, alternative, interstitial story, long neglected and hidden, yet luminous (in this sense, a "light side" of planning), and today, increasingly stronger and more resistant over time.

It is possible to identify two dimensions of this alternative history: on the one hand, the theories, practices, utopias, experiments, and achievements produced by marginal and peripheral figures (or unknown ideas of well-known planning protagonists); on the other hand, the silent history of resistance and the claiming rights struggles of insurgent citizens. This alternative history stems from an impatient and creative strand of planning.

Quoting Paba (2010b), I would call this alternative idea of planning unsettling (disturbing the existing order, dissatisfied, anarchic), unsettling people's heroes, and dissenting the imaginations of their works. Paba was referring to Patrick Abercrombie's definition of Geddes—*a most unsettling person*—which has become the title of a Geddesian biography (Abercrombie 1933, 128). Dissenting imaginations is how an Indian scholar, Chettiyappan Visvanathan, defined Patrick Geddes and Lewis Mumford's visions (Visvanathan 2003, 74).

1.2 A Plural Understanding of "Critical"

The book's contributions are placed on the ridge between two forms of planning which can be described by borrowing John Friedmann's definitions of the planning theory: planning as "societal guidance" that works as conventional planning managed by professional practitioners and planning as "social transformation" which pulls on three oppositional streams of social mobilization: social anarchism, historical materialism, and utopianism (Beard 2003). Friedmann explains that "the operative terms in these definitions are societal guidance and social transformation. Whereas the former is articulated through the state, and is concerned chiefly with systematic change, the latter focuses on the political practices of system transformation. Planners engaged in these two practices are necessarily in conflict. It is conflict between the interests of a bureaucratic state and the interest of the political community [...] (Friedmann 1987, 38)". These two main forms of planning encompass four traditions of planning—social reform, policy analysis, social learning, and social mobilization—which links knowledge to action and overlap.

Standing in this challenging position, the book's authors dig into plural dimensions of a critical perspective of planning and design. They include "radical" as the emancipation of humanity from social oppression but also as an opportunity for a more egalitarian and cognitively realistic self-guiding society; "critical" as understanding the nature of contemporary patterns of urban restructuring under capitalism and the possibility for alternative, radical, progressive or revolutionary response to the crises; and "alternative/insurgent" as opposition to the State.

These three dimensions–*radical, critical, alternative/insurgent*—embed an oppositional element that, respectively, turns to be: in the case of *radical*, conflict strategies (as defined by Friedmann 1987) and capacity for probing problems and possibilities against governmental barriers (according to Charles Lindblom's aspiration to improve social problem solving and challenge the status quo via a self-guiding society, in *Inquiry and Change* 1990); in the case of *critical*, it refers to structural change to be operationalized at multiple scales (in line with David Harvey idea of critical planning in *Spaces of Hope*, 2000); in the case of *alternative* and *insurgent*, it evokes planning practices steered by insurgent citizenship claiming for the right to the city and creating the condition for this right to be possible (as portrayed by Leonie Sandercock in *Towards Cosmopolis* 1998).

These three dimensions embrace the conception of critical urban theory being more or less attuned to its four propositions as defined by Neil Brenner to orientate authors who potion themselves in the universe of critical urban studies:

> they insist on the need for abstract, theoretical arguments regarding the nature of urban processes under capitalism, while rejecting the conception of theory as a "handmaiden" to immediate, practical, or instrumental concerns; they view knowledge of urban questions, including critical perspectives, as being historically specific and mediated through power relations; they reject instrumentalist, technocratic, and market-driven forms of urban analysis that promote the maintenance and reproduction of extant urban formations; and they are concerned to excavate possibilities for alternative, radically emancipatory forms of urbanism that are latent, yet systemically suppressed, within contemporary cities". (Brenner 2012, 19)

In particular, this last proposition describes most of the contributions and represents the inspiring principle of this work.

Critical urban theory has inspired the book that has extended its journey to books and scholars who have circumstantially and more or less implicitly intertwined the propositions of critical urban studies. An intent of the book is precisely to incorporate critical thinking in the approach of critical urban studies through the dialectical comparison with other approaches.

The book is a book about the books of other intellectuals. It aims to propose a journey (one of many possible) through roots, pathways, and frames of critical planning and design.

It has no field-defining or normative ambitions. It only intends to share and deliver to the debate lessons learned by some scholars along the challenging and continually ambitious path of understanding the contemporary urban conditions. Moreover, it aspires to nurture reflections about research and action, with a bias for the critical urban studies to which the book approaches, from which it borrows propositions, with which it enters into critical dialog. Certainly, the book gets its feet wet in an intellectual field "consolidated in the late 1960s and early 1970s through the pioneering interventions of radical scholars such as Lefebvre (2003 [1970]), 1996 [1968]), Castells (1977 [1972]), and Harvey (1976). Despite their theoretical, methodological, and political differences, these authors shared a common concern to understand the ways in which, under capitalism, cities operate as strategic sites for commodification processes" (Brenner 2012, 11). These theoreticians have inspired several scholars that developed the urbanization question as a crucial component of the contemporary critical (urban) theory. Among them is important to mention some influential theorists and their crucial writings such as Soja (2000), who developed critical studies of cities and regions dealing with the dramatically restructured megacities that have emerged worldwide over the last half of the twentieth century, Brenner and Schmid (2015) engaged with a new epistemology of the urban and the theory of planetary urbanization as development of Lefebvre's anticipation of complete urbanization, Robinson (2006) focused on an "urban theory that does not rest on pre-given categories of cities but on a cosmopolitan comparativism that places all cities within the same analytical field […] it is not global cities or third world cities that should be central to academic analysis and policy recommendations, but what she calls "ordinary" cities, in all their complexity, diversity and peculiarity" (Schuermans 2009), Keil (2013), who reconsiders the city from the outside in putting at work an urban political ecology of suburbanisation, and Heynen et al. (2006), who critically explore the re-entry of the ecological agenda into urban theory to understand the contemporary urbanization processes. This field is very rich of contributors that wouldn't be possible to group in a coherent whole. So I apologize in advance for any exclusion I made.

All authors made these inspirational thoughts their own "usable past" and are becoming in turn the usable past of newborn critical planning research which seeks to intertwine critical urban theory with an idea of engaged "planning and design" approach.

1.3 Alternative Planning and Design Histories: Radical and Insurgent

There certainly is a dark side of planning in the history of planning (Yiftachel 1989, 1998; Paba 2003, 2010a; Perrone 2010, 2011; Sandercock 1998). Sandercock (1998) considers it the prevailing character of our discipline, understood in an institutional sense. Arranging bodies and things in space means separating, isolating, segregating, controlling, and repressing: it is urban planning as exclusionary zoning (Healey 1997). This striking critique came out at the end of the twentieth century as a contribution of several efforts to map the postmodern space of cities. While depicting the postmodern "multiple city" (multi-racial, multi-ethnic, multi-gender), Sandercock forges a postmodern age paradigm shift of planning practices. Thanks to this effort, invisible stories of planning are made visible and counterposed to a self-referential, manly masculine, and Eurocentric modernist account of planning. The Enlightenment legacy and its universalistic principles are superseded by an epistemology of multiplicity (see Attili's chapter in this book). Sandercock responded to the urgency of theoretical restructuring in planners' thinking to envision the landscape of postmodernity that is marked by difference. In her flagship book (*Towards Cosmopolis*, 1998, see Attili's chapter in this book), Sandercock depicts the effects of modernist planning practices as follows:

> Planners' historic role has been above all to control the production and use of space. In their state-designated role, they have acted as spatial police, regulators of bodies in space, deciding who can do what and be where, and even, when. They have closed public parks at night so that the homeless can't sleep in them, created ordinances to prevent street vendors, blocked permits for mosques, determined what kinds of housing renovations are permissible, and invented no end of both blunt and subtle ways of keeping certain bodies (marked bodies, marked by colour, by race, by gender, by sexual preference, and by physical ability) out of the sight and out of the way and out of the neighbourhoods of certain other bodies. These discriminations and repressions are the noir history of planning. (Sandercock 1998, 16)

In this description (which is just an exemplification of Sandercock's critique of modernism), Sandercock sheds the foundation of a postmodern practice that relies on appropriate and variegated kinds of knowledge and recognizes the difference as the initial characteristic of the postmodern city. Sandercock started this work by revealing the dark side of planning while bringing out the dark ideas, thoughts, and diversified approaches to the rationality of planning (Perrone 2010). She forged encounters between the theorization of feminism, postmodernism, and the postcolonialism buried by a modernist mainstream.

Sandercock's urgency to change the lenses of planning thinking has inspired the critical planning approach described above in its basic features.

Nevertheless, this approach goes further than Sandercock's goal of rediscovering the planning roots of a postmodern epistemology of multiplicity.

The critical planning angle, suggested in this book, reworks Sandercock's relegating/rehabilitating exercise for reframing planning epistemologies, which relegates modernist planners/thinkers to the dark side of planning on the one hand, while rehabilitating the protagonists of the politics of difference on the other. It widens this effort and brings to the light side of planning even certain roots Sandercock archived as modernist and then outdated, as in the case of the "extreme" Lewis Mumford or the radical Patrick Geddes (see Perrone's chapter in this book). This critical planning approach is radical (not ideological) in this investigation of the roots, which allows the rediscovery of many generative and timely contributions.

What differentiates the critical planning angle from the critical deconstruction work carried out by Sandercock is the conviction that it is possible to reread the same official history of planning, with more detailed attention to contexts in which this knowledge has been produced, to collaborations between theorists, and interactions among ideas and practices. Whether we unify the entire historical tradition of planning in a generic "enlightenment" trend understood as a compact and supportive whole, or we put in the "dark side of planning" not only Robert Moses, but also Patrick Geddes, to give an example based of Sandercock's work, we end up giving up a heritage too rich in theories, practices, and experiences to be thrown away.

A critique from within the official history of planning itself may then be necessary and more productive. The critical planning angle aligns with this endeavor. An example can explain this point of view. In a beautiful essay entitled *Collaborations*, Colomina (1999) reconsidered the work of some great protagonists of the history of architecture by analyzing collaborations, studying the reciprocal influences, and the impact of contexts. The "heroes" of the history of architecture thus appear less isolated, and some other companions appear on the stage, as in the case of women. It is unveiled not only the role of women in famous couplings, in which the female role is recognized and explicit (Ray and Charles Eames, Alison and Peter Smithson, Denise Scott Brown and Robert Venturi, as well as Laura Thermes and Franco Purini, or Gabriella and Massimo Carmassi, if we want to cite some Italian examples), but also the role played by hidden collaborators (wives, lovers, workmates), such as Lilly Reich and Mies van der Rohe, Charlotte Perriand and Le Corbusier, Margaret McDonald and Charles Rennie Mackintosh, Catherine Bauer and Lewis Mumford first and then William Wuester (technology and community, and housing, respectively), Anne Tyng and Louis Kahn (in this case, Anne was a young collaborator, an unfortunate love, an intertwining in the letters between private life and design).

Colomina herself has reinterpreted the work of some of the great architects (in particular, Adolf Loos and Le Corbusier) with a very particular and specialized gaze, revealing hidden aspects that standard critics had not been able to identify until then (i.e., the game of looks inside the houses, "domestic voyeurism", or some minimal yet essential details of projects).

Likewise, the value that some pillars of the "official" history of planning may have depends on our gaze, on our horizon. This can be the case of the anarchist and libertarian movement that has crossed geographic and urbanistic thought since the nineteenth century (from Kropotkin and Reclus, to Geddes and Le Play, from Goodman to Murray Bookchin and Colin Ward, and in Italy, Carlo Doglio and Giancarlo De Carlo); the fundamental work of Patrick Geddes on the co-evolutionary model; a certain Mumford focused on the role of community (see Chap. 2 in this book, written by Perrone); a possible social reinterpretation of Ebenezer Howard and the tradition of the "garden city" (in *Sociable City*, Colin Ward develops an original reading of the Howardian experience, finding many roots of contemporary radical experiences, and expanding it to the concept of a "do-it-yourself new town"; Hall and Ward 1998); the great tradition of American regionalism and naturalism (from George Perkins Marsh, John Muir and Benton Mackaye to the contemporary bioregionalism of Kirkpatrick Sale, Gary Snyder, Peter Berg); and more generally, the theories (including economic ones) that underline the role of "natural capital," or the "biophysical foundations of economic and settlement processes" in planning.

In this way, the history of planning comes back to life, people become people again, with their body, with their lifestyle, the desire to change themselves, maybe even the world, maybe with their sexual preferences.

As mentioned above, we can find many roots and build a "usable past" (a useful tradition) to move forward. Intellectual restlessness, utopian impetus, the desire for a different city, and society characterize many figures and protagonists of the tradition of critical thinking about the city. We (each of us) can build a different tradition and make sure that marginal and excluded figures are the protagonists of this tradition, which is a goal of critical urban thinking. My effort in this sense is again to give an example of this restless and unquiet work of building alternative planning histories useful for discovering new possible futures. To this end, the critical planning "disk" reproduced underneath is designed. It is unfinished, just evoking possible geographies and traditions from a critical planning angle. It is built without any pretense of completeness and, for sure, it is questionable in many ways, but this is the very meaning of a critical planning and design approach!

In short, to conclude, I think we can agree with this final consideration by Leonie Sandercock: "We have moved from planning history—the official story—to planning's contested and multiple histories."

This book itself is a critical walking through roots, pathways, and frames collected among many other possibilities. The next chapter develops the meaning of critical planning and design through an influential Italian "critical planning" thinker (Paba 1998, 2003), whose work has been a tireless journey in search of roots to be discovered (Fig. 1.1).

1.4 The Book

This book does not have a traditional structure; it is not a reading, and it is not built as a systematic collection of essays around a topic or research field. The book is based on the following assumption: every scholar has a favorite book (which is sometimes secret), and every scholar thinks there is an author or a reference figure that is particularly important for a scientific, cultural, or even personal reason. There is (at least) one book that counts for each of us, in the same way as we attribute a specific value to a person (scholar, planner, architect, writer, philosopher, or historian), whose work and writings we consider as particularly alive, current, and stimulating, in a way that could be called necessary for our vision of research and of the world.

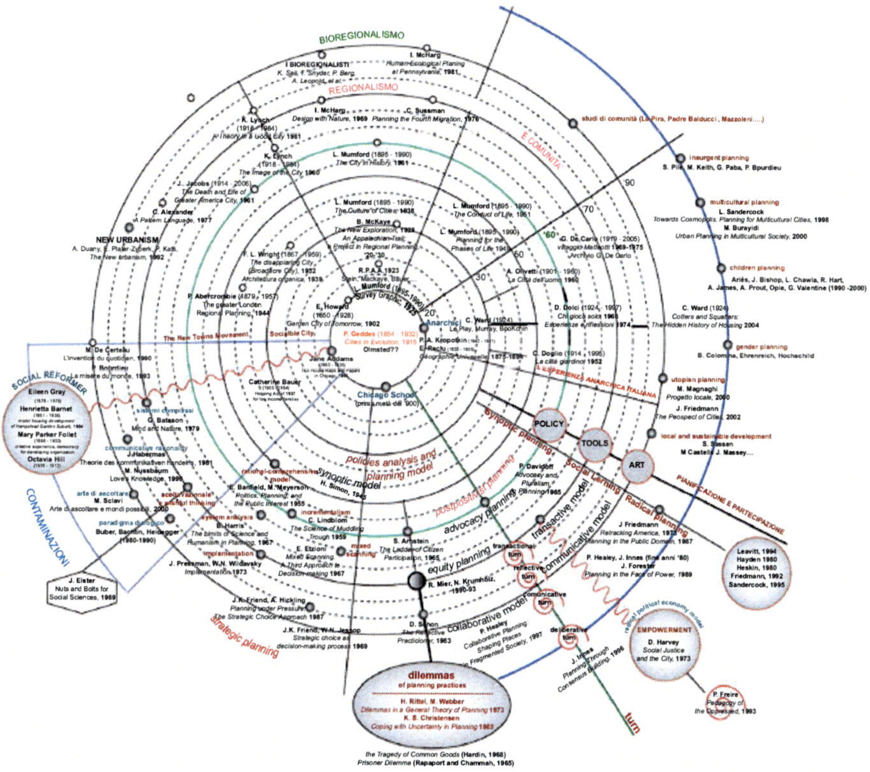

Fig. 1.1 *Critical Planning and Design "Disk."* This mental map suggests many possible "walks through" critical planning roots, pathways, and frames. It is meant to engage the reader in the interaction between planning roots and scholars while inviting them to find one's tradition or planning horizons. Above all, it impels the search for new planning roots. Ideally, the very center of the disk is Patrick Geddes's work and its biosocial program (see Perrone's chapter in this book). The disk is unfinished, in progress, continuously reworked. In fact, this book includes just a few couples of books and authors mentioned in the disk. However, each chapter describes the exercise of tracing trajectories in the intellectual universe of critical planning and design while enriching it with new perspectives. All in all, the disk shows some possibilities for dialog among critical urban scholars, planners, and designers whose research agendas might be antagonistic rather than interconnected and plural

Following this principle, a series of seminars entitled *Lectures on Planning* was organized by the research doctorate in Urban and Regional Planning and the Laboratory of Critical Planning and Design and was held in recent years at the University of Florence. I asked a group of scholars to choose a book that they considered particularly important for their research and to talk about that book and its author. The invited scholars were free to give the lesson, or written contribution, in their preferred tone and style. Some have dealt mainly with their chosen book, reconstructing it analytically, studying it in-depth and intensely; others have valued the book as a relevant testimony of the author's overall work, drawing a more complete portrait of their personality and their achievements; others have placed the book and the author in the historical and cultural context, rediscovering neglected roots and more fruitful ideas; and finally, others have used the book as an occasion, sometimes almost a pretext, to expose their ideas, and a possible philosophy, or genealogy, of their research.

The choice of this method of organization proved to be favorable. The lessons were intense and passionate, and the freedom granted to scholars to choose the book and their style of exposure created differentiated and original readings. The selection of texts was surprising and yet, reasonable. In the introduction to the volume, we return in detail to the readings proposed by the invited scholars and their significance in the contemporary discussion in the field of urban studies, planning, and policies. However, from the choices made by the authors, there are some common traits that we consider important to underline.

The first aspect concerns the need to reconstruct a genealogical continuity of our way of thinking and working. Subsequently, the first section of the book contains a sort of imaginary conversation with some great personalities of the past—Hildefonso Cerdà, Adriano Olivetti, Ian McHarg, and Yona Friedman appear in these conversations of disconcerting usefulness and actuality. More generally, scholars who have chosen to speak of an "ancient" book seem to implicitly affirm that our work requires roots and historical or genealogical continuity—perhaps nourished precisely by discontinuities, revolutions, and paradigm changes that some great protagonists in the history of planning have been able to realize.

A second thread unites scholars who have chosen dialog with some protagonists of the contemporary debate in the field of planning and urban studies. In this case, the scope of choice is potentially infinite, and any selection list would have been insufficient. Scholars have chosen the books of John Friedmann, David Harvey, Doreen Massey, Patsy Healey, Leonie Sandercock, and Thomas Sieverts. The final selection was very significant. Many of the themes arising from globalization and urban planetary processes are reflected in their works: the growth of world cities, post-metropolitan developments, spatial injustice, the insufficiency of traditional planning paradigms, the importance of collaborative and participatory planning, the city of differences, post-colonial studies, the role of in-between spaces, and many others.

The third section brings together the contributions of scholars who have chosen to deal with books of particular relevance in philosophical, economic, and social thought. It includes classics from economics, sociology, and political science (Charles Lindblom and Robert Nozick), as well as the works of great philosophers and thinkers who dealt directly or indirectly with the city and space: Max Weber, Henri Lefebvre, Gilles Deleuze and Félix Guattari, and Georges Didi-Huberman. From the contributions in this chapter, the need for scholars and planners to refer to renowned theoretical and conceptual frameworks of modern and contemporary culture has emerged in defining the orientation of their research and work.

The book is aimed at scholars in the fields of planning and urban studies. They may perhaps find original perspectives from rereading books that they know and, ideally, by participating in a conversation about their current relevance in time and space. The book is also aimed at students, including doctoral students, who can gain confidence in the sometimes-unpredictable network of cultural references inside and outside our discipline. More generally, the book is aimed at all those who think that the destiny of cities, territories, and landscapes is a relevant theme for everyday life and for the future of the planet.

In particular, the book reflects on seminal books and authors of many disciplines (planning, philosophy, sociology, economics, geography, and social sciences) that are very well known and are still assumed to be frameworks of both basic and advanced academic courses and research.

The proposed book, however, has an original construction that makes it different from the others: it highlights the generative value of some books and authors in different fields of planning, under the umbrella of critical urban theory. However, the book is a product that can interest wider fields of reflection, differentiating itself from the typical handbooks or readings of planning theory of a more generalist nature. In particular, our book comes from a sort of "Elective Affinities" between the authors and the beloved and narrated books. This way of construction gives the book a particular acumen and freshness, proposing profound and innovative lectures on seminal books and ideas for planners and scholars.

References

Abercrombie P (1933) Town & country planning. Thornton Butterworth, London

Beard VA (2003) Learning radical planning: the power of collecting action. Plann Theor 2(1):13–35

Blake CN (1990) Beloved community: the cultural criticism of Randolph Bourne, Van Wyck Brooks. The University of North Carolina Press, Chapel Hill, London, Waldo Frank & Lewis Mumford

Brenner N (2012) What is critical urban theory. In: Brenner N, Marcuse P, Mayer M (eds) Cities for people, not for profit: critical urban theory and the right to the city. Routledge, London, New York

Brenner N, Keil R (eds) (2005) The global cities reader. Routledge, New York

Brenner N, Schmid C (2015) Towards a new epistemology of the urban? City 19(2–3):151–182

Castells M (1977) [1972]) The urban question: a Marxist approach. Edward Arnold, London

Colomina B (1999) Sexuality & space. Princeton University Press, New York

Friedmann J (1987) Planning in the Public Domain: From Knowledge to Action, Princeton University Press, Princeton (US)

Hall P, Ward C (1998) Sociable cities. The legacy of Ebenezer Howard. Wiley, New York

Harvey D (1976) Social justice and the city. Blackwell, Cambridge, Mass

Healey P (1997) Collaborative planning: shaping places in fragmented societies. MacMillan, London

Heynen N, Kaika M, Swyngedouw E (eds) (2006) In the nature of cities. Routledge, New York

Harvey D (2000) Spaces of Hope, Edinburgh University Press, Edinburgh

Katznelson I (1993) Marxism and the city. Oxford University Press, New York

Keil R (ed) (2013) Suburban constellation. Jovis, Berlin

Lefebvre H (1968 [1996]) The Right to the City, In: Writings on Cities by Henri Lefebvre, E. Kofman and E. Lebas (Eds.) 63–182 Oxford: Blackwell

Lefebvre H (2003 [1970]) The urban revolution, trans. by R. Bononno. University of Minnesota Press, Minneapolis

Lindblom C (1990) Inquiry and change. A troubled attempt to understand and shape society, Yale University Press and Sage Foundation, New York

Merrifield A (2002) Metro-Marxism. Routledge, New York

Mumford L (1950) Mumford on Geddes, The Architectural Review 108:81–87

Paba G (2003) Movimenti urbani: Pratiche di costruzione sociale della città. FrancoAngeli, Milano

Paba G (2010a) Corpi urbani: differenze, interazioni, politiche. FrancoAngeli, Milano
Paba G (2010b) Radici. Alle origini della pianificazione interattiva. Contesti 1:19–30
Perrone C (2010) DiverCity, Conoscenza, pianificazione, città delle differenze. FrancoAngeli, Milano
Paba G (1998) Luoghi comuni: La città come laboratorio di progetti collettivi, FrancoAngeli, Milano
Perrone C (2011) What would a 'DiverCity' be like? Speculation on difference-sensitive planning and living practices. In: Perrone C, Manella G, Tripodi L (eds) Everyday life in the segmented city—research in urban sociology, 11, United Kingdom, Emerald, pp 1–25
Robinson J (2006) Ordinary cities. Routledge, London
Sandercock L (1998) Towards cosmopolis: planning for multicultural cities. Wiley, New York, London
Schuermans N (2009) Ordinary cities: between modernity and development (J. Robinson). Belgeo, no 1
Soja E (2000) Postmetropolis. Blackwell, Cambridge, Mass
Visvanathan S (2003) Cultural encounters and the orient: a study in the politics of knowledge. Diogenes 50(4):69–81
Yiftachel O (1989) Towards a new typology of urban planning theories. Environ Plann B Plann Des 16:23–39
Yiftachel O (1998) Planning and social control: exploring the dark sides. J Plann Lit 2(4):395–406

Camilla Perrone, PhD in Urban, Regional and Environmental Design (2002), is an Associate Professor of Urban and Regional Planning at the University of Florence (UniFi) Italy. Her research interests include critical planning, interactive design and urban governance, urban diversity, multi-scalar urbanization, urban political ecology and self-organising city. She is a member of the International Network for Urban Research and Action (INURA). Her publications: "DiverCity" (FrancoAngeli 2010); "Space Matters" (with Giancarlo Paba), in A. Balducci et. al., Post-Metropolitan Territories (Routledge 2017); "Città in movimento" (Guerini, forthcoming). Perrone is also the editor (with Giancarlo Paba) of "Critical Planning & design" (Springer, 2020). She is the Founding Director of the Research Laboratory of Critical Planning and Design at the University of Florence.

Part I
Roots

Chapter 2
Giancarlo Paba's Trilogy of *Luoghi comuni* (*Common Places*, 1998), *Movimenti urbani* (*Urban Movements*, 2003), *Corpi urbani* (*Urban Bodies*, 2010): Influential Italian "Critical Planning" Thinking

Camilla Perrone

Abstract This chapter retraces three main bibliographic contributions of the recently deceased Italian scholar Giancarlo Paba, whose critical and radical thinking was very influential in Italian discourse over the past three decades. These contributions are "Luoghi comuni" (Common Places, 1998), "Movimenti urbani" (Urban Movements, 2003), "Corpi urbani" (Urban Bodies, 2010). This chapter proposes a selection of key themes and concepts of Giancarlo Paba's critical approach that the author addresses and recursively develops in the three books (and in some other related contribution). They include the themes of "action", "interaction" and "materiality". In particular, this chapter portrays Paba's critical thinking in planning and design through the key concept of "usable past" associated with that of "roots" of planning (as a discipline), which inspire contemporary actions and, like tree roots, encourage and generate new articulations of thought. In general, this chapter highlights two radical attitudes at the basis of Paba's human and scientific contributions: (1) his tireless drive toward radicalizing problems and solutions, to take another step forward each time, toward unpredictable and fascinating horizons; and (2) the ethical urgency of an ongoing verification of ideas by developing social participation.

Keywords Usable past · Interaction · Roots · Critical urban theory · Critical planning · Radical planning · Regional planning

2.1 Premises

This chapter retraces three main bibliographic contributions (the trilogy) of the recently deceased Italian scholar Giancarlo Paba, whose critical and radical thinking

C. Perrone (✉)
Urban and Regional Planning Divison, Department of Architecture, University of Florence, Florence, Italy
e-mail: camilla.perrone@unifi.it

C. Perrone (ed.), *Critical Planning and Design*, The Urban Book Series,
https://doi.org/10.1007/978-3-030-93107-0_2

was influential in Italian discourse over the past decades. Notably, it is essential to mention that Paba's critical thinking has been nourished by thinkers and activists' ideas and experiences shared within the International Network for Urban Research and Action (INURA, which he co-founded in Salecina, Switzerland in 1991). The trilogy is composed by the following three books: "Luoghi comuni" (Common Places, 1998), "Movimenti urbani" (Urban Movements, 2003), "Corpi urbani" (Urban Bodies, 2010). This chapter proposes a selection of key themes and concepts of Giancarlo Paba's critical approach that the author addresses and recursively evokes and develops in the three books (and in some other related contribution). Therefore, the chapter aims not to describe the content of the three books in a philological way but rather to recount the distinctive features of Paba's critical planning and design thinking (throughout the three books) as an inspirational contribution to the planning discourse. In general, Paba's contributions are not yet well known in the international context, mainly because his writings have not been translated. Moreover, Paba co-authored texts in English only with a limited group of friends. With most scholars, he had a relationship that he called "bookish" (rather than "human" or "personal"). With his closest research companions, he established genuine friendships in order to confuse "planning as friendship" with "friendship as a principle of good planning." This is an aspect that he would write about later on, in a piece dedicated to the friendship between Clarence Stein and Benton MacKaye, which lies at the heart—at least in his view—of a richer, more interesting regional planning model, still inspiring a regional and post-metropolitan approach (Paba 2014).

Paba defined himself as suffering from extreme bookishness (a definition borrowed from Blake (1990), who coined this word for Lewis Mumford), which he practiced in a restless intersectoral and historical "digging." Indeed, his intellectual and personal lives were invaded by books. This wealth has always been generous and generative, has influenced the Italian schools of planning, and contributed to the birth of the Critical Planning and Design research laboratory (directed and co-founded by the editor of this book) at the University of Florence, which today develops critical and radical thinking (in the sense of getting to the root of concepts and thoughts) in the Italian and international context.

This chapter thus attempts to reveal the components of Paba's influence on critical thinking for a wider audience, without the ambition of philologically reconstructing his theoretical stimulus. A striking contribution that reconstructs a synthetic (although exhaustive) map of Paba's thoughts is "Giancarlo Paba's Map: A Remembrance," recently written by Alessandro Balducci, Raymond Lorenzo, and John Forester in *Planning Theory & Practice* (2020) as a tribute to a friend of thinking.

The last section of this chapter recaps the most explicit messages of Paba's beliefs and legacy (as addressed in his three books), accompanied by a biographical note by Paba himself.

2.2 Explorations of Urban Critical Thinking and Action from the Borderlands

The Paba's three books (1998, 2003, 2010) gather the author's reflections on three crucial time thresholds in the debate on urban planning and urban studies in general. The end of the 1990s coincided with the onset of post-modern thought, the spread of multiculturalism, politics of difference, and gender studies in planning (Young 1990; Kymlicka 1996; Qadeer 1997; Valentine 1996, 2002; Sandercock 1998; Parekh 2000; Fenster 1999; Fainstein and Servon 2005), and the diffusion of an anarchic, radical philosophy in areas of research such as the body, childhood, urban inequality, ghettoization, the development of punishment as an institution, and urban marginality (Ward 1973, 1978; 1988; Hall and Ward 1998; Wacquant 1991, 2004, 2008; Haraway 2008). This period also marks the commencement of globalization and the explosion of the debate on sustainability (Appadurai 1996; Sassen 1999, 1996).

The beginning of the new millennium marked the debate around the future of cities and territories by problematizing the mainstream development model (Sachs 1999, 2002; Chambers 2005; Magnaghi 2000, 2020; UN-Habitat 1996). Issues of spatial justice (Fainstein 2011) are addressed in relation to the spatial effects of neoliberal models (Brenner 2004; Marcuse 1982). This is also the moment in which we return to reflect on the role of the planner, on the ways of involving the different components of citizenship, and therefore, on collaborative planning models and forms of civic and community engagement (Healey 1993, 1997; Friedmann 1987; Forester 1989, 1993).

The end of the first decade of the new millennium called into question the Anthropocene and revived the discourse on (urban) political ecology (Latour 2015; Heynen et al. 2006; Heynen 2017; Swyngedouw 1996, 1997, 2004; Keil 2003, 2005, 2011; Kaika 2005). It evoked the humanity needed to deal with climate and environmental crises, and with the changing nature of urbanization processes. The debate on the urban impact of neoliberalism occupies both mainstream and critical/protest thinking (Wilson and Swyngedouw 2015; Metzger 2018). At the same time, a new important reflection on socio-spatial and environmental transformations began (Latour 2017; Keil 2020) that shed light on the need to reformulate an epistemology of the urban (Brenner and Schmid 2015; Soja 2011; Keil 2018) and triggered the need to reload the epistemology of the urban (Brenner and Schmid 2015; Soja 2011; Keil 2018).

Paba's three books capture these strands of the international discourse (which were influential in the formation of Paba's thinking), often opening up a "silent dialog" with articles and books by international scholars, friends, and colleagues frequented in international networks.

The three books have not been translated into English, and in general, there are few contributions by the author translated or written in English (i.e., Paba 2012b, 2013, 2018b, 2019; Paba and Perrone 2017; Paba et al. 2017). However, the contribution of these three books was essential to the formation of a critical, radical research field in Italy, capable of expressing itself today in an international domain. This contribution fuelled the creation of courses (bachelor's and master's degrees in planning at the University of Florence), research and publications, a research laboratory, and an association (see Sect. 2.6 for details). It also inspired scholars and researchers from all over the world (through INURA), as well as the idea of the Critical Planning and Design laboratory, for which this book intends to establish some cultural and critical roots.

Paba's trilogy laid the foundation for critical thinking on planning in the Italian debate in dialog with the international one, even though it was not conceived with this aim.

It is the post-humous reinterpretation of Paba's entire repertoire that makes it possible to attribute to these three books the role of forerunner in a critical, radical culture of planning in Italy.

The idea of writing a trilogy would never have belonged to the profile of the author, who was more interested in exploring new terrains, and building a map of thoughts and notes for those who wanted to read them. The author's thinking is indeed rich and generous, and is expressed in multiple forms and writings in order to investigate, explore, and discuss issues often neglected in the mainstream cultural approach of the discipline (in Italy, it is very anchored to techniques and the technological realm; people are more reluctant to embrace the idea of planning as a branch of the social sciences, as John Friedmann later consolidated).

The three books, therefore, collect scattered reflections developed in many other precious contributions that cannot be recalled here. None of them resolves or concludes a line of thought; rather, they pave the way for interpretative frameworks on planning issues in vital dialog with the roots of the discipline, and the newest and often relegated approaches are confined to the dominant debate of the moment.

Hence, they do not constitute a systemic, orderly contribution to Paba's philosophy. However, it is important to recognize the post-humous, coherent, and generative value of Paba's explorations into urban critical thinking and action from the borderlands of an international debate. Recently, Alessandro Balducci, Raymond Lorenzo, and John Forester used the metaphor of "roots and wings" to describe how Paba "dug deep in search of roots, into the past, into bodies, into territories, into the lives of the dispossessed of the earth, into processes, and then he flew high over the territories of its forms and its inhabitants, just as his inspiration Patrick Geddes did from his Outlook Tower, to give us the vision of a city and a habitable territory" (Balducci et al. 2020, p. 498). In the same text, the authors propose the concept of a map to describe Paba's daily explorations "in literature and in urban practices" such as the social, material, and symbolic periphery of living, claiming the right to the city for everyone.

Perhaps the metaphor of the constellation more adequately represents Paba's explorations and allows us to recognize his contributions: dressing themes with new interpretations, journeying between layers of knowledge and time, enabling past and future relations, connecting the roots of planning thoughts to the needs of urban contemporaneity, often anticipated in its problematic dimensions, up to the "appearing" of striking and clear concepts that suddenly and surprisingly emerge from his reasoning. Constellations are areas in the celestial sphere in which a group of visible stars forms a perceived outline or pattern, typically representing a recognizable animate or inanimate object. Just as it happens in a constellation that a group of stars comes out as sense-making or evoking shape, Paba's beliefs—as outlined in his three main books—form geographies of critical thinking on a multiplicity of themes clustered (in a disorderly manner, but with geographies that can be understood as groups of stars in the constellations) around several fundamental questions. This happens depending on whatever interpretative perspective or scientific posture a reader may adopt.

In the following sections, three of these questions are scrutinized. They are intertwined with each other: the question of the "usable past" (Paba 1998, 2003) in Sect. 2.3, the theme of "interaction" (Paba 2003, 2010) in Sect. 2.4, and the "materiality" and obduracy of territories (1998, 2010) in Sect. 2.5. An outline of Paba's legacy follows (Sect. 2.6). This section portrays Paba's contribution to critical planning (the figures of his constellation) as a result of his explorations of planning roots. Moreover, it prompts us to consider Paba's work as a planning root in itself.

The chapter ends with a brief note on Paba's diversified, branched contribution to Italian critical and radical thinking (Sect. 2.7).

2.3 "Unsettling": A Usable Past for an Alternative (Radical and Critical) History of Planning

Paba spoke of himself as one of the most unsettling people he knew. He loved this expression that he considered tailored to himself, as it had been meant for Patrick Geddes and Lewis Mumford (as explained later). Paba discovered it in his indefatigable work on the roots of planning in search for untold planning histories and dissenting imaginations (another of his favorite words), which he valued as a generative source to deconstruct the modernist epistemologies of planning and to examine alternative planning pathways.

His impatience about what he did not know (yet), his restless and assiduous search for the root (even the most hidden aspect) of a concept, of a biography of scholars, occupied his days. It is this scientific, as well as personal, attitude to which the first essential question posed in Paba's books refers.

It is the idea of an "unsettling" planning, disturbing the existing order of things (disciplines, consolidated or universalistic concepts), and dissenting traditional descriptions of planning issues. It is opposed to the official history of planning, as if to propose a counter-history of planning that embeds the actual transformation of an increasingly rich, branched, diversified, plural city (Paba 2003). It is precisely this idea of planning that orientates Paba's critical and radical gaze toward the rereading of the roots of the discipline and the revision of planning genealogies. He does this in order to execute the complex effort of positioning himself in the scientific debate required of every scholar.

The relationship with the past is not that of a historian who reconstructs history philologically.

Paba's idea of the past refers to a "usable past" (Blake 1990; Mumford 1950; Paba 2003) that inspires contemporary actions and, like roots, encourages and generates new articulations of thought. They are "augmented" by a sense of time and awareness of one's scientific posture (built-in humility and prudence); most of all, they allow for a sense of belonging to a community of thought (Balducci et al. 2020).

2.3.1 *"Usable Past"*

The way to appropriately use this expression, "usable past," is borrowed from Lewis Mumford, who employed it to describe his own attempt to pick out the historical foundation on which to build his vision of American society (Blake 1990). According to Mumford, the construction of a usable past is not a cold, historical reconnaissance or the creation of an objective canon. Instead, it involves the deliberate, often biased choice of a personal pantheon of great intellectual figures. Lewis Mumford also used Geddes' life and practical bearings to clarify his own ideas on the city and planning, perhaps even his own ideas on humankind and society. In short, he built a sort of usable Geddes: a Geddes useful for the present, a tool of self-awareness, and a conceptual and practical device necessary for action (Paba 2003, 2010).

In Paba's view, roots and the "usable past" are essential ingredients of critical, radical thought and looking toward the future. Accordingly, Paba reinterpreted the work of Lewis Mumford, Patrick Geddes, Benton MacKaye, Clarence Stain, and Jane Jacobs (and other relevant planning masters). In doing this, he highlighted new or misinterpreted aspects of their thinking, which today represent helpful inspirations. He also retraced their biographies and readings (often from other disciplines such as botany, history, philosophy, and geography) that inspired their thinking. This is how Paba fed his idea of regional planning (see below).

Not only can the masters of the past be critically reconstructed as part of a usable past for the needs of the present; the same attitude, according to Paba, is needed for territories, history, and the identities of settlements and places. He was strongly convinced that independent thinking could be achieved through a break with tradition and recognizing what can be used from the past as a usable past. To this end, Paba devoted his endless efforts to critically investigating the formation and evolution, over time, of the main planning paradigms. He explored how they are transformed, even the impressive tendency they sometimes have to metamorphose to survive, to reappear, to regain vigor, perhaps in a transformed way, even in contemporary experiences. In this regard, he populated his usable past with three fundamental roots: the *philosophy of regional planning* of MacKaye; the *Fourth Migration* of Mumford and his emphasis on *Planning for the Phases of Life*; and the *biosocial program* of Patrick Geddes. They inspired Paba's thinking about regional planning, interactive planning, the creative management of territorial heritage, and the interactions between bodies, plants, and animals.

2.3.2 *The Philosophy of Regional Planning (Benton MacKaye)*

Paba chose to dig into Benton MacKaye's usable past. He was particularly fascinated by MacKaye's philosophy of regional planning and its flagship experiment, the "Appalachian Trail" (described in the famous 1928 book *The New Exploration*). Benton MacKaye was the great "geotect" of the Regional Planning Association of America (RPAA), and devoted many reflections to the meaning that geophysical features take on in the emerging theory of regional planning (MacKaye 1928, 1968). The word "geotect" (conceived by Mumford to describe MacKaye's contribution to the RPAA) combines George Perkins March's concept of "active geological agent" (1864) with the term "architect." It evokes the idea of a geo-architect. Paba explored some components of this root. In particular, he reused the attention toward the watersheds and physical features that led MacKaye to develop a theory of flows and containments, as well as connections and levees, which would be applied in many planning experiences (from the Appalachian Trail to the Tennessee Valley Authority, from New York's regional plan to the Boston Super By-Pass).

This work of mending between the roots and new planning approaches led him, for example, to bridge the regional planning of the early twentieth century with contemporary bio-regionalism (Magnaghi 2020; Fanfani and Ruiz 2020).

This "usable Benton MacKaye" lies at the origin of Paba's regionalist thinking (2003), which is built around the reinterpretation of the "Appalachian Trail" as a complex system of regional planning alternatives to metropolitan growth, keeping together agriculture (in an anticipatory idea of the relationship between food and planning) and the protection of forests, wilderness and cities, hiking and resettlement, trails, and railways (MacKaye 1921; Paba 2012b, 2014). Paba used Benton MacKaye's book, *The New Exploration* (MacKaye 1928, 57), to grasp the effective characteristics of the transformation of cities and urban spread. He fully embraced MacKaye's conviction that a new exploration (as a sustained investigation) should start precisely from the mere physical, the thing, the original, the primeval aspects of the territories where we live, while rebuilding the interplay between the geosphere and the sociosphere, and the unshakeable specificity of every regional environment.

Paba's regionalist thought is enriched with a second root, which is Lewis Mumford's conscious critique of metropolitan civilization, developed in his writings regarding *The Fourth Migration* (Mumford, Issue #7 of *Suvery Graphics*, 1925, pp. 130–133), and in the experience of the RPAA. Moreover, a great impulse for Paba's regionalist thinking comes from the documentary *The City*, produced in 1939, and which was promoted by the American Institute of Planners with commentary by Lewis Mumford and music by Aaron Copland. Paba was strongly impressed and influenced by the intertwining between the evolution of Mumford's thinking on regional planning and some places that were crucial to his life. These places include the village of Shirley near Boston, Massachusetts (a place of environmental and territorial self-training of the young MacKaye), from which the documentary begins, and Sunnyside in Queens, New York, where Mumford fed his reflections on regional planning and undertook community experimental planning, designing integrated and self-managed neighborhoods. Mumford quoted Sunnyside many times in his work on local communities and minimal settlement units (Mumford 1968), inspiring Papa's radical and engaged participatory approach.

Mumford was crucial to Paba's thinking for another aspect to which we will return in the next section. Briefly, we can anticipate that it refers to Mumford's enhancement of the planning takers' plural profiles, which led Paba to develop his contribution to children's spatial competence (1998, 2003) regarding radical participation and interaction (2010). In reading *Planning for the Phases of Life*—in which Mumford stresses the needs of children and the elderly—specifically about the need to encourage forms of spontaneous cooperation in the construction and management of collective spaces (Mumford 1949), Paba elaborated an important contribution to contemporary Italian discourse.

2.3.3 The Biosocial Program (Patrick Geddes)

The third root is Patrick Geddes, whose philosophy contributed to many areas of Paba's research. This is certainly not the ordinary Geddes, but the university planner, outsider, and deviant academic (Sutherland 2009) that Paba rediscovered in dialog with Geddes's works. Basically, Paba reinterpreted and embraced the biosocial program of Geddes' experiences in Edinburgh (which then developed in India); that is, the open-air laboratory in which he experimented with his theories on the physical and social "body of the city" (the city's materiality): an integrated, cross-sectoral, multi-objective project (housing and services, the economy and culture, training and scientific research); a close intertwining of physical with social recovery; heritage protection and innovation in architectural uses and features; designs capable of understanding the artificial and natural components of the city (stones, earth, plants, and sky); people's direct involvement in the design and implementation of the program; and the ability to move toward a strategic horizon (the identification of goals and actions to achieve them). Finally, he dealt with an incremental, pragmatic attitude, enriched by the awareness that small positive changes can result in important systemic effects. A crucial element of this experience, as well as an important part of Paba's beliefs, is the meaning of coevolution of the Geddesian vision that derives from some of his works (both theoretical and experimental) in the field of biology and natural sciences—not only the most important and systematic (*The Evolution of Sex* or *Chapters of Modern Botany*).

The evolutionary (Geddesian) paradigm allowed Paba to appreciate the city's organic qualities without implying a fixed relationship between the parts and the whole, or a mature optimal form. The city is transformed into a mutual, positive relationship between human beings and the environment, between the diverse and plural manifestations of the population, and the numerous components of the natural and built environment.

There are many other roots of Paba's usable past that he developed in various other writings, including one of the last in which he reinterpreted Jane Jacobs (Paba 2019).

Throughout his career, Paba explored planning thoughts relegated to the margins or obscured by mainstream planning history. In doing so, on the one hand, he helped to represent an alternative history of planning (such as theories, practices, utopias, experiments, and the realizations produced by marginal and peripheral figures); on the other hand, he revealed the silent history of resistance, of struggle, of conditioning from below in urban populations' transformations of cities.

2.4 "Interaction": A Agency of Bodies, Animals, Plants, and Other Terrestrial Elements

The second major question of Paba's work is the concept of interaction, which he extended to human and non-human agents alike, to bodies, to plants, to things, and to the complex domain of interweaving all these things together.

2.4.1 Human and Non-human Bodies

At the center of his urban explorations, Paba established the essential link between bodies (human and non-human) and stones, between forms of social and political organization, and forms of the space that host or determine these organizations. The inhabitants' bodies are always considered the protagonists and victims of the transformation of cities and territories. These bodies are marked, articulated in a plurality of material, cultural, and social differences, often in relation to each other. They are individual and collective, might be in agreement or conflict, united or divided in the increasingly complex game of trajectories, movements, and desires that characterize cities. These are urban bodies, wherever they live or work, forced to deal with the unpredictability of urban interactions, in the proximity of physical relationships, or the virtuality of distant relationships (Paba 2010).

The (urban) bodies are Paba's intriguing investigation object and become "agents" of a project that extends to the house and the city. Paba was a forerunner of this theme (2003). His approach's originality and richness emerged just recently in Italian urban planning discourse, which is sometimes reluctant to welcome new ideas, and is rather stuck on the notion of traditional urbanism. In *Corpi urbani*, Paba talks extensively about "disciplined and contested bodies in the social and political 'game' and in the dispute over the organization of spaces and territories" (2010, p. 43, translation by the author). He rediscovers the body as an object of (bio) power. Bodies are seen as disciplined, trained, mutilated, marked, stigmatized, re-appropriated, proudly exhibited, managed, self-produced, re-signified, and re-identified. Paba speaks of the body as self-architecture, of the body as a house, and of the house as a body. He always underlines the movements of bodies in space and their effects, and by means of this emphasis connotates the meaning of the bodies as urban ones (both variegated and multiple). Paba explored the concept of bodies in the literature in his tireless search for roots of thought to be rediscovered and make them usable. He considered bodies to be agents of the planning process; in particular, he accentuated a heterogeneous, polymorphous multitude of bodies that crowd the city, with an investigative passion for women's bodies and the women's perspective on bodies, from the fluid body of Duden (1994) to the techno-product body of Haraway (2008).

His exercise always goes to the heart of concepts, and challenges mainstream beliefs in "contrast with the standardization so typical of urban planning, which comes from an anonymous and asexual consideration of humans. Our bodies are all different, Paba says, and that diversity is the basis of the richness of life and communities" (Balducci et al. 2020, p. 496).

There is a very long list of bibliographic references to authors who have remained in the shadows of the literature and whose contributions are now known thanks to Paba's radical work. This is true not only with reference to the question of urban bodies. For example, this radical work includes the exploration of women's role in the planning process: a path of thinking that prompted the re-discovery of female activists, social reformers, planners, architects, experts, and scholars of the city (including Octavia Hill, Henrietta Barnett, Jane Addams, Mary Parker Follett, Eileen Gray, Catherine Bauer, Elsie Phillips, Edith Elmer Wood, Mary Symkhovitch, Florence Kelly, Julia Morgan, Eleanor Raymond, Grete Schutte-Lihotsky, Lina Bo Bardi, and countless other figures; see the "Introduction" of the book).

In *Corpi urbani* (2003), Paba also examines extreme positions, such as that of the bio-architect Friedrich Stowasser (known by the pseudonym "Hundertwasser"), who gave—during two performances that were held in Munich and Vienna in 1967/1968—the famous "naked speech" against rationalism in architecture; that is, against the denudation of architecture operated by the functionalist canons. Paba, inspired by the alternative suggested by Hundertwasser, expressed in the theory of five skins, from the body to the cosmos. The first skin is naturally the *epidermis* that surrounds the body, the first house we live in. Our *clothes* are our second skin, considered to be extensions of the body (Hundertwasser also designed clothes and shoes suitable for good health). The third skin is the *house*, and in the Munich and Vienna speeches "the first skin was laid bare to better proclaim the right to the third," the right to a home no longer stripped of all ornaments, a living house, directly transformed by the inhabitants. The fourth skin is called *identity* and is the place in which families and homes are contained (the neighborhood, the city). The fifth skin is the *earth*, the environment, and nature, the ecological framework in which we live. Although Paba was not a true lover of Hundertwasser's projects, and sometimes remained perplexed by some of his architecture, he thought Hundertwasser's message called for renewed consideration. In the old performances and manifestos, there are some relevant and still current themes: The manifestos on the "Right to Window" and on "The Tree Tenant" indicate the possibility of people transforming their own homes and filling them with greenery and trees, the house of Hundertwasser being conceived of as a natural, vegetal machine; the manifestos on the *Humus Toilet* and on *The Sacred Shit* pose the problem of waste recycling and posit a more general idea of our relationship with the environment (the fifth skin) (Paba 2010, 2013). All in all, in the usable Hundertwasser, Paba saw message for a rebalanced, sustainable relationship between human beings, nature, and the environment, which is a very timely message!

2.4.2 Interactive Planning and Design

Digging into these roots, Paba also developed an innovative idea of interactive planning and design that includes, in the interactive game, the values and interests of non-human actors who (called "agents" in Paba's reinterpretation) populate the Geddesian universe. Paba did this by reconstructing a critical, alternative interpretation of Patrick Geddes' figure and of his fruitful contribution to contemporary critical discourse. Paba's usable Geddes is the botanist Geddes, who anticipated the discourse on plant life and the contributions of organisms other than humans (developed herein). He was the creator of the first participatory planning experiments and pedagogical gardens with children and women, in the repairings of *Edinburgh Old Town Tenements, Ramsey Garden*, and the creator of the first open-air zoological garden in Edinburgh (Paba 2010).

This Gedessian root enabled Paba to discover the many forms of agency that help to transform cities and guided him on a small journey among the emerging forms of resistance, perhaps of active citizenship, of animals, and plants. He was passionate about Geddes' research in Naples (Italy) on marine fauna (studying the Dohrn model to reorganize the similar structure of Stonehaven in Scotland), neglected in planning as bizarre and not very influential in the activities of the Geddes planner. By contrast, Paba hypothesized that these Neapolitan studies could deepen the study of the evolutionary basis of the Geddesian vision and explore the possibility of re-establishing an innovative planning model for those intuitions (Batty and Marshall 2009; Paba 2011b). Geddes was drawn by the hybrid character of these living beings that he studied and defined as "disputed organisms": perhaps algae, perhaps animals, perhaps something intermediate, and also vegetating animals that lead a more or less vegetable way of life, worms no doubt, but also a little vegetable (Geddes 1878–1879, p. 450). More generally, Geddes was attracted by the phenomena of coalescence, coagulation, and corpuscular fusion, which manifest in the interactions of living beings and which he carefully analyzed in the work of the Indian scientist Sir Jagadis C. Bose, to whom he dedicated a biography *The Life and Work of Sir Jagadis C. Bose.* Paba attributed to these studies a fundamental role in the Geddesian analysis of the formation of the urban fabric and settlement constellations. Drawing on the ways in which elementary organisms form complex configurations (by aggregation, absorption, and coalescence), Geddes elaborated on his analysis of urban conurbations, addressing his co-evolutionary model. This is not developmental (therefore linear, mechanical, and predictable), but is evolutionary. In the evolutionary paradigm, evolution is open and thus unpredictable in the long term. The city is seen as a collection of interdependent, co-evolving parts: "the evolutionary paradigm allows us to appreciate the organic qualities of the city, without implying that there is a fixed relationship between the parts and the whole, or an optimal mature form" (Batty and Marshall 2009, p. 552).

Paba was fascinated by this co-evolving paradigm, which also includes plants and animals, understood not as objects, but as protagonists of the planning processes. Therefore, he transformed this Geddesian root into his usable Geddes and works out of a *radical, critical, biosocial, and interactive planning model* that attributes value to every manifestation of life on the planet, to every aspect, organic and inorganic, of nature and the landscape (Paba 2010, 2011b). It is grounded in cooperation, mutual adjustment between inhabitants and places, the creative management of heritage sites, and economic interactions between urban animals and plants. The core question is coevolution: The city is transformed into a process of mutual relationship between human beings and the environment, between the plural manifestations of the population and the diverse components of the natural and built environment. Paba then put Geddes' concept of coevolution to work and used it to identify and describe the open, transformative, cooperative interplay between a plurality of actors, animals and plants, human beings and things, and finally performing experiments in the field of (co)evolutionary, biosocial planning. Following this line of reasoning, Paba anticipated the debate on the importance of the non-human and its agency, familiarizing and reinterpreting—before others in urban studies—some philosophical roots, such as Latour's philosophy and the discourse on political ecology (Latour 2015; Swyngedouw 2004; Keil 2020).

2.4.3 Radical Participation

The concept of interaction incorporates the notion of participation, a word that Paba did not like. Therefore, he always felt the need to use adjectives, such as "radical." By radical participation, Paba referred to the involvement of all actorships and citizenships in a city of differences, engaged in claiming their right to the city. Poor people and migrants take center stage in Paba's participatory practices. Peripheries and neglected places are the laboratories of research. The forms of appropriation of intermediate and residual spaces—scraps of the dominant urban planning processes, usually—together with informal practices of living, compose the object of Paba's research action. The concept of interaction is deeply rooted in Paba's radical participatory approach. His research is in fact always engaged and becomes "action," often on the edge of the city, on the boundaries of discipline, on the threshold between bodies, things, animals, plants, and not humans. Paba's radical participatory practices aim to overcome poverty and spatial injustice while engaging with human and non-human agents and delegating to places' specificities a fundamental role in the pursuit of spatial justice. In Paba's view, individuals' well-being (people's prosperity) is related to the quality of places (place prosperity). Thus, by pursuing the prosperity of places through radical participation, poverty can be effectively contained (otherwise, it is reproduced by the inequalities of individuals' economic incomes). The contrast is between the ideal of improving people's well-being, regardless of where they live, or to improve the well-being of groups of individuals, defined by their spatial proximity to places (Paba 2014).

For Paba, participation became radical and critical, and developed in three fields of intervention:

(1) *Expanding citizenship rights and the right to the city, and to pursue a city of gender, age, and cultural differences.* At the heart of Paba's participation are children and migrants, whose rights and needs he wanted to bring to the fore in urban policy agendas and practices: "Giancarlo's particular approach to exploring, reading and interpreting urban paths and places, as well as his critical knowledge of planning history and contemporary instruments, enriched professionals' understanding of children's spatial marginalization" (Balducci et al. 2020, p. 497).

Paba introduced the question of children in and for the city into the Italian debate on planning from a multidisciplinary angle (Paba 1998, 2003, 2010; Paba and Perrone 2004; Paba and Pecoriello 2006) and with the crucial approach of going back to the roots that characterize his gaze, reminding people that the masters of planning (e.g., Geddes, Jacobs, etc.) "had much to say regarding the role and well-being of children in city spaces. [...]. [Paba's] passion for activism led him to work together with, and learn from, children—whom he once called builders of tree-houses, diggers of caves and diverters of streams—as he engaged with innumerable small-scale local projects aimed at changing the city, including the ways children might inhabit and transform it" (Balducci et al. 2020, p. 497). Paba became one of the national coordinators of the Italian Ministry of Environment and the UNICEF Program 'Sustainable Cities for/with Children.'.

(2) *Encouraging the construction of local democracy and radical participation at different levels of the city's transformative action.* The local dimension constitutes the strength of participation and, in general, the strength of local government. Participation in Paba's radical perspective plays a political role and entails antagonism, contestation and reconstruction from below the processes of globalization (Innes 1995). It is a local, strategic alternative to the domination of neoliberal globalization and the destruction of territories, the environment, humanity, and social wealth (globalization). Cities and territories, partially freed from globalization's external constraints, can become the nodes of an alternative "neoglobal" network comprised of collective actors who are free and dynamic subjects. In this vision, participation is horizontal, plural, decentralized, federated, non-hierarchical, and non-competitive; it adds bonds, densifies relationships, and increases communication. Paba explored its many dimensions, methods, and gradients, both in research and in action in the city, which led to his foundation of practices and learning alongside literature and scientific explorations. Three main aspects characterize this kind of participation. The first aspect is concerned with the extension of the participatory process, the awareness that the entire cycle of the architecture and city projects can be intertwined with citizens' protagonism. Participation not only deals with the characteristics of a single project, but also with everything that comes first: the decisions made to establish the new structure, where to locate it, on the resources that must be allocated to it, and on the needs that it must fulfill. The second aspect relates to the composition and role of the local community involved in the architecture and city projects. In trying to

summarize, in a single sentence, the basic requirement of Paba's idea of participation—namely, what must still happen in any slot, even a small one, of the participatory process—is that one leaves a participation experience feeling different from how one entered it. The message is that radical participation: changes people who enter the process through social interactions; reframes problems; produces interactive knowledge, ideas, and expectations; and changes the interplay between the protagonists of the process (the inhabitants, administrators, technicians). Participation turns out to be a positive-sum game, as it allows for a collectively more advantageous solution than the one that would emerge in its absence. Moreover, it enables hidden energies and mobilizes underutilized collective resources, what Lewis Mumford defined as the "margin of energy," another concept of the roots of the discipline that Paba rediscovered and put to work in the actuality of participatory practices and the debate on participation. However, in his more in-depth critical work.

Paba dwells on the most extreme dimension of interaction, the insurgent and conflictual one.

(3) ***Paba finally spoke of "the background noise of participation," consisting of the thousand forms of resistance and social conflict, acted out through self-organized*** *urban practices* (Paba 2003, 2010). This is what Danilo Dolci, another usable past of Paba, defined as "fighting exactly" (Dolci 1962, p. 2) to intervene in a complex reality that requires precise attacks at certain points and nodes, with suitable tools. This noise helps to reveal positive transformative actions in the struggle for the survival of the marginal and excluded layers of the population, or even in the most radical manifestations of antagonism and struggle. This is what can be defined as the "insurgent" characteristic (which is alternative, emerging, and constructive) of the individual or collective, spontaneous social practices (Paba 2002). Once again, with a lucid and insightful rereading of the roots of planning, Paba found such characteristics in the works of Patrick Geddes and Lewis Mumford, although they had been relegated to the dark side of planning by Sandercock (1998), one of the authors who rehabilitated the insurgent planning practices in post-modern planning discourse. Indeed, she labeled them as "enlighteners" whose influence is necessary to ignore in order to re-establish a planning discipline based on the emergence of a plural world of emerging citizenships. The same cancelation was made by Friedmann (1999), who spoke about insurgent citizenships, and Holston (1996), who was the first to reuse the term at the end of the 1990s. On the other hand, "insurgent" is a keyword in the original Geddesian vocabulary, which Paba retrieved and took as the root of a contemporary approach. It is peculiar that Paba, Holston, Sandercock, and Friedmann never remembered the Geddesian origin of the term. Conversely, Paba loved the word "insurgent," and used it exactly because it is deeply Geddesian-Mumfordian. Geddes life (mind and body) is insurgent: Geddes added to the usual list of organic characteristics a quality that he himself possessed at the highest level, at least in Mumford's description (1950). This quality is the insurgence itself; that is, the ability to overcome—through power or cunning, a plan or a dream—the forces that threaten the organism. The most important point (and perhaps useful even today) is that insurgence is not only a generic quality of life and of the person; it is also a specific quality of the community and places, and indeed constitutes the foundation of

collective life. Geddes's forward movement of life, his insurgence, and his expectance also belong to the community: it has "an underused margin of energy" (see the next paragraph). Community development depends on the use of this resource for new goals and new destinations. The image that summarizes this interpretation by Paba (2002) is the *Insurgent City* (2002, 2003). The insurgent city is not pure and simple antagonism: the mechanical reversal of good into evil and evil into good. On the contrary, the city, made of new emerging citizenships, is a constructive laboratory of new sociality and new urban and environmental quality.

The insurgent city is a meticulous, hopeful recognition of individual and collective positive energies emerging in cities. It is a renewed trust in the social bond, in the "new community practices" in the nascent forms of brotherhood and collective action.

2.5 "Tenacious Cities": The Materiality of Territories, Obduracy, the Margin of Energy, Walking Through, and the Militant University

In Paba's books, there is a recurring notation on the concept of the territory: "places, cities, territories matter more, not less." This expression summarizes his contribution to the debate around the urban question, characterized by the awareness that contemporary urbanization processes have caused a profound crisis in the relations between space and community, sociality and proximity, placed-based and place-neutral activities, the fixedness of places, and the fluidity of communication.

2.5.1 Obduracy (the Materiality of the Territory)

Two aspects are crucial in Paba's perspective on the meaning of "territory." The first is the materiality of the territory, traced back to two roots: Benton MacKaye's work in the Boston region and the latest evolution of Melvin Webber's philosophy on the inertia of cities (rediscovered thanks to a contribution on the concept of the tenacious city, neglected by the post-humous literature) (Balducci et al. 2020). The second aspect is concerned with the role local communities play in producing territories to live in, developed in dialog with his "usable Mumford" (the "extreme" Mumford, who is an anarchist, utopian and insurgent, and relies on the community's margin of energy as an essential resource per the city's social construction) (Paba 2003). This second aspect evokes Paba's insurgent city and the Mumfordian ideas of insurgence, mentioned in the previous paragraph and elaborated on below.

For the first aspect, Paba accentuated the role of geohistory. Territories have formed in a long process of reciprocal adaptation between geographical, environmental, urban, and cultural factors. This is something that Paba defined somewhere else, quoting Saverio Muratori (1967) and De Landa (2000), "a kind of efficacy in the spatial arrangement, which tends to remain, resist, reproduce in time (also through an open and not linear game of variations, modifications, adaptations). Urban and territorial settlements could be considered a sort of mineralization of humanity: 'human-made structures are very much like mountains and rocks: the accumulation of materials hardened and shaped by historical processes'" (De Landa 2000, p. 55 as cited in Paba et al. 2017, p. 98).

The persistence of the material structures of the territory over time (physicality) and resistance of the geo-historical matrices to the change of the built environment are always probed in dialectics with the opposite (flows of the change). In particular, the concept of obduracy synthesises and describes the resistance of the territory to processes of transformation and urbanisation (Paba 2010; Hommels 2005). Paba identified the interplay between "distensive forces" that connect, unite and link, and "tectonic forces," which regulate, delimit, and protect the places of urban and social life. Later in his contributions on the post-metropolitan transition (anticipated in the texts of 2003 and 2010), he would specify this concept by relying once again on one of his beloved masters, Benton MacKaye. In this case, Paba referred to the concept of "streams and levees" that MacKaye used for the region of Boston to describe the evolution of major regional units "as a game between structured and complex flows (water, natural resources, people and goods) on the one hand, and levees (geophysical, environmental, cultural) on the other hand" (Paba and Perrone 2017, p. 263).

According to Paba, territories are coarse and corrugated (in their resistant material configuration). He described the territory obstinacy not to dissolve completely in the forms imposed by urbanization processes, with reference to the transformation in the time of the famous Malevich's Black Square painting (oil on canvas, painted from 1914–1915, depicting a geometric figure centrally placed within a white area that extends to the edges of the non-framed, perfectly quadratic canvas of 80.01 × 80.01 cm). Made to represent an undifferentiated and uniform nothingness (and "not even nothing"), the picture embodies something, showing a figure, conveying a meaning (as described by Olson 2007). The original black square came to be a square less black, not uniform, and corrugated, despite resisting the change. From Paba's perspective, this autonomous existence (independent from the artwork's creators) is metaphorically associated with the resistance and transformation of the territory in time. "The picture's transformation is important because it represents the matter's rebellion, the emersion of traces, the plots' resistance, the depth and three-dimensionality of the things of the world (material layouts, places, cities, urban filaments). Then, the picture's transformation—if *Black Square*'s agency can be used as an example of the territory's agency—represents the resilience of territories, the persistence of matrices and textures, the obduracy of technologies and infrastructures, the thickness and stratification of the territorial palimpsests also live and transform in time" (Paba and Perrone 2017, p. 259).

2.5.2 *Tenacious Cities*

In his reaction to the idea of the dissolution of places, Paba borrowed a crucial (not very well known) writing from Melvin Webber, who effectively pinpointed the image of the dissolution of places, the assertion of "community without propinquity," the city's disappearance (the Post-City Age), the domination of individual means of transport (The Joy of the Automobile), and the joyful triumph of the sprawling metropolis (*The Joy of Spread-City*) (Paba 2003, pp. 20–28).

This is an article with a surprising title, *Tenacious Cities*, a contribution from the end of the 1990s (Webbcr 1996). Weber does not deny his vision of the city, in particular, his positive judgment of the spread of settlements in space. Instead, he underlines how the importance of places and the inertia of cities have not disappeared. It is precisely to this aspect that he links his idea of a dialectical relationship between places and people, between stocks and flows, between streams and levees. Drawing on this dialectic between the tenacity of the city and the irrepressible development of urbanization processes, Paba grafted his notion of the territory: It is a place where collective action patiently builds the conditions to live in and produces community heritage while turning the interplay between the resistance of geo-historical matrices (on the one hand) and the persistent change of urbanization (on the other hand) in opportunities for a new territorial organization (Paba 2003).

2.5.3 The "Margin of Energy" and "Walking Through"

From this perspective, Paba relied once again on his usable extreme, Mumford, this time unearthing *The Conduct of Life* (1952), a book that according to Paba, reveals Mumford's thoughts on the human condition (according to Mumford, the human being is a self-fabricating, unfinished animal). Paba was intellectually intrigued by the Mumfordian concept of the "margin of energy." It is the creative, existential (underutilized) drive that belongs to human beings and defines the human condition and its potential for transformation. Geddes described it as the forward movement of life. This concept explains Mumford's notion of an active, reflective society that continually reconstructs its foundations, as human existence is an ongoing process of self-construction (which in the book is supported by a significant number of expressions such as self-examination, self-justifying, self-appraisal, self-fabrication, self-transcendence, self-understanding, self-willed, self-love, self-consciousness, self-regulation, self-direction, self-confidence, self-respect, self-deception, self-discovery, self-revelation, self-transformation, self-fulfillment, and self-reliance) (Paba 2003, pp. 9–15; Mumford 1952).

Paba conceived of his interpretation of the "territory" through this Mumfordian lens of the "self" and the "margin of energy": The territory is the result of an incessant emergence of energy from the communities that build relationships with the materiality of the places in which they live (Paba 2003, pp. 15–19).

Like Mumford, Paba made the journey from the neighborhood unit to the region and back. The neighborhood represents the active social presence of a micro-community in its settlement shell. The region becomes a polycentric system of self-sufficient communities. Paba fell in love with Mumford's neighborhood (1954, 1968), and constantly tried to bring his experiences of radical, engaged participation closer to the idea of the neighborhood developed by Mumford, a community made of the intertwining of social and physical aspects that daily reinvent collective life bases. Unlike Mumford, however, Paba hybridized the concept of the region with the Geddessian co-evolutionary model and described the region (as the most representative unit of the territory) as a polycentric, polynucleated system of interconnected, interdependent communities persistently engaged in a continuous process of self-construction. However, the elective territory of Paba has always been the city, especially his suburbs (peripheral neighborhoods), of which he constructs an exemplary unconventional description—*the second non-conventional description*—in his first book, *Luoghi comuni* (1998, p. 79). He describes the margin of energy of its inhabitants, who are often poor, marginalized migrants, and he identifies the local, situated potential for transformation, capable of generating a new balance between the resistance of places and the needs of new communities. This tension constantly accompanies Paba's research, actions, and experiences, and inspired his political and institutional commitment (see Sect. 2.6).

Paba was passionate about and refined a real "art of looking at the city" as mental training before exploring the notion of the territory. In this exercise, he relief on a "usable Geddes" (unlike the cumbersome character of literature), built this time by Ferraro (1998), an important Italian scholar who devoted his research to Patrick Geddes' work in India. Paba borrowed Ferraro's reinterpretation of the Geddesian exercise of looking at and walking through the city as a survey tool of places. Building on this, Paba began to work on the idea of planning as a continuous, creative reflection on evolution, able to put back into circulation the otherwise oppressive heritage of the past. The past then becomes a tradition voluntarily chosen and made available for further journeys, and to feed thoughts with the hope of the future (Ferraro 1998, 117; Paba 1998).

2.5.4 The Militant University

The very idea of the materiality of the territory and the legacy of a "usable" Patrick Geddes (the biosocial program, the "walking through" and the agency of non-human organisms) form the basis on which Paba operated a retrospective reconstruction of the cultural background of the Italian "Territorialist School" of Magnaghi (2000, 2010).

Paba himself was a protagonist of this school. He helped to develop the concept of the "urban village," which took the center stage of territorialist philosophy (2010). In *Luoghi comuni* (1998), *Movimenti urbani* (2003), and *Corpi urbani* (2010), Paba laid the foundations for an articulated reflection on the notion of the "militant university." This is the definition Paba attributed to the Territorialist School (see also Paba 2011a), emphasizing common intersections and traditions between the Italian Territorialist School, Mumford's regional thinking, and the Geddesian approach, being fundamental in encouraging the subsequent encounter between Magnaghian and Geddesian thought, which helped to consolidate the explicit link between Geddes and Magnaghi (in conferences and writings under publication). In particular, Paba identified the following aspects of the Mumfordian-Geddesian influence on the Territorialist School: identity and self-sustainable local development; a new local/global relationship (as an extension of the *Think Global, Act Local* formulation, attributed to Geddes); interactive planning; the relationships between identity and differences; and the bioregional vision (as the city/territory relationship). Paba's search for roots and intersections between Magnaghian philosophy and the "masters of planning" continued in a game of cross-references that can only be hinted at here (very incompletely and merely evoking): the polycentric urban bioregion recalls Benton MacKaye's regionalism; the ecomuseums develop the tradition of the Outlook Tower, the Regional Museum, and the Civic Galleries of Geddes; the ecological and solidarity gardens proposed on the outskirts of the city evoke the recovery of agricultural terraces on the slopes of Edinburgh; attention to the role of women, migrants, children, and in general the theme of the city of differences recalls the motto "adapting the city to child life" and "Sunday talks with my children" by Geddes; planning for different phases of Mumford's life; the revival of these themes in the works of Colin Ward; the Magnaghi's urban village, which takes up some suggestions of the neighborhood unit of Clarence Perry and Henry Wright; and the tradition of community planning (Paba 2011a, p. 12).

Above all, Paba underlined the creation of a school in the proper sense, embodied in university and non-university teaching courses, as a crucial dimension of territorialist thought, just like Geddesian philosophy and its founding activities of schools and universities (i.e., the Collège des Écossais of Montpellier in 1924, and the University Extension Movement, launched in 1887).

Territorialist thought embraces the school (whether institutional or not) as an instrument of a militant thinking in which it recognizes itself. In this sense, Paba defined the Territorialist School as a militant university made of laboratories and universities (such as the Laboratory of Ecological Design of Settlements, established in 1994, and the University of Empoli—a branch of the University of Florence, Italy—in 2001). Both Geddes and Magnaghi share the attempt—not realized (or only partially in the case of the Territorialist School)—to create a polycentric, regional university system, based on local funding and on the active role of cities and communities.

In this sense, Paba considered the Geddesian conception of the relationship between the university and the territory as an important part of the usable past that can be put back to work: a root, which has already produced a small plant. At the same time, it is a horizon, which can guide the path of the school in the future: "the seed, whose period of germination is followed in due season by leafage, flower and fruit" (Paba 2011a, p. 14).

2.6 Paba's Legacy and Contribution to Critical Planning

Paba's "message"—expressed in his books and numerous other writings published throughout his life (about 200 in total)—is varied, sometimes complicated, and not always exposed with a linear coherence. However, the nature of his contributions reflects the urgency to approach the discipline of planning with critical thought, always alert, never too convinced of the mission of thought, always challenged and challenging, involving the debtor and curious interpreter of the roots of the discipline.

This chapter is not able to comprehend the richness of Paba's contributions, nested in the humility of his work, never imposed, always offered, and generously disseminated, even when this did not suit his academic career. Each reader will be able to grasp the most appropriate starting point for their questions; in this sense, the constellation metaphor is useful. Each individual will identify his/her own figure in the dust of ideas and suggestions scattered across Paba's writings.

However, it is important to conclude this text, with some particularly identifying contents of Paba's radical, critical approach (the *usable* Paba) to the city as a material "body" and laboratory of collective projects in which the streets, in its concrete and political dimension (streets as democratic streets), assume perhaps the most important role in Paba's re-education to look at the city (Paba 2016).

At the core of Paba's thought is the idea of a radical interactive planning, difference-sensitive, shaped based on the plural urban bodies' needs and desires. Therefore, it recognizes diversity (gender, religion, lifestyle and consumption, sexual orientation, conditions of physical and mental health, geographical and cultural origins), and stresses the social practices of self-organized, creative insurgency, which aim toward happiness (place prosperity) and overcoming poverty and inequality.

From this angle, Paba's most important contribution is recognizing children as holders of rights in the proper sense. This implies also to overcome a concept of childhood seen as a social construction of the adults, and as a constraint, a forced reduction of humanity, competence, and responsibility. Paba helped to animate the "urban revolt of children," silent but powerful, which participates in the struggle for the rights of all "the different," "the diminished," and "the defective" of complete citizenship (Paba, 1998, 2003, 2010; Paba and Pecoriello 2006). He dealt with children as one of the thousand forms of subversive insurgent citizenship, which in the abstract circuits of the widespread metropolis, try to impose their collective bodies' materiality and radicalism.

Paba's radical interactive planning is also anticipatory and inclusive of non-human and other human bodies. In this opening, Paba anticipated the critique of the Anthropocene and built a dialog with the late Latour, who recognized in "the Terrestrial (in Paba's broad sense, the territory) the place for the reconstruction of a new relationship between territorial roots and planetary dynamics, therefore, the only way out of the rebellion of the earth and the effects of the climate and environmental crises" (Latour 2015; Paba 2018a).

Paba's thesis is that only planning models based on critical and radical interactions, "social conversation," and support for collective experiences of self-organization are today able to interpret the intertwining of trajectories and aspirations of the multitude of urban bodies (both human and non-human).

For this reason, Paba became passionate about the disquieted, creative strand of planning (his usable past), which he considered anarchic and dissatisfied, unsettling, open to social interactions, and sensitive to the materiality of territory and bodies; in one phrase: *dissenting imaginations of planning*. Paba borrowed this definition from the Indian scholar Chettiyappan Visvanathan, who used this term to define Mumford's and Geddes' more radical works (2003, p. 74).

The theme of involving local inhabitants in the construction of the city is, therefore, crucial to Paba's reflection, in dialog with the dissenting Mumford, who theorized the reabsorption of government by citizens as a way to pursue well-being and the fulfillment of the right to housing. He referred to self-managed neighborhoods and social community experiments (Mumford 1952).

Paba reworked this Mumfordian vision. In his view, the realization of the right to housing and the achievement of well-being depend on *place prosperity* (not just on *people prosperity*). He clarified this vision by contrasting two forms of happiness: context-free and context-specific (Paba 2010, 2012a). The first depends on ourselves, perhaps on our relationship with the cosmos or with god. Consequently, collective happiness is simply the sum of individual happiness. The second depends on the context, even on the qualities of the places. Paba stressed the second one, which he considered urban and collective. He intertwined this idea of happiness with his notion of radical planning, whose objective cannot fail to coincide with the well-being of people in their own places in which to live.

Finally, Paba's relationship with regionalist thought is radical and critical. He fell in love with it, not so much as a dictate of mainstream planning with particular reference to some Italian experiences of the 1980s and 1990s—which he rejected in favor of the restless, dissenting vein of his usable Geddes, Mumford, and MacKaye—but as a tangible way to link diverse parts, places, cultural and natural landscapes, and physically joining the city to the wilderness (thus rethinking the metropolitan and post-metropolitan model) (Paba and Perrone 2017). From this perspective, he reconceptualized the relationship between agriculture and community, planning and spontaneity, civilization and wilderness, culture, and solidarity. Benton MacKaye's regional philosophy, expressed in the Appalachian Trail and the Wilderness Society, inspired Paba's thinking and practices in many ways as a planner (not just as a scholar).

Paba reworked the regionalist philosophy of these masters of planning as a tangible way to connect diverse parts of places and to physically join the city to the wilderness (thus establishing the foundation for rethinking the metropolitan and post-metropolitan model) (Paba and Perrone 2017). From this perspective, he rethought the relationship between agriculture and community, planning and spontaneity, civilization and wilderness, culture, and solidarity. Benton MacKaye's regional philosophy, expressed in the Appalachian Trail and the Wilderness Society, inspired Paba's thinking and practices in many ways as a planner (not just as a scholar).

Overall, Paba's legacy stands on a particularly dense, innovative meaning of critical planning. It includes a number of adjectives that denote this approach as radical, dissenting, interactive, insurgent, biosocial, and multi-agent (human and non-human). It transforms the tenacity and obduracy of cities and territories into resources for new communities to find/produce their own place to live in.

In general, at the root of Paba's human and scientific contributions, there are two tensions: (1) his tireless drive toward radicalizing problems and solutions, to take another step forward each time, toward unpredictable and fascinating horizons; and

(2) the ethical urgency of an ongoing verification of ideas by developing social participation. These radical tensions represent the conceptual and practical background of his perennial pursuit of happiness.

2.7 Unveiling a Radical Planner's Biography

Constantly and restlessly fostering radical change in society and cities, throughout his lifetime Giancarlo Paba, Professor of Town Planning at the University of Florence, since 1994 (where he began his employment in 1973), was, first of all, an insurgent, energetic and mindful intellectual, tormented by the search for happiness (of others, the city, the wider area).

He expressed his cultural and social endeavor with keen intelligence, elegance, and efficacy, at times translating it into uncomfortable truths.

His intellectual life mixed radically and critically with political engagement. In the same way, his academic career was entirely filled by a necessary and virtuous combination of research and action; action that he carried out in a militant, convinced, and courageous manner.

In this tangle of life and research, his inevitable aim was to get to the root of the problems and the solutions, to discover new terrains of "exact" action (as he loved to say while remembering Danilo Dolci), through social participation. His favored terrain in the field of research-action was participatory planning (or interactive planning as he himself would define it later on).

His experiences of interactive planning and community building—with university research and as personal civic engagement—made him a pioneer, as did his research on the condition of childhood in cities which he carried out in the role of consultant at the Istituto degli Innocenti in Florence, and then as a promoter of the Associazione "Città Bambina" (as of 2006).

The initial adventure with the political group "Potere Operaio", which he shared with many friends, gave rise to his first attempt to build a scientific interpretation of the city and regional transformations as the upshot of the conflict between capital and work. This effort led to his work with the journal "Quaderni del Territorio". Later on, in the mid-1980s, Giancarlo helped to build the "Territorialist School", consolidated in the research-action work of the 1990s, with the foundation of LaPEI (University of Florence Laboratory for the Ecological Design of Settlements) which became the platform for numerous research projects of national interest (CNR, MIUR, PRIN) from 1985 to 2011. In 2009, this ongoing work led to the foundation of the "Società dei Territorialisti e delle Territorialiste" (Territorialist Society), of whose advisory board Giancarlo was part, and lastly, in 2014, the idea for the University of Florence Critical Planning and Design Laboratory, inspired by critical social theory.

During his academic career, Giancarlo was the founder, promoter, and member of the academic advisory boards of institutions that combined an academic thrust, research activities, and civic engagement. Among these, he played a convinced and "transformative" role in the International Network for Urban Research and

Action (INURA), which formed in 1991, giving rise to field investigations and study seminars in different cities around the world in the following decades.

The vision of a militant university inspired Giancarlo to create the academic and teaching structure of the Planning courses in Empoli, alongside the Rete del Nuovo Municipio experience in the 2000s.

His institutional career was marked by other important governance and management roles, which he always carried out untiringly and creatively.

At the same time, Giancarlo Paba carried out important research and teaching activities at an international level (fellow scholar, visiting professor, lectures, and conferences in many universities around the world). In 1986/87 he was called to Tokyo-no-daigaku, Kyoyo Gakubu (University of Tokyo, College of Arts and Sciences), with funding from the Nihon Gakujutsu Shinko-ka (Japan Society for the Promotion of Science). This was the start of a thirty-year partnership with Japanese research institutions and universities which resulted in numerous publications on the Japanese landscape and urban culture. Not only that, he also gained an uncommon intellectual, human, and visual sensitivity for "other cultures", which enabled him to contribute to the dialog with the Orient of some important Florentine institutions such as the "Gabinetto Vieusseux".

Other international experiences saw his involvement in European and international projects from 1998 to 2003.

Giancarlo Paba also played an important role in promoting the divulgation of academic research results. He was director of the journal Contesti, published by FUP, and a member of the editorial boards of academic journals and book series (Quaderni del territorio, Plurimondi, the FUP "Territori" series, the SdT journal of Studi Territoriali, Fondazione Michelucci's La Nuova Città).

His engagement with the Fondazione Michelucci was another important strand in the life of Giancarlo Paba, who, first a researcher at the Fondazione at the end of the 1990s and then a member of the academic advisory board, became its chair in 2012. He was engaged on that crest where the Foundation was balanced, relating society and built space (as he himself loved to remember): "one front involves the activities linked to space (houses, architecture, towns), and the other is directed toward society (people, communities, institutions)".

This was the crest of all of Giancarlo's human, political and academic engagement in the three overarching dimensions of his substantive life, which sometimes shunned the limelight, and was always accompanied by an intelligent degree of understatement. On the academic front, he cultivated the dialectical nature and complexity of the relationship between physical space and urban movements, social ecologies and "insurgent corporalities" in the city, in dialog with the legacy of Patrick Geddes and the American regionalist movement; he remained loyal and at the same time critical in his political engagement; he was a pioneering innovator in the field of social action and associationism.

References

Appadurai A (1996) Modernity at large, cultural dimensions of globalization. Minnesota Press, Minneapolis, London

Balducci A, Lorenzo R, Forester J (2020) Giancarlo Paba's map: a remembrance. Plann Theor Pract 21(3):495–499
Batty M, Marshall S (2009) The evolution of cities: Geddes, Abercrombie and the new physicalism. Town Plann Rev 80(6):551–574
Blake CN (1990) Beloved community: the cultural criticism of Randolph Bourne, Van Wyck Brooks. The University of North Carolina Press, Chapel Hill, London, Waldo Frank & Lewis Mumford
Brenner N (2004) New state spaces: urban governance and the rescaling statehood. Oxford University Press, Oxford
Brenner N, Schmid C (2015) Towards a new epistemology of the urban? City 19(2–3):151–182
Chambers R (2005) Ideas for development. Earthscan, London
De Landa M (2000) A thousand years of nonlinear history. Swerve Edition, New York
Dolci D (1962) Spreco. Documenti e inchieste su alcuni aspetti dello spreco nella Sicilia occidentale. Einaudi, Torino
Duden B (1994) Il corpo della donna come luogo pubblico. Bollati Boringhieri, Torino
Fainstein S (2011) The just city. Cornell University Press, Ithaca and London
Fainstein S, Servon LJ (eds) (2005) Gender planning. Rutgers University Press, New Brunswick, New Jersey, London, A reader
Fanfani D, Ruiz AM (eds) (2020) Bioregional planning and design: volume I—perspectives on a transitional century. Springer, Cham (Suisse)
Fenster T (1999) Gender, planning and human rights. Routledge, London
Ferraro G (1998) Rieducazione alla speranza. Patrick Geddes in India, 1914–1924. Jaca Book, Milano, 1998
Forester J (1989) Planning in the face of power. University of California Press, Berkeley
Forester J (1993) Critical theory, public policy and planning practice: toward a critical pragmatism. State University of New York Press, Albany (US)
Friedmann J (1987) Planning in the public domain: from knowledge to action. Princeton University Press, Princeton (US)
Friedmann J (1999) Claiming rights: citizenship and the spaces of democracy. Plurimondi 1:287–303
Hall P, Ward C (1998) Sociable cities: the legacy of Ebenezer Howard. Wiley, Chichester, New York
Haraway DJ (2008) When species meet. University of Minnesota Press, Minnapolis
Healey P (1993) The communicative turn in planning theory. In: Fischer F, Forester J (eds) The argumentative turn in policy analysis and planning. Duke University Press, Durham & London, pp 233–253
Healey P (1997) Collaborative planning in fragmented societies. MacMillan Press, London
Heynen N (2017) Urban political ecology. People, the Earth, Environment and Technology, American Cancer Society, International Encyclopedia of Geography, pp 1–9
Heynen N, Kaika M, Erik Swyngedouw E (2006) (eds) In the nature of cities: urban political ecology and the politics of urban metabolism. Taylor & Francis
Holston J (1996) Spaces of insurgent citizenship. Archit Des Profile 124:54–59
Hommels A (2005) Studying obduracy in the city: toward a productive fusion between technology studies and urban studies. Sci Technol Human Values 30(3):323–351. https://doi.org/10.1177/0162243904271759
Innes JE (1995) Planning theory's emerging paradigm: communicative action and interactive practice. J Plann Educ Res 14(3):183–189
Kaika M (2005) City of flows: modernity, nature, and the city. Routledge
Keil R (2003) Urban political ecology 1. Urban Geogr 24(8):723–738
Keil R (2005) Progress report (2): urban political ecology. Urban Geogr 26(7):640–651
Keil R (2011) Transnational urban political ecology: health and infrastructure in the unbounded city. The new Blackwell companion to the city. Wiley-Blackwell, London, pp 713–725
Keil R (2018) Suburban planet: making the world urban from the outside. Polity Press, Cambridge-Milford, USA
Keil R (2020) Urban political ecology for a world of cities. Urban Stud 5(11):2357–237

Kymlicka W (1996) Multicultural citizenship: a libera theory of minority rights. Oxford University Press, Oxford
Latour B (2015) Face à Gaïa: Huit conférences sur le Nouveau Régime Climatique, La Découverte, Paris
Latour B (2017) Down to earth. Wiley, New York
MacKaye B (1921) An Appalachian trail: a project in regional planning. J Am Inst Architects 9:325–330
MacKaye B (1928) The new exploration: a philosophy of regional planning. Harcourt, Brace and Company, New York
MacKaye B (1968) From geography to geotechnics. University of Illinois Press, Urbana
Mumford L (1949) Planning for the phases of life. Town Plann Rev 20(1):6–16
Magnaghi A (2000) Il progetto locale, Bollati Boringhieri, Torino, trad. en The urban village: a charter for democracy and local self-sustainable development. Zed Books, London
Magnaghi A (2020) Il principio territorial. Bollati Boringhieri, Torino
Marcuse P (1982) Building housing theory: notes on some recent work. Int J Urban Reg Res 6(1):115–121
Metzger J (2018) Postpolitics and planning. In: Gunder M, Madanipur A, Watson S (eds) The Routledge handbook of planning theory. Routledge, New York and London, pp 180–193
Mumford L (1950) Mumford on Geddes. Archit Rev 108:83
Mumford L (1952) The conduct of life. Secker & Warburg, Londra
Muratori S (1967) Civiltà e territorio. Centro Studi di Storia Urbanistica, Rome
Mumford L (1968) The urban prospect. Secker & Warburg, London
Olsson G (2007) Abysmal: a critique of cartographic reason. University of Chicago Press, Chicago
Paba G (1998) Luoghi comuni: la città come laboratorio di progetti collettivi. FrancoAngeli, Milano
Paba G (2002) (ed) Insurgent city. Racconti e geografie di un'altra Firenze, Mediaprint, Livorno
Paba G (2003) Movimenti urbani: pratiche di costruzione sociale della città. FrancoAngeli, Milano
Paba G (2010) Corpi urbani: differenze, interazioni, politiche, FrancoAngeli
Paba G (2011a) "Militan university": tradizioni e intersezioni della scuola territorialista. Contesti 2:7–14
Paba G (2011b) Le cose (che) contano: nuovi orizzonti di *agency* nella pianificazione del territorio. Crios. Critica Degli Ordinamenti Spaziali 1:67–80
Paba G (2012a) Felicità e territorio. Benessere e qualità della vita nella città e nell'ambiente. In: Magnaghi A (ed) Il territorio bene commune. Firenze University Press, Firenze, pp 33–56
Paba G (2012b) Democratic water: from the "Watershed democracy" to the participatory river planning. In: Takeshi I (ed) Conference proceedings agenda for sustainable asian cities and regions. Graduate School of Engineering, pp 3–11
Paba G (2013) Body, architecture and city. In: Perrone C (ed) Conference proceedings living landscape landscape for living: policies, practices, images. Planum. J Urbanism 2(27):1–7 (Sect. 6)
Paba G (2014) Amicizia e pianificazione. I sentieri di Clarence Stein e Benton MacKaye. (ibidem) le letture di Planum 3/2104. Planum 2(29):18–22
Paba G (2016) Democratic Streets. Contesti 1–2:6–17
Paba G (2018a) La ribellione della terra e il terrestre come orizzonte: osservazioni e letture. In: Perrone C, Paba G (eds) Confini, movimenti, luoghi: politiche e progetti per città e territori in transizione. Donzelli, Roma, pp 105–116
Paba G (2018b) How planning practices have changed (some) Italian urban planners. In: De Leo D, Forester J (eds) Reimagining planning: how Italian urban planners are changing planning practices. INU Edizioni, pp 250–257
Paba G (2019) Geddes, Patrick. In: Orum A (ed) Wiley Blackwell encyclopedia of urban and regional studies (15). Wiley. https://doi.org/10.1002/9781118568446.eurs0113
Paba G, Perrone C (2004) I bambini e la città: esploratori, progettisti, costruttori. Rassegna Bibliografica. Infanzia e Adolescenza, vol 5, pp 3–28

Paba G, Pecoriello AL (eds) (2006) La Città Bambina. Esperienze di progettazione partecipata nelle scuole, Masso delle Fate, Signa, Firenze
Paba G, Perrone C (2017) Place matters: spatial implications of post-metropolitan transition. In: Balducci A, Fedeli V, Curci F (eds) Post-metropolitan territories: looking for a new urbanity. Routledge, New York, pp 256–265
Paba G, Perrone C, Lucchesi F, Zetti I (2017) Territory matters: a regional portrait of florence and tuscany. In: Balducci A, Fedeli V, Curci F (eds) Post-metropolitan territories: looking for a new urbanity. Routledge, New York, pp 95–116
Parekh B (2000) Rethinking multiculturalism, cultural diversity and political theory. Harvard University Press, Cambridge, Massachusetts
Qadeer M (1997) Pluralistic planning for multicultural cities: Canadian practice. J Am Plann Assoc 63(4):481–494
Sachs W (1999) Planet dialectics: explorations in environment and development. Zed Books, London
Sachs W (2002) Development and global environmental change. In: Timmerman P (ed) Encyclopedia of global environmental change, vol 5. Wiley, Chichester [u.a.], pp 150–162
Sandercock L (1998) Towards Cosmopolis: planning for multicultural cities. John Wiley and Sons, Chichester, New York
Sassen S (1996) Losing control?: sovereignty in the age of globalization. Columbia University Press, New York
Sassen S (1999) Globalization and its discontents. Essays on the new mobility of people and money. New Press, New York
Soja EW (2011) Regional urbanization and the end of the metropolis era. In: Bridge G, Watson S (eds) New companion to the city. Wiley-Blackwell, Chichester
Sutherland D (2009) Education as an agent of social evolution: the educational projects of Patrick Geddes in Late-Victorian Scotland. Hist Educ 38(3):349–365
Swyngedouw E (1996) The city as a hybrid: on nature, society and Cyborg urbanization. Capitalism Nat Socialism 2(7):65–80
Swyngedouw E (1997) Power, nature, and the city. The conquest of water and the political ecology of urbanization in Guayaquil, Ecuador: 1880–1990. Environ Plann A Econ Space 29(2):311–22
Swyngedouw E (2004) Social power and the urbanization of water: flows of power. Oxford University Press, Oxford
UN-Habitat (United Nations Human Settlement Programme) (1996) An urbanizing world: global report on human settlements. Oxford University Press for UN-Habitat, Oxford
Valentine G (1996) Children should be seen and not heard: the production and transgression of adults' public space. Urban Geogr 17:205–220
Valentine G (2002) Social geographies: space & society. Prentice Hall, Edinburgh
Visvanathan S (2003) Cultural encounters and the orient: a study in the politics of knowledge. Diogenes 50(4):69–81
Wacquant L (1991) Influences: voices of creative dissent: quiet voices of dissent. Green Books
Wacquant L (2004) Body and soul: ethnographic notebooks of an apprentice-boxer. Oxford University Press, New York
Wacquant L (2008) The body, The Ghetto. The Penal State. Qual Sociol 32(1):101–129
Ward C (1973) Anarchy in action George Allen & Unwin and Freedom Press (UK). Harper & Row (US)
Ward C (1978) The child in the city
Ward C (1988) The child in the country (society today). Robert Hale, London
Webber M (1996) Tenacious cities. In: Proceedings of the NCGIA sponsored research conference on spatial technologies, geographic information, and the city. Baltimore, pp 214–218
Wilson J, Swyngedouw E (ed) (2015) The post-political and its discontents. Spaces of Depoliticisation, Spectres of Radical Politics, Edinburgh University Press, Edinburgh
Young IM (1990) Justice and politics of difference. Princeton University Press, Princeton

Camilla Perrone, PhD in Urban, Regional and Environmental Design (2002), is an Associate Professor of Urban and Regional Planning at the University of Florence (UniFi) Italy. Her research interests include critical planning, interactive design and urban governance, urban diversity, multi-scalar urbanization, urban political ecology and self-organising city. She is a member of the International Network for Urban Research and Action (INURA). Her publications: "DiverCity" (FrancoAngeli 2010); "Space Matters" (with Giancarlo Paba), in A. Balducci et. al., Post-Metropolitan Territories (Routledge 2017), "Città in movimento" (Guerini, forthcoming). Perrone is also the editor (with Giancarlo Paba) of "Critical Planning & design" (Springer, 2020). She is the Founding Director of the Research Laboratory of Critical Planning and Design at the University of Florence.

Chapter 3
Ildefonso Cerdà, *Teoría General de la Urbanización*, 1867: An Innovative Approach

Luigi Mazza

Abstract This contribution critically re-reads *Teoría General de la Urbanización* written by Ildefonso Cerdà in 1867, admitting admit that the great lexicon of this "theory" has not left any identifiable trace in urban studies and planning. However, this approach was innovative and still useful for planning that aspires to respond to a social objective of equity and spatial justice. This contribution focuses in particular on two dimensions of this theory; the continuous underlining of the social dimension and the insistence on the relationship between the ways of live and move, and the continuity established between plan and construction. An originality which appears again in the dynamic is consideration of urban phenomenon. Above all, this contribution tries to critically respond to the crucial question whether Cerda's planning was comprehensive or not, exploring the continuous relation general/particular. In conclusion, drawing on other scholar's comments, the contribution suggests that may be the continuous tension between individual freedom and general coordination is the most significant character of Cerdà's theory and work.

Keywords Ildefonso Cerdà · Squared Network · Equity · Justice · Urban theory · Rule · Barcelona · Madrid

3.1 Diffusion of Cerdà's Works

Cerdà is usually known for two reasons: for being the author of Barcelona plan and for having written a big volume entitled the *Theory of Urbanization, Teoría General de la Urbanizacion, y aplicación de sus principios y doctrinas a la Reforma y Ensanche de Barcelona*, published in fac-simile in 1971. In 1971 many unpublished works are found, and we uncover that Cerdà worked a lot in particular: the "Teoría de la Construcción de las Ciudades", and the *Teoria de la Construcción de las Ciudades. Cerdà y Barcelona* and the *Teoría de la Viabilidad Urbana. Cerdà y Madrid*, originally published in 1859 and in 1861. In short, Barcelona and Madrid come before

L. Mazza (✉)
Department of Architecture and Urban Studies, Politecnico di Milano, Milan, Italy
e-mail: luigi.mazza@polimi.it

C. Perrone (ed.), *Critical Planning and Design*, The Urban Book Series,
https://doi.org/10.1007/978-3-030-93107-0_3

of eight and five years, compared to the Teoria General, but only twenty years later scholars can read them. It is proper to think that this lack influenced cerdian studies, not only the Spanish ones.

For instance, a selection of the *Teoria General de la Urbanizacion* is translated into French in 1979, edited by Antonio Lopez de Aberasturi, with a preface by Francoise Choay. In 1980 her book "*The rule and the Model*, a large part of Chapter 6, The theory of planning" is focused on Cerdà, despite she could not read the books on Barcelona and Madrid.

We are today in condition of analysing Cerdà's thought in the temporal order of the original publications and possibly of the development of his thought. I will consider Cerdà's works according to the original order of publication.

In the meantime, a note: the Spanish term "reforma" has at least three corresponding terms: reform, restoration, alteration. Even if restoration is the technical term, I preferred reform which seemed to be the more inclusive of the most inclusive of the theoretical and practical objectives of Cerdà (and, I admit, the most pleasant from a phonetic point of view).

3.2 The Theory of Cities Construction

The first volume to consider is the Teoría de la Construcción de las Ciudades *y Teoría General de la Urbanizacion y aplicación de sus principios y doctrinas a la reforma y ensanche de Barcelona*, published in 1859.

The volume on Barcelona is organized in three main documents: *Memoria, Teoria de la construction de las ciudades*, and the *Economic Plan*. In the three documents almost all themes distinctive of Cerdà appear, some of them will be deepened in the Theory of viability which will add some new ones.

The *Memoria*, descriptive of the preparatory project of Barcelona expansion, collects the results of the studies on the geographical characters (climatic, demographic, etc.) of the town and is opened by the description of the regrettable sanitary conditions of the people, of the epidemic and infections suffered since XIV century. A significative beginning because it marks the importance which has and will have for Cerdà the hygienic question; a classic question for the reform processes of '800 and it is here exalted by the particular conditions of Barcelona, strangled by the walls. The quality and circulation of air, the exposure to the sun, the hot and cold, the humidity of streets and dwellings, the poor supply of water, all problems heightened by the excessive densities and crowding of the town.

The hygienic problems are a thread running through all cerdian work and are linked to the form of the street network, to the wide of the streets and their ventilation. About the street network and the problem of the healthiness, the problem of functionality is added. Cerdà does not fail to mark that the project of the network cannot be considered an artistic problem of "variety and whim" but has to be traced back to the directions of the healthiest winds in addition to the directions according to which do exist and are expected the largest circulation of goods and people.

To these problems the dwelling theme is added, not only a hygienic one, but also a social and economic one. The lack of the offer of urban land and the building costs affect the final price and the lease, producing hard trouble for the poorest population. Round the dwelling issue face conflicting interests which we need to identify to maintain the balance between any types of interest in the game. This is another question recurring in the whole work of Cerdà.

These main themes of the plan project are continuously associated to the questions of equity and justice.

The theory of cities construction, which follows the *Memoria*, opens with the known question of the lack of books on the city themes, which press Cerdà to devote himself only to these studies.

The project is composed of two parts: the analytic and the synthetic one.

After a review of the walls and their destruction questions, it develops and deepens the studies on Barcelona already started in the Memoria. In particular Cerdà expatiates upon the many technical, economic, social aspects of the dwelling question. Problems of air, light and space which prevent a life in conditions of healthiness, morality, decency and independency come back. Every year because of the conditions of artificial monopoly the number of owners of urban land decreases and rents increase, unlike other monopolies, urban property is not taxable or subject to other restrictions; more over to avoid the need of destruction of the wall, it is allowed to augment the number of storeys creating a second artificial monopoly increasing densities and a further worsening of hygienic conditions.

In the synthetic part we find again the themes of the analytic one in the perspective of a project which does not hesitate to treat the technical details even the most detailed. For instance, among the four criteria to respect about the street orientation, together with the direction of healthier winds, the direction of existent and foreseen movements, there are also the major or minor slope of the road, and in the end the easy discharging of the manholes. In all questions it proceeds from general to particular and vice versa, and at times it is not content with particulars, but it is looking for major details.

About street orientation the preference for N-E and E-O is justified through a very detailed analysis of the character of the winds and their daily frequency in the different seasons. On this basis it is concluded that the preferable street network is the orthogonal one, beside it happens in most American cities. According to the Theory of construction of cities, this is the origin of the famous board of Barcelona.

Against the objection of the excessive and presumed monotony of the board, Cerdà refutes that benefits gained when the streets follow the most convenient direction, blocks can have a good orientation, and houses which form them can have a good exposition to air and sunlight.

About the width of the streets Cerdà analyses the examples of numerous European and American, foreign towns. To satisfy the owners' rights, the most suitable solution would be an equal width of the streets, because it would guarantee to everybody a fair treatment and condition.

At this point Cerdà summarises the principles to be followed for the restructuration and expansion.

Principles for the restructuration. Restructuration of the town has the precondition of the wall demolition and of the link by main streets of action centres such as port and railway station. Other important streets must connect the districts together and with the main action centres. In this way it is remedied to the casual process of construction of the old city. The construction without rules has produced narrow and winding routes unable to satisfy the contemporary needs.

Principles for expansion. Expansion must be boundless preventing mistakes made in Paris for the opening of the Boulevard Rivoli, Strasbourg, Sebastopol. Present space has to be multiplied fivefold, and expansion will have no limits except near villages. Streets will consider winds and future circulation.

Expansion has to be considered as a whole from the viewpoint of local interests (agricultural, commercial, industrial) and of national interest to make easier the relations of the town with the whole nation and abroad, and therefore the port is a theme of upmost importance.

About the blocks, to have a good exposition would be rectangular, or better, squared, with the side not too long to avoid that a large part of surface cannot be built on. And because is not possible to build isolated houses, blocks have to be opened on two sides.

In the *Economic Plan* Cerdà considers implementation problems and makes a comparison between Barcelona and Paris where there is only one decision-maker and the public treasury anticipates allowing a rapid realization of projects, and where costs have to be paid according to equity, therefore who has a gain has to pay; a rule Cerdà quotes in many occasions and assumes for his project.

3.3 La Teoría de la Viabilidad Urbana y Reforma de la de Madrid

The theory of urban viability appears only two years later than the theory of cities construction. Note that the *Memory* 1855 concerns Barcelona expansion, the project for Barcelona 1859 is a restructuration and expansion project, while the Madrid project 1861 is a restructuration one, because the expansion plan was already prepared by Carlos Maria de Castro and approved in 1860.

The theory of urban viability consists of three parts: the first one devoted to technical and scientific studies, the second to the Economic Plan and the last one to the restructuration proposal for Madrid.

Reason for proceeding with the proposal are important: lack public spaces and piazzas, dwellings are old and poor, and viability is comfortless. But the public opinion is divided, and a conflict occurs between who is in favour of the expansion and who is in favour of a major use of the city within the walls.

Cerdà offers a solution arguing that there is no restructuration without expansion: expansion is needed for many reasons and in particular to offer residential spaces to people removed by demolitions.

According to Cerdà to give rise to discontent and a widespread uneasiness among the population derive mainly from the conditions in which are and are the houses and the cities.

In comparison with Barcelona studies, Madrid studies have some innovative characters.

Importance of historic knowledge that explains nature and reasons of the "geroglific" is underlined. Cerdà uses this term for describing the imprint of history on the urban land. In particular is advanced a reading of facts and events of construction linked to the street network for understanding the life and construction of city since the foundation.

The expert charged with intervening on the city was likened to a doctor caring with a very silk man (TVU, § 1055), and Cerdà offers to him a method for working. The first thing to do is to study the natural topography of the land to design the urban viability in the better conditions for its functioning. Before drawing the first line it is necessary to have a topographic study with contour lines, a hydrographic and geologic study.

To understand which results have been achieved and which mistakes have been done, Cerdà suggests doing a comparison between the natural and artificial topography of the city.

It is then necessary to have an exact idea of the centre or of the centres of movement and the life of people; and to know statistics, habits and practices and the consequent needs. Only then it will be possible to draw the geometric network of urban streets to form the skeleton of the city. The empty spaces of the skeleton will then be filled with the constructions.

Cerdà underlines values and role of the viability which is an essential element for the life of population. In the meantime, the viability "exerts a decisive, incalculable, immense influence on the public and public hygiene, on the administration, on the order and security till to the domestic economy. The viability is not certainly the life of a population, but it is the only one to show and make this life always working" (TVU, § 646).

In a theoretical section several geometric alternatives of network are offered to the expert. Radial and annular are the roots from which all the others derive from, in particular the rectangular and quadrangular model. It is curious how to construct them, an example of the recurrent attempt of Cerdà of giving a 'theoretical' foundation to the elements of the town he is analysing.

The general convenience demands that the means needed for an easy functioning, without too many difficulties, are left to the social and political, commercial and economic, industrial life. The quadrangular model has the inestimable profit of not creating hateful artificial preferences for no one of the streets, distributing the benefits of viability and edification among all the streets and blocks with equality and perfect justice.

It is sad that the quadrangular is guilty of a heavy uniformity and of an irritating equality, in a word of monotony. But the fundamental basis of proceeding of whatever expert has to be always and above all the justice, and justice demands, requires, imposes uniformity and equality, what fools call monotony.

Justice is always and above all equal and uniform; and in this sense there is in the world no major monotony than equality in front of the law, which no doubt nobody went so far as fighting on the basis of the concept of monotony (TVU, § 701).

Viability demands that restructuration and expansion join up to link the old street network to the new one, and the two harmonize among the internal and external interests of the town and link to the system of the great external viability. So, a plan of only expansion has to be considered uncomplete. Restructuration and expansion are indissoluble.

In conclusion, Madrid needs a restructuration and an expansion, to harmonize all interests internal and external to the town. After having investigated and analysed the centres of activity principal and lower have boldly proposed the viability internal system which in association with the external we believe to be the most adequate, within the limits of the topographically possible and economically implementable, to realize a high and great objective, a city capital worth of its future.

In the *Economic Plan* Cerdà underlines two themes. Even if respecting private property, we must not reach sacrifice to it social and humanitarian needs of urban reforms. It is administration concern to make use of compulsory purchase, justified more by the public utility, than by the public need. And because the compulsory purchase is a change of values that always favours private property, its resistance cannot be justified nor encouraged on the pretext of sacrificing that strictly speaking is not imposed on it.

No owner can rationally oppose the reform of Madrid. Based on experience three cases and three methods to solve them. 1° case: opening of a road in land without edification aimed to facilitate the way in and way out, light, air and view of the one or more owners' house or houses of the owner. 2° case: one or more owners with the administration approval open a road in their block and pay the costs. 3°case: is opened a road throughout different blocks aimed to a more comfortable viability and a minor density of a large town. In this case the principle is that who has benefits and utilities pay the costs.

3.4 The General Theory of Urbanization

The two volumes on Barcelona and Madrid collect all what Cerdà knows on the problems of cities, on methods and techniques to know and solve them, on principles and ethic and political convictions from which draws inspiration and on goals he wants pursue. In one word all that Cerdà knows and wants to communicate about urban planning for new towns and the old ones.

But Cerdà's route is not finished yet, on the contrary it is the most difficult task, the construction of a science capable of describing and motivating the characters of urbanization processes and identifying the laws for producing a perfect urbanization. These are the aims that Cerdà pursues in his last work the *Teoria General de la Urbanización y aplicación de sus principios y doctrinas a la reforma y ensanche de Barcelona.*

The theory is designed as a treatise divided in four volumes. The first one is focused on the study of urbanization "considered as a done". The second volume is "The urbanization as a concret done. The urban statistics of Barcelona".

The two volumes have respectively 800 and 700 pages of a large size, a true challenge to the reader, but it has to be said that the second volume is mainly a collection of statistic tables on the most different characters of the city of Barcelona. In the general plan of the work the third volume is devoted to the general principles for producing a perfect urbanization, and the fourth one to the restructuration of Barcelona, a concrete example of application of the theory. Of the four volumes, only the first and the second are available, the third and the fourth have not been drawn up.

3.4.1 Structure and Articulation of the First Volume

The first volume "the urbanization considered as a done" has an internal title: "First Part. The urbanization in general". This first part is divided in *books*, the books in sections, and when necessary the sections or the books are divided in *chapters*, and the chapters in paragraphs. An architecture rather complicated which the initial *Notice* complicates furtherly adding an alfa-numeric criterium of classification of the paragraphs which is not of easy interpretation (T25). The reader may also be confused from the fact that not always the titles of books and sections in the general index correspond exactly to the titles in the text; and by some discrepancies too, for example, in the two conclusive pages firstly writes that "the Second Part is devoted to the exposition of the theory, and the Third to the technicalities, to reserve the fourth to the development of the practical example applied to Barcelona (T 814), but in the last page writes that the foundation of an urban theory is covered in the Third Volume, in the Second the statistic which is a complement of the treatise of urbanization (T 815). It is possible that these and other contradictions, and a wordy and prolix, repetitive style are caused by Cerdà's wish of giving to his treatise the mark of a 'scientific' treatise.

The first volume is subdivided in four books. The book I. "Origins of urbanization" is shortest of the volume, its content is summarized by the titles of the two chapters that form it: "The need of dwelling is at the origins of urbanization" and "the sociability of man is the cause of development of the urbanization". In the book 2. "Development of urbanization" different forms of urbanization, elementary and combined, are proposed. The book 3 developed an analysis of the present urbanization. In the book 4, the reasons of the present urbanization are described.

The first page of the volume has an epigraph which seems summarise Cerdà's programme: "Independence of the individual in the home; independence of the home in the urb, independence of different type of movement in the urban streets. Ruralize the urban; urbanize the rural. … *Replete terram*".

The 'liberal' idea of independence is central, it is a tool of construction and articulation of the urban universe; to it is associated the general rule "ruralize the urban and urbanize the rural" applied to Barcelona; the biblical conclusion is a *captatio benevolentiae*, a recall to a superior authority in confirmation of the goodness of the program.

The volume opens addressing to "The reader" which begins with a short autobiografical outline: "Born in the first third of this century, when the Spanish society revealed itself rather linked to its old practices of inaction, I remember the strong impression I received from the application of steam to the manufacturing industry which, being still very young, I saw for the first time tested in Barcelona".

In this way breaks into the text the novelty of those years: steam and electricity offer unforeseen opportunities to a society suffering a widespread uneasiness without until now an adequate remedy has been proposed. Cerdà is fascinated by the opportunities that the application of steam and electricity offer to manufacturing industries and transports, but he is worried too because cities are not suited to host a new period of development. The uneasiness is lurking in cities and manifests itself in a continuous fight turned in particular to the obligations that cities oppose to the new demands of life and development. The great cities will be the place of development of this fight, no doubt exists about its end. A new civilization will transform radically the functioning of humanity. The problem is "eminently social", it is to manage in a way that the society knows and appraises the current changes and accept it.

Cerdà is surprised to find no trace of all this in Spanish and foreign books and then decides of leaving any other commitment and dedicate himself completely to the study of what is happening. Because of the social uneasiness emerges mainly in the cities, his research begins from the cities and in a more general way from the phenomenon of urbanization. Cerdà studies the urbanization in a historical perspective which allows him to analyse its characters by a work of "anatomical dissection".

Cerdà acknowledges the appearance of a new civilization, obstructed in its development by the characters of the great cities which instead of ease it, they shut and obstacle it. To remove these obstacles, he thinks necessary to investigate the nature of cities in the double dimension urban and social, to understand the machinery of historical and present urbanization to find principles and rules of a perfect urbanization and show the possibility and the benefits with a concrete application to Barcelona.

The *First Part* is closed underlining the importance of statistics to compensate for the lack of adequate studies on the urbanization; but before that Cerdà underlines that there are two tools of intervention; the first one more suitable for new foundations and expansions is the science, and the second more suitable for restructuration and urban transformation is based on intelligence and art, without forgetting the scientific principles.

3.4.2 Structure and Articulation of the Second Volume

The second volume is subdivided in five parts which are not numbered: the container, the content, the functioning, epilogue, and appendix. It stands out the principal distinction between the container, the physical city, and the content, the society, hosted by the container. Analytical data refer to Barcelona.

Of the container are distinguished the topographical and atmospheric circumstances; the component elements of the plan and elevation, street, blocks, buildings; the component elements of the town.

Of the content are indicated the characters of population composition in different years, and its dynamic (the law of continuity).

The functioning is expressed by the relationships between content and container, between population and viability, population density, mechanical motors and population.

The Epilogue contains a synthesis of the essence and value of the container and of the existence and burdens of the content. In the end, the Appendix, is the statistical monography on the working class of Barcelona.

3.4.3 New Words

The study of a subject completely new implies the use of concepts and new words which the author feels obliged to justify and qualify. To this aim he dedicates some pages of the Introduction (T 28–32), with the advertence that he will be back to them (Liber III, Chapter III, § VII), to explain "that with very detailed, etymological, analytical and philosophic" all the words used to define the term *urbanizacion*.

The word *ciudad* seems to him inadequate to express beyond the materiality of the city the life which develops inside it. He is struck by the Roman term *urbs* "syncope of *urbum* or plough with whom was traced out the external furrow" the opening of the plough "urbanized the enclosure and all what was inside it. That is the opening of the plough was a true *urbanizacion*; i.e. the act of converting a free and open field in *urbs* (T 30)."

These are the "philological reasons" which persuade Cerdà to adopt the word *urbanizacion* not only to denote whatever action aimed to group building and to rule its functioning in the group just formed, but also the set of principles, doctrines and rules which has to be applied so that building and its grouping far from repressing, perverting and corrupting physical, moral and intellectual faculties of the social man, serve to stimulate its development and strength and increase the individual welfare whose sum form the public happiness. For the same reasons I adopted the words *urbanizar, urbanizador* ... (T 30).

And a little further he confirms that the *urbanizacion* is "a set of knowledge, principles, doctrines and rules aimed to teach in which way has to be ordered a group of buildings so that can respond to the aim of make the inhabitants live comfortably and can help themselves mutually, contributing to common welfare (T 31).

Cerdà notes moreover that "the numerous words we have—ciudad, villa, pueblo, *lugar, aldea, feligresia, caseo, alqueria, quinta*—focus on the different hierarchy that form according to the number, size and extension, and in ancient times, according to the statute, privileges, prerogatives, etc.". Therefore "to state simply and clearly a group of buildings without any concrete relationship with the size and almost at all indifferent to the applications of the fundamental principles of the urbanizacion, ..." introduces a new term, the new term is *urbe*.

At this point I may attempt to sketch a synthesis.

3.5 A Synthesis' Attempt

Looking back, I think that we have to admit that the great lexicon of the General Theory of Urbanization has not left any identifiable trace in urban studies and planning.

But its approach was surely innovative at least for two reasons.

The *first* is the continuous underlining of the social dimension, "Para la urbanización, la forma es nada, la satisfacción cumplida y adecuada de las necesidades humanas es todo". (T 50). Social dimension is underlined too by the couple container/content with which analyses urbanization, and insistence on the report block/viability to which corresponds insistence on the relationship between the ways of live and move. An originality which appears again in the dynamic consideration of urban phenomenon.

However, more than results of his analytical project, it is his activity as a designer which recalls and incites our attention. Cerdà projects following a method which provide that at the beginning a systematic survey of information on the natural and artificial topography of the city. The comparison between the two topographies and a historical analysis of the city development allows to understand which has been successes and failures and what remains to do.

Then Cerdà projects according to a rule which has in the squared and in the squared network the definitive support of the urban project. The rule is coming from the conviction many times confirmed according to which, the network or viability system, is the first decisive choice of the plan, because the network structures through the blocks all building.

Not only but the quadrangular network has those spatial distinguishing features that guarantee equity and justice. The plan must draw inspiration from this rule, beside that expansion plan, the restructuration plan too has to be traced back to the rule in the way of joining to the expansion plan.

The rule has some exceptions identified in the need of direct link between the centres of action and between the suburbs.

Cerdà indicates a rule that we can consider still valid, and that we can translate into a disposition more general: the redaction of a plan demands the supply of a rule to which conform structure choices and then, ordinances to which conform the building choices.

Cerdà's rule is contemporarily an analytic and project tool. The application of the quadrangular network allows the interpretation of the existent city, noting the deviations from the structure considered optimal and suggesting the necessary transformations to reduce the deviations and if possible, to annul them. To the rule Cerdà admits exceptions only for the main streets.

Cerdà's rule is built on two basic convictions. The first one does exist a privileged orientation of the streets and blocks, which only the orthogonality gives the possibility of arranging the best micro climatic. The second conviction is that type of network which allows the most equality of treatment of the owners. Put it with Cerdà, hygiene and justice are at the basis of the choice.

It is not bizarre to think of this rule which, underline is, at the meantime, analytic and planning, still relevant. A century after Cerdà, another important analytic and planning contribution to the culture of planning, is that of Buchanan, *Traffic in Towns*, 1963. Buchanan states a method to plan the viability and through it, indirectly the town. For Buchanan too the concept of environmental area is an analytic and planning tool. Buchanan's rule is aimed to reduce times of routes and to put in contact, to use Cerdà' words, action centres. In Buchanan there is a segregative dimension which is extraneous to Cerdà, but the cares for the smooth flow of traffic are for some aspects common.

The *second dimension* of Cerdà still very present is the continuity established between plan and construction. The plan is articulated in blocks and these in buildings and open spaces; the buildings in plans and dwellings, and of dwellings are important distributive and constructive details: windows, doors, internal distribution, and services.

Cerdà's observations on this subject integrate the observations of the dwelling sociologist with those of the construction expert. The open spaces are courtyard and patios and communicate with dwellings through the stairs. A continuity that makes inseparable viability structure and constructive elements.

The plan is the unity of these ensemble, a project that moves continuously from the general to particular and vice versa. Shortly, the construction characters are not less important than the infrastructural frame. Cerdà would not understand neither acknowledge a distinction and a break between a general frame and construction details. His texts are a continuous mix of distributive and technical characters, sociology of family, building engineering and urban geography.

A comprehensive planning that of Cerdà? Yes and no, within the continuous relation general/particular, there are still many spaces of compositional freedom that founded on the structure can get different features.

Barcelona is a concrete example of the opportunity and of the results that is possible to get organizing a general plan and details. May be that the continuous tension between individual freedom and general coordination the most significant character of Cerdà.

These characters recall the almost contemporary experience in Paris. Cerdà quotes Paris even to regret that there is not in Spain a man alone as Napoleon III who can decide and implement the decision thanks to the support of Treasury and the financial tools organized by Haussmann, who however, do not forget it, in the end, caused his dismissal.

The reading of Haussmann reintroduces the essence of planning as land use control and as a comprehensive vision of the control at the urban and metropolitan scale. What is important is to regain a connection network that cover the whole town. The road condition of the town and the excellence points as the goal of road perspectives. Besides the capacity of moving from this network level to the details of the common parts, a singularity of design that strengthen the network, gives it individuality and strengthen and is strengthened by the construction control of buildings.

In other words, the control operation is tridimensional: the communication network, the form of public and common spaces, and the external aspect of buildings. This tridimensional control gives unity and individuality to the urban texture in Barcelona and to a certain extent in Paris.

3.6 Soria y Puig and Choay on Cerdà

About Cerdà have been written many essays, not a book on him. For the sake of brevity, I will consider two authors who, for different reasons, I think they are important—Arturo Soria y Puig and Françoise Choay—omitting others as, for example, Fernando de Teran, who made any way many interesting considerations.

In Spain Arturo Soria y Puig is considered the guardian of Cerdian orthodoxy; he edited the two volumes on Barcelona and Madrid and the one volume on Cerdà written in English, published in 1999, *Cerdà, The five base of the General Theory of Urbanization.*

I admit that Soria is not liked, even too because for years he developed a severe criticism to prevent that doctoral thesis were developed on Cerdà. Who wanted to write about Cerdà, to avoid his criticism, had to do it indirectly, treating lateral issues. Soria wrote the theory developed in the books on Barcelona and Madrid are the partial development and revisions of a same theory that definitive of The General Theory of Urbanization. This sentence implies that Cerdà had a general theory before the studies on Madrid and Barcelona. We have no evidence of it, while it is possible that the general theory, for its aspects properly "theoretical", has been the result of the previous experience and reflections.

Soria makes another operation rather questionable. In 1861 Cerdà published a short text: "About two words that Don Pascual de Uhagon sad to the owners of land included in the expansion zone of Madrid, still four words to the same owners by Don Ildefonso Cerdà". Soria extracts from the text the five base that Cerdà show as necessary to develop the work of expansion which are the technical, legal administrative and economic, and the base aimed to "harmonize the being with the must be". Less problematic is the essay published by Soria in 1955 in the TPR, where the base are four and are the necessary reference that a plan may become operational.

Much more interesting the contribution by Françoise Choay. Despite she is conditioned by the target of reducing Cerdà's contribution to the theoretical frame of her book, Choay offers many interesting critical ideas and it is a pity that she cannot read the materials published in 1991.

Choay's sentence that Cerdà's planning discourse "is normative and cannot belong if not in the average to any scientific practice" is certainly true, even if it can be noticed that Cerdà concludes the *General Theory* underlining that "two possible intervention tools; the first one, rightest for new foundations and expansions, is the science; the second, rightest for urban restorations and transformations, is based on intelligence and art, without forgetting scientific principles" (T 814). It may be noted that for Cerdà science is above all the notions he borrows from biology and anthropology completed by robust injections of good sense.

Notice again Choay that the "idea of a building or town structure may depend on a set of rational considerations which have a proper logic, marks the decisive break that demands the study of contemporary planning pass through the study of treatise of architecture, and consider these two categories of text as part of an equal set belonging to a common denomination" (1986, 20). A central remark to focus on the multiform contribution of Cerdà who, as we said, moves from small constructive details to metropolitan and regional scale.

Again Choay" no specific term does exist today for the planning studies aimed to offer a theory of organization of space (1986, 303)" and complains that the "founding character of the *Theory* has not been recognized ... this text did not have a direct posterity (1986, 305)" and historians like Lavedan ignore it.

In the end Choay offers numerous other traces, from Patte to Haussmann, to whom Cerdà himself refers, but on these themes another essay would be necessary.

I stop here, not before a revengeful notation: Cerdà is the most tedious author I ever read. To read him is a true penance. The pages flowing and engaging are very few: the short biographical outline I quoted, the notes on steam and electricity. For the remainder a wordy pompous repetitive. To summarize it is not easy, one have the impression of repeating always the same things, even when it is not so.

Post scriptum. In the picture by Ramon Martì Alsina, realized two years after the death of Cerdà (1878), Cerdà has a glance ironic and allusive so that rethinking to his history, suggested me an 'exchange' with the character of Don Chisciotte.

The Dulcinea of Cerdà is the science, pursued without a break and with renewed hope of arriving in the end to get it. In this exchange the loyal Sancho Panza is the reader who follow without any break Cerdà-Don Chisciotte in the long inexhaustible route of his texts.

References

Cerdà I (1859) Teoría de la Construcción de Ciudades (Vol I: Cerd° Y Barcelona; Vol II: Cerdà y Madrid), republished by Ministerío para las administraciones públicas, 1991, Madrid

Cerdà I (1861) Teoría de la Viabilidad Urbana y Reforma de la de Madrid (original edition)

Cerdà I (1968) Teoría general de la urbanización y aplicación de sus principios y doctrinas a la reforma y ensanche de Barcelona. Instituto de estudios fiscales, Barcelona (riproduzione in copia anastatica dell'edizione originale stampata a Madrid nel 1867, Imprenta Española, Torija, 14, Bajo)

Choay F (1997) The rule and the model. MIT Press, London, UK, Cambridge MA (original ed. 1986, La regolae il modello, Officina, Roma)
Soria Y, Puig A (1999) Cerda: The five bases of the general theory of urbanization. Gingko Pr Inc., Berkeley, CA

Luigi Mazza is a Professor Emeritus at Politecnico di Milano. He has been an active planner mainly in the urban planning field. He is also the founding editor of the *Planning Theory* journal, and a member of the board of the *Town Planning Review, European Planning Studies, DISP,* the *Planning Review, Spatium, International Review,* and *Planning Theory*, all of which have published his articles.

Chapter 4
Concrete Community and Territorial Principle in Adriano Olivetti's Thought

Alberto Magnaghi

Abstract Reflecting on two fundamental concepts of "concrete community" and "territorial principle" and their connections, the chapter traces back Adriano Olivetti's progress from the purely philanthropic culture of an "enlightened" entrepreneur to a practice of community self-government of territories which advocates for a general reform of the institutional hierarchy. Such progress, founded on a "holistic" conception of territory as a historical-social subject pre-existent to production relationships, and resulting in a political project in which the primary decision-making level resides in local territories, was interrupted by its defeat in 1958 elections; but this did not diminish its impressive anticipating power, covering fields like production governance, corporate responsibility, urban and regional planning, and theory and practice of democracy. And culminating in a territorial project which anticipates the spirit of integrated multi-sector districts and even the environmental care relationships typical of the "urban bioregion" perspective, interpreting the "life worlds" of populations as foundations of the entire socio-economic development model.

Keywords Adriano Olivetti · Concrete community · Territorial principle · Territorial self-government · Territorial project · Urban bioregion · Life worlds

Reference text for this essay is Adriano Olivetti's book *L'Ordine politico della Comunità* (The Political Order of Community) in its 2014 edition[1] (Olivetti 2014a, or. 1945).

Such work, completed during the Swiss exile (1942–1945), represents the philosophical, political, religious "*summa*" of Adriano Olivetti's thought and, around a communitarian hypothesis, proposes a complex institutional political system (with a detailed proposal for a constitutional reform) which proceeds from local community to state through all the intermediate institutional levels typical of a regional federal

[1] Edited by Davide Cadeddu and with no subtitle (unlike following editions). The essay is a concise re-elaboration of my reflections about Olivetti contained in: Magnaghi 2015; 2019; 2020.

A. Magnaghi (✉)
Department of Urban and Regional Planning, University of Florence, Florence, Italy
e-mail: alberto.magnaghi@unifi.it

C. Perrone (ed.), *Critical Planning and Design*, The Urban Book Series,
https://doi.org/10.1007/978-3-030-93107-0_4

state. In this system the "concrete community" (including the territories of several municipalities) is introduced as the political ground of the federal state.

With respect to the extensive literature developed around this text,[2] I will focus my comment on the importance given, within its political construction, to two founding concepts: the *concrete community* and the *territorial principle.*

Both concepts involve a centrality of the "territorial"[3] issue.

For Adriano Olivetti (at the same time a theorist of community federalism and one of Italy's foremost entrepreneurs) the "territorial" issue does not concern the "private" context of the big enterprise system, like in the historical experiences of production communities represented by the *workers' villages* of the early twentieth century[4]; territories of such villages are marked by a philanthropic culture and a policy of social integration (claimed e.g. by the *Werkbund* movement) in a life model for workers completely managed by the company through the villages, whose houses, in some cases, are equipped with spaces for domestic agriculture and breeding as well as of services like schools, swimming pools, libraries, parks, etc.; the *company town* is a model opposed to urban massification, but closed within the corporate system (even if sometimes, as in Valdagno, with the ambition of proposing urban models alternative to the relevant cities). Although Adriano's father, Camillo, founder of Olivetti, was closer to such models, for Adriano there is no link with them: the Olivettian model recognizes the territory belonging to the company's workers in its *pre-existence* as a historical community, recognizes its cultural and identity complexity and socio-productive autonomy from which the dialectical relationship with the company emanates. More: realizing the disproportion between the attractive power of the big Olivetti company and the decision-making abilities of the local territories, assumed as a foundation in his political model, Adriano goes so far as to activate territorial actions and projects to diminish the enterprise centrality and weight: to give substance to his "concrete community" project, founded on the primacy of the "territorial principle", he invests a lot of money, projects and energy to found agricultural cooperatives and many other autonomous business initiatives in many small business sectors throughout the Canavese, to counter the abandonment of marginal areas and strengthen the relationship between inhabitants and producers in territorial self-government.[5] The productive investment in the area to ensure the

[2] For the complete publication of Olivetti's works and the conspicuous literature on them, please refer to "Edizioni di Comunità", publishing house founded by Olivetti himself in 1946, whose website is: http://www.edizionidicomunita.it (12/2020).

[3] [The Italian word "*territorio*", as used by the author, is almost impossible to translate into English, since it refers to the complex higher-order living organism, resulting from co-evolutionary relationships between settled communities and "natural" environment, which represents "the environment of humans"—whereas the English "territory" is far from having such a meaning. Therefore, all terms related to "territory" throughout the text should be read in this key (*Translator's Note*).].

[4] Such as, for example, the workers' villages of Saltaire in England, Mulhouse in France, Crespi d'Adda, Lanerossi (Schio), Marzotto (Valdagno) in Italy.

[5] Anna Marson (2019) details Adriano's actions on the Canavese production system, which formed a production fabric of small and micro-enterprises similar to a multi-sector district model.

autonomy of the "life worlds" is therefore definitely antithetical to the intent of organizing the reproductive life of workers under the aegis of the large company welfare. Territories, therefore, in the concrete community practice of corporate management, represent *the epicentre of the social governance of production*, which substantially affects the political project set out in *L'Ordine Politico della Comunità* and whose principles, in 1947, will be the fundament of the Community political movement. The extreme attention to architecture and urban planning (the Val d'Aosta Plan, the Canavese Territorial Plan), the role attributed to industrial architecture in relation to the quality of the work environment (see in particular the industrial settlement of Pozzuoli), the mobilization for these plans and projects of the best architects and urban planners of the time, the presidency of INU (National Institute of Urban Planning), witness in Olivetti a conception of "territories" that commits him to treat them as a "generative" element, in integrated and holistic form, of the productive, social, environmental, governance aspects of the "concrete community" model.

It is no coincidence that, in his presidency of INU, Olivetti substantiated his conception of territories considering urban planning a primary category of territorial political action: «an idea of urban planning at the service of super-individual and therefore ethical purposes, which unifies economic and territorial planning: attributing superiority to the "territorial principle" over the "functional" one, it integrates economy and territories into the urban planning project» (Olivetti 2014a: 187).

Such integration requires the urban planning project to be founded *on multidisciplinarity*—consider e.g., the workgroup of the Ivrea Plan, later become technical group for the urban planning coordination in Canavese, made up of urban planners, geographers, statisticians, agricultural economists—and *on integration*, as stated in the 1956 INU Congress:

> through this methodological complexity a project is formulated which, with the aim of raising the well-being of the entire territory of the communities, radically rejects the current vision of a gap between two types of design: one, in dwarf proportions, for the uses of Saturdays, dedicated to beauty, ideals, goodness, truth; the other of vast proportions for its alleged practical usefulness, mixed with ugliness, squalor, new barbarism (Olivetti 2014b, or. 1956).

The *ante litteram* reference is here to "dual regime" territorial policies between protected areas (governed by protection restrictions) and non-protected areas (governed by economic laws); conception (Council of Europe, UNESCO) which is being overcome only with the recent interpretations of landscape plans in compliance with the Code of cultural heritage and landscape (2004), which are extended to the entire regional territory and refer to the "life worlds" of populations, as the object of the policies proposed by the 2000 European Landscape Convention; interpretations developed by the bioregionalist approaches to territories, whose "constructive elements" holistically concern the integration of hydro-geo-morphological and environmental balances with polycentric urban systems, the multifunctionality of agro-forestry systems, local production and energy systems, within federalist forms of bioregional self-government.

For Adriano Olivetti entrepreneur, the word "territory" in its use in "territorial principle" assumes the founding importance of a historical-social subject capable of substantiating the idea of "concrete community". This can take shape in a specific geographic space in which nature and history, city and countryside, factory and local society can be integrated within proximity relationships in which it becomes possible to exercise community self-government of administrative, economic, cultural and trade union factors: self-government that, being founded on the territorial community, becomes the foundation of a grassroots federalism—"to participate", as Silvio Trentin (1987) would say.

In the territory of concrete community, against gigantism and metropolitan centralization, *decentralization* takes place as cultural, economic and social enhancement of small town networks, boroughs and villages: a supportive and non-hierarchical territorial system, managed by "community centres",[6] multi-sector systems of small businesses, and directed by an integrated economic-urban planning. *Measure* is the guideline of urban relations in the concrete community: bond.

> a certain measure suits the city as well as everything else in the world, animals, plants, tools: since each of these beings, either too small or too large, cannot keep its strength, but now will lose its nature, now will get corrupted, as a ship a span or two stages long will no longer be a ship, and with its dimensions either excessively large or excessively small will render itself useless for navigation; similarly, a city made up of a few inhabitants cannot be self-sufficient as a natural agglomeration; while self-sufficiency is the main character of the city; the one made up of too many inhabitants will be sufficient for itself from the natural point of view, not as a city, in its genuine political meaning, since it is not easy to found in it a genuine constitution. (Olivetti 2014c, or. 1953)[7]

In Olivetti's vision, territorial decentralization is not interpreted as a fight against urbanism or as an anti-urban idea (in 1959, Olivetti—2014d—expressly cites, as a negative example, the fascism policies that prevented southern populations from leaving their province); nor is it assumed as a simple function of reducing labour costs (as will actually happen, starting from the 1970s, in the "widespread factory" of productive decentralization, resulting from the territorial breakdown of the Fordist factory city); on the contrary, decentralization (which we today call "reterritorialization") is assumed as an *opposition to urban and metropolitan gigantism* («the great monsters of our time»), to affirm another idea of the city and the urban world, through the revitalization of villages, hamlets, small towns which constitute the Italian urban historical framework, close to nature and landscape; decentralization is therefore assumed as *a function of polycentrism*, of equipping each centre with

---

[6] This "federative" movement of small towns and villages in the peripheral communities is activated by Community centres (at that time "community operators", today we would say local development agencies, LAGs, etc.); centres that, among other things, «improve the link that binds workers of the factories and the countryside» (growth of "relational goods", Zamagni would say) and enhance the resistant social and territorial structures.

[7] Here Olivetti echoes Alberti (1989: 153): «the ancients affirm that the city is like a ship which must have such dimensions as not to sway when empty, nor to be insufficient if full»; and also Botero (1973: 72): «it is indeed worthy that the cities, having reached a certain limit of greatness and power, do not pass beyond, but either stop in that limit or go back [...]. The nourishment is obtained either from the countryside of our city or from the countries of others, and if the city has to grow it must be brought to it from afar [...] but the greatness that depends on remote causes or awkward means does not last long».

all the socio-economic functions, for «the material, cultural and social elevation of places» (Olivetti 2014b).[8]

In this sense, the Olivetti model represents, in those mature years of Fordism, an anomaly planted in the heart of its maximum Italian expression, the Piedmont of FIAT's "factory city".

I will briefly recall the three Piedmontese models of production systems of the 1950s that well exemplify the different types of relationships between factory and territories[9]:

1. *Turin/FIAT, the Fordist model of factory city*

In this model, dominant at that time in social production relations, local territories are treated as a mere technical support of the mass production system: the enhancement of capital implodes in the man/machinery relationship, implementing a process of subordination of territorial organization, functional to the division of work of the large factory system and its reproductive system: large transport systems, large dormitory districts, large leisure sites and so on (Le Corbusier's "*machine à habiter*"), along a process of productive and reproductive massification in the central-peripheral structure of metropolis. Territories, cities, villages over which the factory city expands in the metropolitan area are buried, homologated and with them local cultures as well as cultural, morpho-typological, landscape and environmental identities. The factory city model is pervasive: the three shifts of Ottana, Marghera, Gela factories require the transformation of a Sardinian shepherd, a fisherman from the lagoon, a Sicilian farmer into three identical chemical workers, who sleep in public houses and eat salami in plastic bags. The Turin factory city is pervasive throughout the territory: it bases its production concentration on a great exodus from the mountains and the countryside, first in Piedmont, then in Veneto, then in the South.

2. *Langhe/Ferrero, the model of the peasant worker*

Also in this Alba model (which would have a multinational development), business profit is the core of territorial relations (in this case with Langhe) and defines the subordination of territories for the purposes of the production system; but the identity peculiarities of local territories, instead of being brutally denied, here become opportunities for business development; the territory is "put to work" with its "farmhouses" for the production of hazelnuts; the Langhe farmhouse is therefore kept alive (which places a limit on emigration to Turin/FIAT): the peasant worker carries out seasonal work both in the factory and in the fields; every evening a minibus picks him up from the farmhouse up to the upper Langa and brings him back in the morning to start working the fields after the night shift in Alba. This initiates an investment on some peculiarities of territorial and social heritage, with an "anticipation" of the recognition of the economic value of place peculiarities (in environmental, productive, social, community terms) that will characterize the industrial districts of "made in Italy".

3. *Canavese/Olivetti, the concrete community model*

In *L'Ordine politico della Comunità*, Adriano Olivetti states the passage from the *functional* principle adopted by the other two models (which recalls the analysis, the breakdown into parts, the action by sectors and separate functions) to the *territorial* one (which recalls the synthesis, the holistic principle, in a synoptic view). By reversing the relationship

[8] Olivetti goes on: «technology and culture lead towards decentralization, towards the federation of small cities with an intense life where there is harmony, peace, greenery, silence, far from the current state of overcrowded metropolises, as from the dismay and isolation of the lonely man».

[9] The description of the three models takes up the one used in Magnaghi (2017).

between factory and territory, principles of reciprocity between development of technology and community arise: the *self-government* of local territories guides production and technology. From profit, to innovation, to social well-being, the social and political project is focused on orienting the company development according to the territorial community development, bringing out the spirit of the place, the life worlds of population, of its social life, the complexity of local production system, first of all agriculture, thus stating the importance of local identity. In this vision, territory is interpreted as *the place* for the company social development, through the rejection of the metropolitan model and the enhancement of networks of small towns and villages, which constitute the main Italian urban framework. It follows from this the maintenance of complex and integrated economies, of the city-countryside relationship, applied in particular to the enhancement of the structure – mainly agricultural and made of small properties, small towns and resistant social structures – of the Canavese territory.[10]

Compared to other models, the most important aspect of Olivetti's is that the first and crucial level of the direct democracy system set up by concrete community is placed in local territories: by integrating factors of life and work, the inhabitants/producers govern the institutions of primary decision, which are *territorial*, with respect to higher levels of government. It is from the concrete community in its local (territorial) dimension that political representation emanates "from below" towards the higher levels (functional to the achievement of the concrete communities goals) of the federal state.

This Olivettian "inversion" of the political representation system exists in many other "secular" cultures that refer, albeit in different forms, to the orientation of the decision-making system "from the bottom up".[11]

Therefore, it is from territorial self-government, as the matrix of concrete community, that not only the political system, but also the orientation of production and technology towards the well-being of the community can develop *and not* vice versa.

---

[10] The territory of "concrete community" cannot ignore the inseparability of city and countryside since it is composed of «units that have their foundation in nature and their limits in man»: The Community is therefore necessarily governed by a single administration among urban centres and vast agricultural territories made available to social life, producing a virtuous symbiosis between agricultural and industrial economy. The community territory is a «"*life areal*" where fields, factories, nature and life, brought back to unity, find the harmony of peace and freedom» (Olivetti 2014c).

[11] «To cite just a few examples: from Pëtr Kropotkin, who describes the networks of small territorial units and groups of medieval cities as structures of solidarity self-government antagonistic to feudal powers, to Carlo Cattaneo, who attributes centrality to cities and their synergies with their territories in the federalist political system, up to the libertarian thought of Murray Bookchin, who sees in his eco-community municipalism (neighbourhood, borough, city assemblies, networks of cities and confederations of municipalities as higher levels of decision) the base of "politics" which the state seizes expropriating it from citizens; from the proposals of political ecology of André Gorz, who entrusts to productive and mutualistic relations of neighbourhood the primary level of local democracy for the autonomous satisfaction of the needs of dwellers, to the self-planning hypotheses of poor communities by Yona Friedman; up to radical examples like the autonomous Zapatista communities of Chiapas (organized in Municipalities and Regions), to the "stateless territories" of the Mapuche Communities in Chile, to the democratic confederalism based on autonomous municipalism by the Kurdish communities of Rojava in Syria» (Magnaghi 2020: 11–12).

In this primacy of the "territorial principle" lays therefore a problem that transcends the Olivettian temporal context, projecting into today's alternatives between a world that, based on technological optimism (neuroscience, digital economy, hydroponic crops, cybernetic communities, etc.), results in the fulfilment of an artificial second nature without land, environment and history; and a future that, consistently with the recognition of the importance of the co-evolutionary processes between human settlement and nature that have characterized all human civilizations, proposes as vital, for the future of humanity, to measure technological innovation with the "return to territories"(Becattini 2009); assuming more generally the territory as a «life place» where it is possible «to create a common moral and material interest among humans who carry out their social and economic life in a convenient geographical space determined by nature and history» (Olivetti 2014a: 33, where the concept of concrete, solidarity, personalist community is expressed); which brings out the spirit of places (character, personality, soul, genius, as Vidal de la Blache, Hillman, Norberg-Schulz would say); interpreting the "life worlds" of populations as foundations of the socio-economic development model.

As a political movement, the territorialized model of local community development aimed at social well-being was defeated in the 1950s (1958 elections), although it was not culturally isolated at the time (Rossi Doria, Sereni, Ceriani Sebregondi, Dolci, etc.); it returned as a "background" in the 1980s, first with the economic culture of the "productive atmosphere" of districts (linked to the peculiarities of long-term territorial cultures), then with the social emergency of environmental issue and local development. It has later matured, in recent cultural acquisitions, in the form of multiple local experimental responses to the global structural crisis: all united by the profound link of innovative productive, social and cultural systems with local knowledge, environments and lifestyles, "excavated" by living communities in the history of places and reinterpreted as patrimonial wealth; an anchor that can be seen as a sort of antidote to the financial crisis of economic globalization, indicating ways to overcome the crisis itself which, precisely from the "return to territories" and to their heritage redesigned for the future, draw the strength of innovation. This is the topic extensively dealt with by the Territorialist Society, both with the Observatory of good local development practices[12] and the journal *Scienze del Territorio* (Territorial Sciences).[13]

The "holistic" conception of territory, first overwhelmed in the 1950s by the functions of the Fordist "factory city", then buried by the deterritorializing abstraction of economic-financial globalization, shows today many signs of rebirth: from the new farmers who, recovering the knowledge of historical rural landscapes, rebuild city-countryside relationships producing food and ecosystem services for cities, repopulating villages and mountain hamlets; to the inhabitants of metropolitan suburbs who weave sharing networks, rebuild public spaces and new solidarity forms of living and producing; to local communities that reinterpret and care for local patrimonial values to produce and exchange goods which are unique in the world; to new forms

[12] See http://www.societadeiterritorialisti.it/2019/01/22/schede-gia-elaborate/ (12/2020).

[13] See https://oajournals.fupress.net/index.php/sdt (12/2020).

of integrated and participatory planning which, on the growth of active citizenship and "place consciousness", found visions of the future anchored to the return to territories: river, lake, landscape contracts; ecomuseums, local landscape observatories, multifunctional agricultural parks, biodistricts and so on.

I will give just one example on the topicality of Olivetti's "territorialist" thought, recalling the latest elaborations by the economist Giacomo Becattini.

In Becattini's latest book, *La coscienza dei luoghi* (The Consciousness of Places, 2015), the link between this historical evolution of the concept of "local" with the "return to territories" is evident. Becattini's territory is not a geographic space but a set of places endowed with historical depth and identity; it is the *place* that educates the community that inhabits it: the place understood as a set of patrimonial deposits, both of *environmental knowledge and resources* ("the metaphor of the lake" in the last essay of the book, or the relationship between "stocks" and "flows", in reference to Georgescu-Roegen), and of cultures, lifestyles, characters, cultural flairs, social models. Referring directly to Olivetti, Becattini carries out a radical role reversal between the concepts of economy, production and territory; in the essay "The long march of economic studies towards territories" the place, as the primary agent of productive transformation, is synthetically and metaphorically defined by Becattini as "a spring wound over the centuries":

> many apparently anodyne places were springs wound over the centuries; which, if the conditions were created for their release, could change the face of a country. In short: the people of Prato, Biella, Carpi and many other local populations have done something that seemed impossible to most economics professors: the water of their artisan know-how and their local cultures has been transformed into the wine of export and in the *joie de vivre* of social groups, even those of modest extraction. Here, this is the territory – something deeply different from the space of the theorists of location and transport, in which cultural distances cannot be measured, and where the potential for development is hidden in the most unexpected folds of local societies. It is to these "springs wound over the centuries" that we anchor a return to the Cattanean concept of territory as a "reality built by humans"; a summary term allowing to deepen through comparative analysis the differentiated development of places. (Becattini 2015: 95)

To such idea of territory, to this "spring wound over the centuries", Becattini attributes the strength to produce not only wealth but also happiness, which leads him to propose an economy of happiness aimed at satisfying the interests of dwellers and not profit.

Another aspect of the "topicality" of Olivetti's thought on the centrality of territories in socio-productive transformations is found in the recent thrusts of innovative companies to move from social to socio-territorial corporate responsibility (De La Pierre 2018), in the transition from industrial districts to multi-sector production districts and in attributing to *a multiplicity of actors* in territories (and therefore to the power of social demand) the definition of the goals of industrial production, aimed at social well-being. And this process applies even more in a territory with a complex and multi-sector production system, compared to the situation in Ivrea or in the Canavese area where, as I said, the territorial principle was applied in a "hard" situation, with a big company playing on the international market, and on which Adriano intervenes with a territorial project anticipating the spirit of integrated multi-sector districts and even the environmental care relationships typical of "urban bioregion" models (Magnaghi 2014, 2020, Chaps. 5 and 6).

In any case we find analogies on the *new social form of enterprise*, in the relationship between enterprise and territory, if we compare the model recalled in a conference of the "Adriano Olivetti" Foundation—«Community owns a part of the share capital of large and middle factories, appoints some of the main managers [...]; buys and sells land and properties in relation to the technical development needs of the Community; [...] assists the development of crafts and tourism...» (Zagrebelsky 2014: 29)—with Becattini's most recent hypotheses on the company socio-territorial management—«we could think e.g. of a "bicameral" system of the industrial cluster or district that provides for the presence of local representatives in the board of directors of the company and of those of the company in the local political council» (Becattini 2015: 46)—or with experiences of "foundations" that simultaneously address social and productive aims in the management of "evolved social districts", a holistic experience inspired by the theory of complexity with the expansion of social capital as a constraint to the logic of profit; this is the case of the Messina Community Foundation, whose cluster[14] includes: profit enterprises; social enterprises, ethical finance, solidarity economy networks ("Sole" Consortium, Ecos Med); support for energy micro-productions; social housing, focus on environmental technologies (Horcynus Orca Foundation); cultural park of youth cooperatives of weak people; network on sustainable engineering and architecture; eco-gastronomy; rehabilitation of slums and refugee camps; community welfare models, rehabilitation of areas of archaeological and environmental value.

Where, on the other hand, the model of concrete community is (unfortunately) not at all topical, is in the consequences of the Olivetti model with respect to federalist forms of territorial (and social) government: the overwhelming wave of neo-centralism—which involves at the regional level decision-making concentration and territorial centralization of services in cities and metropolitan areas, at the state level megaregions like in France, at the global level the concentrations of megacities and megaregions (up to the "Greater Beijing" project with 140 million inhabitants) and the globally interconnected growth of a neo-state-cities—has long since wiped out in Italy, also for the Lega (which in fact remains the only inadequate guardian of Carlo Cattaneo), any federalist hypothesis, not only those of solidarity federalism "from below" that took up the hypothesis of "federalism to participate" by our "New Municipium Network" in the 2000s.[15]

---

[14] As described by Gaetano Giunta, President of the Foundation, in his speech at Focus Adriano Olivetti, Milan 2016 (https://focusadrianoolivetti.it/2016/11/08/gaetano-giunta/—12/2020).

[15] The New Municipium Network (2000–2011) was a national association, promoted at the first World Social Forum of Porto Alegre (2001), made up of municipalities, academic research-action laboratories, local movements and associations of active citizenship, aimed at promoting innovative policies of self-government of territories as common goods, under the banner of a participatory and

Currently, with dishevelled power conflicts related to the management of pandemic emergency, all the "left" oppositions make no mention of federalist alternatives but, even with the justified criticisms for the alleged "differentiated regional autonomies", praise the unity of Nation guaranteed only by the central state (up to propose the abolition of Regions).

Thus, even Olivetti's idea of *community*, still today, clashes with the sociological thought of modernity in its academic expressions, which substantially considers the very word "community" a concept to be avoided. Yet, for the most part, the experiences that address concrete territorial initiatives for the defence of the environment, the territory, the landscape, which regain urban and agricultural spaces for the self-organization of experiences of life, cultural promotion, self-managed productions, which promote productive and exchange activities with ethical, social and mutual aid value, "community cooperatives",[16] "energy communities" (Bolognesi and Magnaghi 2020) and so on, ***define themselves*** as "communities" to underline that the seek for alternatives to social pulverization and the subtraction of powers, from dwellers and producers, by the global flows of financial capital, pass precisely through the reconstruction of bonds of proximity, solidarity, re-identification with places; characters that are re-attributed to the word "community" as an emerging political dimension with respect to that of society, whose outlines are increasingly blurred and atomized.

To underline the emergence of such community cultures, the Territorialist Society entitled "The democracy of places: actions and forms of community self-government" its 2018 annual national conference, held in Castel del Monte (Apulia) from 15 to 17 November.[17]

The signs of this return to Olivetti's territorial principle in a new context of globalization are visible in transparency in the innovations of the Italian socio-territorial fabric, which we brought out in the laboratories with the social actors of this process at the same Castel del Monte conference (Gisotti and Rossi 2020).

These signs allude to a new generation of *self-sustainable local development models*, based on the enhancement of territorial, environmental, landscape and socio-cultural heritage, in which productivity does not depend on a single sector but on the local environment, on the maturation of *place consciousness* and on technical and merchandise features related to complex and integrated supply chains, from multifunctional neo-agriculture to the advanced tertiary sector, which substantiate the generative processes of community self-government institutions.

---

dolidarity federalism for a "globalization from below"; a "historical" collection of the documentation produced by it can be found on http://www.nuovomunicipio.net/documenti.htm (12/2020).

[16] Even institutional initiatives such as the "community cooperatives" promoted by Legacoop and financed by various Regions and Municipalities use the term "community" to denote the common presence, within a socio-productive enterprise, of citizens, economic operators, local institutions.

[17] And whose themes are the subject of issue n. 8 of the journal *Scienze del Territorio*, see https://oajournals.fupress.net/index.php/sdt/issue/view/484 (12/2020).

Along these lines, the reference to Olivetti's idea of Community, an open idea that contains the recognition of the potential for self-government of territories by their dwellers/producers, is constantly present in our initiatives, conducted with the multiform innovative experiences of active citizenship and with the new forms of participatory planning and agreements that "streak" our territory in an increasingly dense (but still very weak, institutionally) way.[18]

References

Alberti LB (1989) L'architettura. Il Polifilo, Milano (or. 1452)
Becattini G (2009) Ritorno al territorio. Il Mulino, Bologna
Becattini G (2015) La coscienza dei luoghi. Il territorio come soggetto corale. Donzelli, Roma
Bolognesi M, Magnaghi A (2020) Verso le comunità energetiche. Scienze del Territorio, special issue "Abitare il territorio al tempo del Covid", pp 142–150
Botero G (1973) Delle cause della grandezza delle città. Adesso suppl. 83 (or. 1588)
De La Pierre S (2018) Dalla responsabilità sociale alla responsabilità socio-territoriale d'impresa. Scienze Del Territorio 6:95–102
Gisotti MR, Rossi M (eds) (2020) Territori e comunità. Le sfide dell'autogoverno comunitario. SdT Edizioni, Firenze
Magnaghi A (2014) Il progetto della bioregione urbana. Regole statutarie e elementi costruttivi. In: Id (ed) La regola e il progetto. Un approccio bioregionalista alla pianificazione territoriale. Firenze University Press, Firenze
Magnaghi A (2015) Dal territorio della Comunità concreta alla globalizzazione economica e ritorno. In: Id, Bonomi A, Revelli M (eds) Il vento di Adriano. La comunità concreta di Olivetti fra non più e non ancora. DeriveApprodi, Roma
Magnaghi A (2017) Le fondamenta di un'utopia concreta. In: Id, Bellandi M (eds) La coscienza di luogo nel recente pensiero di Giacomo Becattini. Firenze University Press, Firenze
Magnaghi A (2019) Ripartire dal territorio. Dalla comunità concreta alle bioregioni. In: risposta a 11 domande su Olivetti. La Rivista di Engramma 166:131–141. http://www.engramma.it/eOS/index.php?id_articolo=3691 (12/2020)
Magnaghi A (2020) Il principio territoriale. Bollati Boringhieri, Torino
Marson A (2019) Breve riflessione sul rapporto tra Adriano Olivetti e il territorio eporediese. In: risposta a 11 domande su Olivetti. La Rivista di Engramma 166:143–147 (12/2020). http://www.engramma.it/eOS/index.php?id_articolo=3689
Olivetti A (2014a) L'Ordine politico della Comunità (edited by D. Cadeddu). Edizioni di Comunità, Roma (or. 1945)
Olivetti A (2014b) Noi sogniamo il silenzio (paper presented at the 6th INU Congress, 1956). In: Saibene A (ed) Il mondo che nasce. Dieci scritti per la cultura, la politica, la società. Edizioni di Comunità, Roma

[18] «The *Concrete communities* advocated by Olivetti», after all, «are not far from the territorialist idea of urban bioregion, within the framework of a territorial "urban planning" that was a constant concern of Adriano together with the revitalization of the city-countryside relationship. This ideal meaning of "concrete community", coinciding with the centrality of the "territorial principle", shows how all today's multiform expressions of alternative societies and economies (bio-districts, bioregions, civil economy, circular economy, community cooperatives, "participatory" revival of villages, neighbourhoods, valleys, marginal areas and so on) are indebted to Olivetti's fundamental intuitions» (De La Pierre 2018: 95).

Olivetti A (2014c) Dalla fabbrica alla comunità (Perugia 1953). In: Saibene A (ed) Il mondo che nasce. Dieci scritti per la cultura, la politica, la società. Edizioni di Comunità, Roma
Olivetti A (2014d) Il mondo che nasce (1959). In: Saibene A (ed). Il mondo che nasce. Dieci scritti per la cultura, la politica, la società. Edizioni di Comunità, Roma
Trentin S (1987) Federalismo e libertà. Scritti teorici 1935–1943 (edited by N. Bobbio). Marsilio, Venezia
Zagrebelsky G (2014) Presentazione. In: Olivetti A (ed) Le fabbriche di bene. Edizioni di Comunità, Roma

Alberto Magnaghi, architect and planner, is a Professor Emeritus of Regional planning at the University of Florence and the President of the Territorialist Society (www.societadeiterritorialisti.it). He has coordinated national and international research projects on self-sustainable local development, identity representation of territories and the urban bioregion, as well as urban, territorial and landscape projects and plans for social production of territory and landscape. His recent works include La conscience du lieu (Paris 2017) and *Il principio territoriale* (Turin 2020).

Chapter 5
All the Layers of an Ecological Commitment at the Frontier: Ian McHarg, *Design with Nature*, 1969

Francesco Domenico Moccia

Abstract In McHaig biography there are events, relations, and profiles that let us better understand his book *Design with Nature*: the love for nature cultivated in the childhood hiking, the passion of rhetoric speech learned by the missionary father, the committeement to reconstruction of a better world after the war, his popularity in TV broadcast, and his prestige at top government officials. Design with Nature concentrate the best the author can teach as a reflective practitioner and leading academic. It is also a pivot of a cultural movement and a powerful fuel of progressive politics. In modernism urbanism we found two trends: the rational, with anthropocentric bias; the organic, coming from Geddes (1915), and developed by McHarg. After the book was written, and the urban change in progressed, two critical thinking were borne in the University of Pennsylvania, one claiming for social equity starting from Davidoff and the other for ecological sustainability from McHarg. Because now each of them compete for primacy, the suggestion of this paper is that their cooperation should be better fruitful because they could—as is needed—integrate ecology and equity in the next urban and regional change. Put in this track that book will still be a central knowledge in planners training.

Keywords Landscape · Regional planning · CBD · Overlay mapping · Planning theory · Nature

5.1 Introduction

Design with Nature was published 50 years ago and will be celebrated by the University of Pennsylvania McHarg Center. Penn Department of Landscape Architecture is honoring one of his most famous head and promising that his lesson may still lead a new wave of sustainable planning and also a movement for ecological purposes. Meeting around such mythic figure the best thinking and profession of our time will not only reaffirm the pivot rule of the ivy league University, but also to foster

F. D. Moccia (✉)
Department of Architecture, Federico II University, Naples, Italy
e-mail: fdmoccia@unina.it

C. Perrone (ed.), *Critical Planning and Design*, The Urban Book Series,
https://doi.org/10.1007/978-3-030-93107-0_5

some central principle still projected to a future envisaged half a century ago but no more achieved. The choice of supporting ideas and proposals clearly opposed to the majority trend in the United States and Bush Presidency (an alumni of the Penn Warton School) is shown as a sharp positioning in the political arena in addition to academic ambitions.

In this contribution, at distance, to that celebrations, as a post-doc researcher to the close Penn Department of Planning, I will try to select what seems to me the most promising ideas still able to stimulate and address the work of landscape, regional and urban planners.

I will start recalling few information of the McHarg biography that may explain why his commitment to environmental issues was so challenging and may be still fertile. Then I will outline some features of the planning culture of the time the book was written having special attention to the closer colleagues and theories growing in the Philadelphia melting pot. I find also of great interest to point to the way the book was born not only as the output of the complex character of its author, but as a measure of the deep construction of proposals whose aim was to face new and hard problems in a fresh vision and practical efficiency.

Many critics, better that I can do now, forwarded many and sometime contrasting appraisal of the values of this book together to the whole McHarg production. Some stated that in *Design with Nature* is condensed all McHarg had to say or at least the best he could. That is why, discussion over faults and merits of his contribution to landscape design could be concentrate there.

My effort is more oriented by the interest of a reflexive practitioner, always in search of effective tools for practice. In the list, many things may be added: new perspective to look at complex word phenomena and be able to understand them; how to take advantage of the knowledge coming from science and technology and use it to human purpose; why deep beliefs spread in our culture and lead our actions, belonging to decision processes; how the knowledge bag of planners benefitted of his teaching, and still do.

5.2 Biographical Premises

Ian L. McHarg was born in 1920 in Scotland. In his autobiography (1996) he explains why since his childhood he fell in love with landscape. Living in a suburb of Clydebank near Glasgow he did not had the opportunity to experience the urbanity of cities like gothic and neoclassical Edinburgh, but only a smoky and ugly industrial space at his adolescent eyes; on the contrary he explored the magnificent countryside in the most intense and joyful times of freedom from study and other obligations. Such passion addressed his education to landscape architecture whose competence sited him in the Corps of Engineers of the British Army where he served during the Second World War first in the Africa campaign and then he landed in Italy crossing almost the whole country from Apulia to Cassino and farther (McHarg 1966).

One more mark over his character should have been what he recalls as a father inclination toward the function of a minister of the Presbyterian Church, though he preferred to marry and to perform many businesses in the economic hardship of the Scottish crisis before the Word War. A mystic familiar atmosphere fueled the vigor of his deep greed and the warm of his speeches in a time where fighting for environment was a pioneering effort.

After the war he studied at Harvard, having as a mentor Lewis Mumford, with the intent to rebuild his war-ravaged homeland working on housing and new town in Scotland. From this effort, just undertaken after his graduation and return at home, was diverted by Dean G. Holmes Perkins who enticed him to build a new graduate program in landscape architecture at the University of Pennsylvania in the Department of Landscape Architecture and Regional Planning established in 1924. For many years both the academic department and the firm Wallace, McHarg, Roberts and Todd engaged in action research, advancing in both disciplines and professions.

Design with Nature advances a new theory for design and a new mandate for public policy while presents insightful case studies.

In addition to academic research and teaching in close linkage with his professional practice, McHarg was also a very popular figure. In twenty-six Sunday, CBS television broadcasted the program The House We Live In where he interviews leading theologians and scientists to discuss man and environment relationship and to give a sense to the place of mankind on the hearth. The acquired popularity gained many important friendships as the first lady Lady Bird Johnson—founder of the *Society for a More Beautiful National Capital* in behalf of the city and region of Washington DC, Steward Udall—the Secretary of Interiors promotor of an environmental policy whose outcome was the establishment of many Nation Parks and Laws for the protection of nature, and Laurence Rochfeller—the owner of the environmentally focused hotel chain.

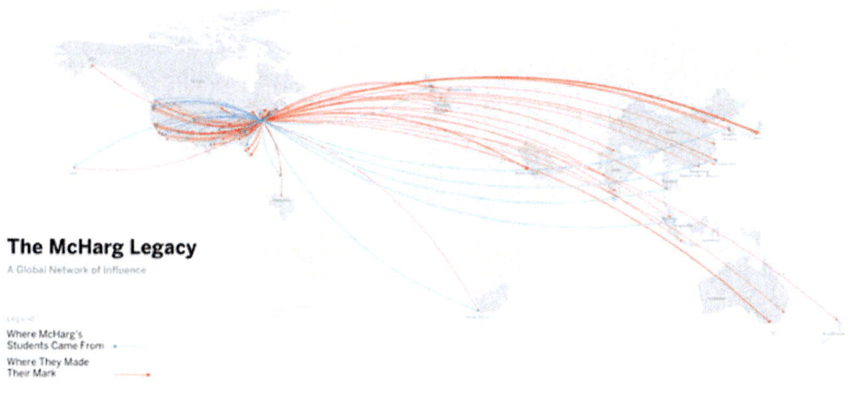

Fig. 5.1 Geographical diffusion of McHarg yeaching (PENN source)

5.3 How the Book Was Born

His excellent reputation gained with the prestigious academic career in one of the most important USA Universities, his forerunners ideas on an emergent policy challenge, the wisdom of the proposals he carried on and the practical demonstrations of the feasibility of agreed on solutions, all scientific and professional attitudes joined to a wide popularity given by TV broadcast. All these put McHarg in the forefront of a policy change in the environmental issue and let him to serve in very important panels and commission as the White House Commission on Conservation and Natural Beauty. In 1965, May 24–25, the Commission held a Conference in Washington opened by a President Johnson message, demonstrating McHarg leading role.

In PENN he started a course named "Man and Nature" in 1959 and went on in the '60s and '70s as one of the most popular and crowded. Because University praise herself of brilliant lecturer, there is a habit among its students to follow whom they feel of more interest and attraction independently of their program, and the McHarg class was awarded of one of the most successful. Again, the reason why we recall it is for his syncretism in mixing academy with outside word, more discipline and skills together. The format of Man and Environment was some of the TV Broadcast and reflect itself in the *Design with Nature* book too (Steiner 2006).

All the above reported events give the book commented here a very special profile. It is not, as generally people think to literature, the solitary effort of a thinker closed in his intimacy or reflecting over his readings. Neither this book may be considered the simple account of a successful practice, nor a research report. It is a demonstration of how the focalization on one single issue may convey in that point a multiplicity of interest. But it is to consider not only the way the book was written but also the effect it had on his time. Being a pioneering effort, it had a leading role in collecting around McHarg and the University of Pennsylvania students angry to understand natural forces and the way to design sustainable human spaces (Fig. 5.1). This movement grow out of academic borders and had influential effects over the beginning of an ecological policy of the USA federal government demonstrating how knowledge may be a powerful fuel of progressive politics.

5.4 City and Country

Man and nature is one of the main topics of Design with Nature. McHarg shows the opposition in shining words that evoke images of his life in Scotland but can be understood by any people living in modern industrial cities as they took shape in the after-war time. As it seems a period of time far from our days, we can assimilate it to the critical accounts of modern urbanism on problems it was obliged to face. Pollution, crowding, lack of hygiene, and all the worst living conditions of slums tenements for workers in Europe and America generated with the industrial development in cities, as pointed out from the Engels inquire on through philanthropist movement until city planner at the beginning of the XX century.

Modernist answer to the problem was to escape from the crowded city and refuge in the country. Both Olmsted's garden city that German socialist democrat siedlungen propose to colonize countryside. As well as McHarg declare, they do not choose for one of the two opponents, but try to mix them together to obtain the best from each of them. In the famous circle drawing, Olmstead try to visualize the joining effect of these conflicting settings in his proposal of housing in the rural atmosphere able to structure a new metropolis. In this narrative, the discourse speaks of houses only, of the land cost and real estate investment. Budget are studied to let houses to be affordable for workers and responding to their family needs. Some concern is given to assure communication with working place while trying to strengthen the autonomy of a self-administrating community. Of this social and special project, the environment is only a background, surely healthier that the former escaped from. The colonization proposal is focused on the social innovation with its economic and technical nuances. CIAM congresses, committed to the modernizing endeavor architectural thinking, made one more step on the anthropocentric account of city-country swinging.

At the side of the mainstream city planning culture, in British seminal regional studies, a different root may be found, whose contribution is essential for the McHarg combination of the dyad so central to modern planning. In contrast with the British, the Scottish school rooting in Patrick Geddes legacy grants more relevance to the natural side of the balance. As a biologist, his comprehension of the industrial conurbations has to take in account geographical, geological, botanic and hydrological factors. According to his natural sciences approach, also conurbations should be considered living organism integrated in the general flux of natural processes. At the wider scale, city and region; the former expression of human endeavor, the latter domain of natural forces, are more balanced. At list, nature can teach a better method to understand urban dynamics commanding environmental conditions to its development.

In modern urbanism to the machinist rash ideas of the functionally efficient metropolis was opposed an *organic* bias where Geddes sensibilities survived and developed. One of the champions of organic urbanism was Lewis Munford, a teacher of McHarg and the author of introduction chapter of Design with Nature. While Mumford was attracted more by the social organization of metropolis, his image of cities was close to a metaphor of a living organism where neighborhood may be similar to the cells of a body of the size proportional to the performing functions and, primary, that of nurturing the social nexus of the local community.

My search of McHarg roots does not reduce the novelty of his seminal work. The stream of thinking I am connecting him is more an orientation of feelings and perspective. Benefitting of that energy and attention, there were still the task to elaborate theory and method, to show haw beneficial could be to the society of his time that ideas, and to move them into practice.

5.5 The Book Geo-Historic Setting

Before moving to the utility of McHarg in our time, let me consider the years when to book was written. In 1969 United Sates were in the post-war renaissance, the name adopted by many cities in the process of renewal, slum clearance, CBD creation. While central cities entered in the service era, the development of global corporations, and the communication network extension, a pool of forces concentrated in key metropolis wealth and command; suburbanization was spreading housing in scattered neighborhoods under the exuberance of inexpensive fossil fuel powered culture. War victory give people trust they could dominate the world and the nature, supported by the unprecedent technology advances and scientific discovery. All problems should be solved, even the wickedest as poverty.

Criticism to uncontrolled growth will come only some decade after and could not be understood at the time when economic and environmental inconvenience were not evident. The whole picture gives the measure of the difficulty of advancing ecological sensitivity. As the effect of the development boom over displaced minorities erupted in social protests, city planning found some champions that reconsidered the role of the discipline and found it too much aligned with government, capitalism interests. To contrast growth machines and its devastating effects over society, advocacy planning was born in some Myerson Building where McHarg had his departmental offices (Davidoff 1965).

In this impetus of deep planning reform McHarg had to elaborate over a long, and well-established American tradition: Park movement. In the *laissez fare* city, that movement monopolized public intervention and was the most important urban public policy (Dal Co 1973). Sometime urban planning was identified with park design and give to landscape the status of urban planning. We can say more. Landscape was the only practical American urbanism. Due to this historic legacy, no surprise if again in the higher-ranking universities, the most successful landscape architect proposes a seminal discipline named landscape urbanism.

In American urban history parks are a central component of cities when they offer the opportunity of civic representation of government symbols in State Capitols, municipal institutions centers or structure, as connecting neighborhoods spread in the suburbs. In the former role, celebration of the nation values evokes the virtues of the original continental wilderness celebrated by poets, father of the nation, as Thoreau (1906) and Emerson (1836). They are meaningfully implanted in congested cities to contrast the symbolic representation of vice, profit and success research, of urban socio-economic environment. In the latter, parks were asked to function as integrating machines of immigration waves through sport exercise where competing needs group cooperation, while educating bodies to toil and stress. Prime public space should be considered also a health environment thanks to the nature benefits profiting of these resources to shape citizens (Cranz 1989).

Also, if based over that tradition, whose main contents he sharply confirmed, McHarg contribution was innovative. His strategic move was from the design of pieces of nature or protection of some tract of land inside the industrial metropolis—a task so well performed by excellent professional of the stature of both Olmsted—to the territorial or regional scale. He inverted the elements considered and made the natural setting of the urbanization come at the forefront of design instead of the background.

5.6 A Multiple Track Narrative

Design with Nature develop many parallel tracks quite independent at the first glance, but metaphorically or causally related each other's, during the development of the story. McHarg share with some other famous colleague of his University a concern of deep thinking. Also, if he is developing a technical proposal for regional planning, he takes care to explore fundamental values of cultures with the attempt to let his message to be accepted by the wider audience. This means to involve culture and, as Luis Kahn in architecture, Edmund Bacon in Urban Planning, all teaching in PENN in some time, McHarg is willingly to made changes in the way people think to the world, and, as a consequence, he has to discuss deep values. This approach went on with the following teaching of Seymour Mandelbaum.

There is some discussion if McHarg was or not a scientist, opposing his main competence in planning. Notwithstanding, he was involved in epistemic discussion to find a reference anchor, a path among scientific paradigm explaining natural process dynamics. So, he faced different naturalist approach (competitive vs. collaborative evolution of leaving species) or criticized the second principle of thermodynamics, opposing to entropy the more positive negentropy preferring the more optimistic solution to many of discussed dilemmas.

One of the thickest tracks is ecology, where he takes advantage of the Odum's (1983) work and feed his progressivism with the regenerative effects of metabolic processes. Focusing on such natural devices he nurtures the hope that humans may address nature to better future if design comprehends directions of its dynamics and leads, with proper action, toward the better choices. In the ecologic field, an original McHarg contribution is regional analysis. In his studies of areas close to Philadelphia, New Jersey, Baltimore, and Washington (what Gottman will identify as the West Coast Megalopolis) he perfected progressively a tassology of land mosaics to capture natural and cultural values. In subsequent times this effort proved to be very fertile, and many more studies developed the tool and consolidated the theory framework (Forman 1995). Though the book collected some case study whose selection comes from opportunities of jobs or appointment, they cover the main landscapes: river shores in New Jersey, the plain of Philadelphia, river basin in the study on Washington; large Delta areas as in New York, hilly landscape in the Baltimore region.

Design methodology became known as overlay mapping and is the most used in landscape planning. It is a collection of values and hazards on different thematic maps. Their overlaying shows areas of risk and areas to preserve for their supply of ecosystem services (to use a definition not yet existing at the book time, but that categorize exactly what McHarg wanted to preserve). The criterion of intervention should be suitability. Other world, neither of the identified issue should be, for opposite reasons, hurt. The best location of development should be chosen in the empty areas of the overlaid maps. This methodology of data collection, data bank, thematic mapping and layers moved to GIS technology and become the widespread tool every planner use today (Steiner 2000). One more effect of that methodology was impact assessment and strategic environmental assessment because *Design with Nature* is not only a system to decide development locations, but, building a knowledge of natural processes, define also development as a process that interact with the natural one of the sites, addressing their trajectories.

The last track of the book is a group of real plans made for real sites and aimed to regulate the use of land and steer development. Here an original contribution to regional and landscape planning should be registered. After a period when Landscape Architecture was devoted to design of parks and open spaces, the McHarg holistic approach turn to comprehensive planning at regional scale, without losing attention to detailed site solutions, especially in most sensible environmental nodes. However, the general picture is not only the framework for the landscaping of riverbanks or shorelines, nor for deciding institutions of new parks over natural areas to protect. It tries to give order to the sprawling metropolis before criticism against the sprawl had gained public recognition.

The strength of McHarg proposals is evaluated by Heavers (2019) in the Washington waterfront. McHarg collaborated on two important planning reports in the mid-1960s: the *Potomac Report* and *Toward a Comprehensive Landscape Plan for Washington DC*. Both studies were the basis for several chapters of *Design with Nature*. Heavers demonstrate that in the different plans issued after the book, suggestions presented in it are left alive and steered ecological policy and design. The author added some issues were not considered by McHarg, and they fall in the social policy field.

5.7 Landscape and Planning

Design with Nature was born in years of booming capitalism. The nation after the triumph of the second world war, when, defeating Nazi empire ambition and empowering his leadership, in negotiation with Soviet Union over the planet, was deeply restructuring his economic regime and, subsequently, the whole urban system. Urban redevelopment started under the flagship of a social enterprise with at the top the blighted neighborhood cleaning. What was initiated as a housing betterment public program, implemented in a public private partnership, turned to be the tool international companies could use to install in the city cores their top management offices. Despite objectives as poverty reduction and social integration were main efforts of federal government and inspired critical revisions of former programs failures, suggesting new proposals and experimentations, American cities seem be shaped more by economic forces of the real estate market. Expansion of America companies over larger and larger markets, supported by communication technologies development, engineered their organization in more centralized management and permitted fusions to advantage of scale economies. Such processes are at the root of Central Business Districts dissemination in any large America city. Housing demand came from blue and white collars (while the former where declining, the latter supplemented them) families in the time of baby boom with a majority preference for cottages and single home in car connected suburbs. Only the '90s a reurbanization process was fueled by single high technology or finance workers.

At the face of the so stylistically sketched process of city change critical thinking departed in two divergent directions both represented in the University of Pennsylvania with the main national representative. The planning stream, led by Davidoff, attacked the social side of urban policy stressing failures of the solution of the housing blight. Democratic arena, where social and economic parties negotiated for their interests, claimed for equity. While other authors developed other critical approaches inspired by Marxism analysis, Frankfurt School of critical rationalism, Habermas claim of social conversation; planning theory domain has been completely inscribed inside a sort of critic of social impact of urban development.

On the other side, one more critical thinking was aware of the nature degradation determined by infrastructure and general urban development of city sprawl. This second stream may be considered lead by McHarg and grounded in the Landscape Department. While the other criticism was involved in values and rights, the second one was aware of all the things combine in the environment. They are physical things, albeit non-urban, but part of the nature as rivers, beaches, woods, hills, trees. His claim is developed from a nature preservation to a promise of a new type of future city where landscape is its ordering and frame public space system. In fact, a long tradition gives to landscape architect the task to design the main public space of America cities: parks and parks chain.

As a result, separateness of the afore mentioned thinking streams generated a dead treatment of urban problems whose eco we can perceive easily when we hear about claims of ecological policy able to not generate social injustice. On the contrary, support of social integration does not propose a vision of public space and a physical structure of a just city. According to this perspective, the best way of recovering McHarg lesson is also to understand his limits and try to overcome the dual development of urban critical knowledge, avoiding that it become a conflict.

5.8 Celebrations

While I am writing this short presentation, in June 21–22 2019 the 50th anniversary of landmark McHarg book will be celebrated by a two-day conference and an exhibition of landscape projects by the Department of the University of Pennsylvania he worked in. In October, many of the lectured and design will be conveyed in the book Design with Nature now, edited by Frederick Steiner, Richard Weller, Karen M'Closkey, and Billy Fleming, published by the Lincoln Institute of Land Policy in association with the University of Pennsylvania Stuart Weitzman School of Design and the McHarg Center. Celebrations will reaffirm the inspiring legacy of *Design with Nature*, how its teaching is necessary for a better relationship of mankind and nature and its theories and technology can still lead regional planning.

References

Cranz G (1989) The politics of park design. A history of urban park in America. MIT Press, Cambridge, Mass

Dal Co F (1973) Dai parchi alla regione. L'ideologia progressista e la riforma della città Americana. Ciucci, Dal Co, Manieri Elia, Tafuri (eds) La città Americana dalla guerra civile al "New Deal". Bari, Laterza, pp 149–314

Davidof P (1965) Advocacy and pluralism in planning. J Am Inst Planners 31(4):331–338

Emerson RW (1836) Nature. James Munroe and Company, Boston, p 5f

Forman Richard TT (1995) Land mosaics. Cambridge University Press, The ecology of landscape and regions

Geddes P (1915) Cities in evolution. An introduction to the town planning movement and the study of civics. Williams & Norgate, London

Heavers N (2019) Ian McHarg enduring influence on the ecological planning and design of Washington's waterfront. Socio Ecol Pract Res. https://doi.org/10.1007/s42532-019-00015-5

McHarg IL (1996) A quest for life. An autobiography. Wiley, New York

Odum Eugene P (1983) Basic ecology. CBS College Publishing (1988, Padova, Piccin)

Thoreau HD (1906) Journal. Houghton, II. Boston, pp 341–342

Steiner FR (2000) The living landscape. An ecologic approach to landscape. McGraw Hills Companies, New York

Steiner FR (ed) (2006) The essential Ian McHarg writings on design and nature. Washington Island Press

Francesco Domenico Moccia, held a master degree in Architecture in 1973. He is now a retired Professor of planning after teaching in the Architecture Department of Federico II University, serving as City Planning Department Chair, Coordinator of the planning program and many long learning programs. Former AESOP CoRep member and Metrex vice-president, he is the General Secretary of INU. Visited Penn, University of Colorado at Boulder, University of Porto. His research and publications involve planning theory, strategic planning, ecologic transition and green infrastructures.

Chapter 6
L'*Architecture de survie* (1978) is Back Talking to EU Cities in Crisis. The Provocative Message by Yona Friedman as a Key for the Present and Future Urban Agenda

Paolo Pileri

Abstract Thinking of an essential architecture or a city with less dependence on energy, cars, and concrete seemed utopian when Yona Friedman published *Architecture of Survival*. Today they are urgent and necessary challenges. His book continues to be a resource in the overview of town planning literature, although it is much less considered than it should. *Architecture of Survival* is a very current prism to review our way of thinking and designing the city. Living with less and generating more benefits is the bet we face. For Yona, welcoming poverty in the urban and architectural projects is an indispensable condition for everyone (the ordinary man). It is a tool to give space to simplicity, which would help us live better, have a deeper relationship with the environment around us, re-inhabit urban spaces, appreciate the little things, and understand and respect nature. No architecture can afford to trample man and the planet by impoverishing both. These pillars of 'good living' always need new narrative projects to keep them alive and current in the urban project. Yona Friedman tried it, and his utopia is not only current, but today it is still a precious and dense material to train the young architects and urban planners of tomorrow.

Keywords Survival · Simplicity in architecture · Utopia · Regeneration · Nature

6.1 Starting from a Fact: Architecture of Survival is a Best Seller

"*Architecture of Survival*" was published between 1977 and 1978. The book has been translated in over a hundred languages and, although it is forty years old, it has been, and still remains, a best seller. Its success was not just ephemeral, as may often occur but, on the contrary, enjoys greater critical success today than when it first came out.

P. Pileri (✉)
Department of Architecture and Urban Studies, Politecnico di Milano, Milan, Italy
e-mail: paolo.pileri@polimi.it

C. Perrone (ed.), *Critical Planning and Design*, The Urban Book Series,
https://doi.org/10.1007/978-3-030-93107-0_6

All those who write about urban planning, architecture or landscape design dream of a success even twenty times less remarkable. The book's success is the first inevitable historical fact regarding this work. Whether we like it or not, in the face of this combination of success and longevity, we must come to terms with the fact that its message still survives and still has something to communicate.

Although negative critic reviews persevered for years, they did not manage to overwhelm Yona Friedman. Such criticism was supported by the opinions of experts who, legitimately, distanced themselves from Yona. We should today have the clarity of mind to reinstate this book which is still left out of our bibliographies and which we hardly introduce to our students from the schools of Architecture and Urban Planning.

6.2 Ambition: For the Next World

In its first edition the book was published with a very ambitious subtitle, *«ou s'invente aujourd'hui le monde de demain»*. I believe this is an important key to understanding the primigenial sense of this work and perhaps also to understand—and reject—part of the bad reviews received in the wake of the current subtitle, *«une philosophie de la pauvreté»*.

We know full well: nobody likes the word 'poverty', least of all within Architecture. I feel that this second subtitle may have favoured a hasty and unfair classification of the book among utopian literature or, in the minority, in a section of the Architecture that wanted to dedicate itself to the cause of the poor and destitute. As though dealing with the poor were a disgraceful matter.

In fact, it is not rare that anyone reading the book gather just that, even reducing Yona's message to some sort of list of recommendations on minimum Architecture for those who have no other possible prospect. Who reads the book in this light often decides that it makes for inadequate reading for the western architects in that, he feels, it is anachronistic.

Unfortunately, the power of the present subtitle has contaminated the sense of the book, broadening the possibility of a negative interpretation that prevents the reader from getting involved, without preconceptions, with the message and deep sense of Yona's poetry.

So we will endeavour to follow a more hidden path between the pages of this book. A path which will not make us slip into clichés and which does not linger on inevitable failures that western culture has placed on countries of the global south, of which Yona has been an extraordinary critical observer.

All this may be found in the book and is a fundamental part of it but we will attempt to go beyond that and try to capture passages, meeting points and variants with which we can shape a more general philosophy. A philosophy that I believe is still very contemporary and useful to Architecture and Urban Planning.

A path which will try to give to the concept of poverty a wider and more complex meaning of which we still have a strong need in order to "adjust" our understanding of reality and go forward into the future. This too is part of both Yona's ambition and that of ourselves.

6.3 To the Ordinary Man

In order to avoid trivial preconceptions it is not necessary to wait for the last chapter but it is suffice to focus on some key clarifications in Yona's introduction that immediately reveals that this book is not a manual for First-Aid Architecture in developing countries, even though Yona has spent a great part of his life there.

> «This book is written for the ordinary man; my aim is to make him think, not to shock him». (p. 11)[1]

And here is Yona's key statement: addressing everyone and not only professionals. An ambition which was probably not understood or not welcome. Yona chooses to go beyond the boundaries that keep the holders of the Architectural Code on one side and everybody else on the other.

He determines that the things he has to say on Architecture concern everyone and that they should make everyone think and—something important which he will then say repeatedly—everyone should be able to understand the language of Architecture or that of Urban Planning so as to be free of an alleged holder of the truth on Architecture.

Bold statements which some consider ideological. But once again we must not take this for ideological impetuousness and disregard everything. We can better understand what Yona wants to tell us and find out how his ambition is not only still ongoing but should, and can, become our ambition as well as shows us something which is still a weak point in the role of Architecture and Urban Planning.

6.4 Communication is Part of the Architecture and Town Planning Project

> «Any organisation, no matter which, is based primarily on communication». (p. 63)
>
> «...a fact often overlooked: in a group of human beings the communication mechanism works only within a highly restricted number of people». (p. 76)
>
> «How does one impart such worthy advice to four billion bipeds? And, above all, how does one impart them rapidly and with the certainty that there are no misunderstandings on how to apply them?». (p. 77)

---

[1] The text and all the quotes have been translated by Ludovica Sechi, architect and english mother tongue, and by me. Into the brackets the reference of the page of the Italian version: Friedman Y. (2009), *L'architettura di sopravvivenza. Una filosofia della povertà*, Bollati Boringhieri.

In these three sentences, Yona makes it very clear what he is concerned about and, consequently, what is to be addressed. Speaking to everyone and, more so, succeeding in making everyone understand are fundamental aims for an Architecture that wishes to address things of the world, namely, which wants to improve the world we live in and our way of living the world.

Failing to invest in communication means isolating Architecture in a field available to few and others either look on, endure or, worse, become convinced they are taking part in some improvement, because they are told so, but that they are unable to recognize as they are not provided with critical knowledge.

The subject grows complicated. Not being able to understand what is happening because one does not have the access codes to certain language is perhaps the first form of unavoidable poverty. Many are forced to endure such a condition from which they are unable to free themselves or—worse still—do not even suspect they should free themselves of.

The forced urban drift that millions of people have experienced in the past, deceived by a development and wealth model which conditioned them into leaving their small villages and moving to large towns, created millions of poor people that had not hitherto been poor.

According to Yona, Architecture has not been able to supply those tools of critical analysis which would have been advantageous for many. It just limited itself to support an ill-fated development model, distorting language to the extent that nobody was able to understand and, perhaps, not even Architecture itself.

The forced urban drift has by no means ceased today. It has merely taken a new turn. The lure of the big towns still applies, depleting inland areas and urging millions of citizens to become puppets in a metropolitan game that needs their money, their debts and their firm belief that taking out a mortgage for a flat in the suburbs is right and proper.

Communication in these town projects has not been at all lacking. Rather, it has been strong and convincing. That which was missing and is still missing—and Yona notices back in the 70s—is a good tool box with which another part of Architecture should react giving those 'millions of bipeds' the necessary tools to understand and object to, if the case may be.

6.5 Becoming Aware/Knowledgeable

Hence, for Yona communicating is a great and earnest project that has the ambition of making everyone aware through a plan for accessing knowledge. It is in this manner that inequality is abated.

Upon accessing knowledge, one is immediately able to acquire an 'independent' conscience, that is to say a conscience that is not blind and dependent upon the thoughts and will of others. Self-determination is central in Yona's thinking and Architecture cannot be a discipline that stays on the outside and compels people to perform acts of faith.

People have the right to understand and ask and Architecture has the duty to explain and attend to the legitimate requests of everyone. Yona's self-determination is not specious nor trivial. It is not even an ideological claim for supremacy, just for the sake of it or just to go against authority. It is a status of a society that experiences fully its right not to be poor due to its being dependent on other people or on some other matter. This position is still a fully current topic now.

For instance, an Architecture that actually obliges people not to be able to go without a substantial amount of energy is an Architecture that makes people poorer in that it forces them to live a life of increasing conditioning.

Yona is interested in Architecture, in fact, he adores it and I believe it is for this reason that he wrote this book. He gave it the great dignity and responsibility it has and that it must not lose. Yona is not interested in instigating a meaningless protest nor in being the leader of committees or of the deprived.

Yona is interested in diminishing inequality, giving great dignity to people in the face of an urban planning decision and, to counterbalance, make Architecture understand the enormous challenge it faces and that millions of people depend on its outcome for a happy and wealthy life.

> «It is better to be poor and independent than rich and dependent upon others [...] He who wishes to be independent must accept the fact that he is poor. [...] Being independent means accepting not to become rich». (Ch. 4, pp. 91–93)

This dilemma is, right now, a basic topic for ethics in architecture and in urban planning, especially for those who have a public and independent position, like academics. When they accept to hold a task or a job as professional, this makes them richer but less independent. In the case of academics, the reduction of independence is a social problem as the whole society loses an independent voice. This is a great damage that puts on risk the hope for a better future.

6.6 Understanding Nature

Nature is the first to be greatly misunderstood by Architecture. And this is incredible according to Yona because, as a matter of fact, the natural environment is the inevitable stage on which the urban and habitability project is performed and should therefore be the place which has been given the most careful thought and is closest to one's heart. But it is not so. Even the solution to this old-time contradiction may be found in knowledge.

> «Nature is habitable as long as we are aware of how to inhabit it and are able to behave according to its needs». (p. 87)

But Yona knows full well that the contradiction is detrimental. Knowledge is a necessary condition, but it is not enough. We need to distance ourselves from our selfishness, our pride in not being economically dependent (according to Yona, this is an ontological condition of political power), and awareness that «resources are exhaustible». (introduction, p. 12)

Since 1977 Yona saw in all this awareness the driving force of the expected "ecological revolution". This passage in which he highlights how urgent it is to understand that resources are limited and therefore to plan Architecture according to this limitation is Yona's chance to explain to the reader just what he means by 'philosophy of poverty':

> «I did not intend to idealize slums, nor the poor, nor the people. Rather, my hope is that through this book the idea of rationing may be revived. That is to say, the conviction that resources are limited and the idea of a just rationing, which must thus be decided by those who will suffer the consequences». (introduction, p. 14)

If resources are few, why take away decision-making supremacy from those who will suffer the consequences of the squandering due to a short-sighted selfish Architecture which based its principle on the denial of limitations? The question is an extremely contemporary one.

6.7 Habit and Anti-habit

The path set by an Architecture that impoverishes Man and the Planet goes back hundreds of years but never as in recent times has the increased pace produced such consequences and intensified so many circumstances of inequality and poverty. Ironically, this attitude has become a habit.

The habit of using one's car even if only for a journey of a few hundred metres. The habit of depending on ever-increasing allocation of energy. The habit of being accustomed to a housing market that produces newly poor people and new exclusions. The habit of being accustomed to towns without green space. The habit to towns lacking in high-quality facilities. The habit of looking upon the population outflow from inland areas to cities as a common and acceptable fact. The habit of convincing ourselves that the major projects or the major infrastructures for high-speed mobility will free us from this predicament. The habit of accepting the fact that land and soil be used up profusely as long as whoever is planning promises a few jobs, no matter if in a dark warehouse, no matter whether they vanish within a few years, no matter if it is for an economical model of pure extortion. The habits, within the urban project and that have blurred our idea of the town, are many. Moreover, habits are deceitful and harmful and Yona is extremely clear-headed in realizing how this is a disadvantage rather than an added value as:

> «Custom is very powerful and the human animal constantly endeavours to keep most of its habits». (p. 31). And
>
> «Habit is the worst flaw of all planning which is always founded on a hypothesis». (p. 141)

And here is Yona's reply: clear, intense, precise and, as always in this book, challenging and never banal.

«Today an education that prepares for anti-habit is essential. [...] Anti-habit is a quest for new ways of solving new problems. Not in the way that an artist – who feels in a class of his own – would do but, rather, as a vigilant alpinist who is aware that just one mistake on his part could be fatal». (p. 141)

Poverty abides here. It may be discerned in the blindness in perceiving a threat and not always an opportunity in something that has always been done. Yona turns the tide and, frankly, we are not surprised at the reticence on behalf of most Architecture and Urban Planning in the face of these stances as it is difficult to eradicate conservatism. Yet after forty years his appeal to education—a special education, one that is critical, independent and aware—is all the more contemporary and necessary. Rather, the power of illusion of augmented reality and the seductive charm of technology are extremely insidious and do not help people to think in an unconventional way and distance themselves from the new habits that those tools are marking and towards which much Architecture is moving. The best antidotes for poverty are schooling, culture, reading, dialogue and art. Politics should aim at anti-habit.

6.8 Survival

We approach here the key concept of Yona's thinking and his originality which still exists and which is very helpful in asking ourselves which Architecture and Urban Planning we need to learn to claim in the future. Yona, very clearly, states that,

«An architecture may be considered an architecture of survival if it does not complicate (or, rather, if it facilitates) the production of food, water supply, climate protection, the safeguarding of public and private goods and the commons, the organization of social relationships and the fulfilment of one's aesthetic sense». (p. 103)

The concept of survival takes full form and power in these words. But also responsibility. In Yona's view, concepts are never descriptions but always acts of responsibility. In this he teaches us how activism is a crucial ingredient in the architect who cannot be a dull "aye aye, sir", but only an observer provided with keen critical thinking. Respect for nature and the philosophy of the limit causes Yona to also say that

«...Architecture of survival consists of seeking to limit transformations and preserving only that which are most necessary so that Man may survive in conditions adequately favorable». (p. 88)

It is a manifesto of the idea of a sustainable town *ante litteram* and, today, the authentic version of urban regeneration. The pillars of Yona's thought support one another and are summarized in the concept of survival.

«Architecture of survival is therefore the search for an architecture and life 'plan' which reduces an individual's dependence upon the others. The poor have behaved in this manner since long before such conduct was ever analyzed». (p. 89)

Not only respect for the environment but also reduction of dependencies. Pure genius. If we were to paraphrase Yona, today we would have to say that it is not suffice to 'convert' to the electric car in order to define oneself as ecological and that the new hybrid traffic is not a survival choice. It does not help us in any way to get away from some severe conditioning but, rather, continues to keep us enslaved to a specific way of moving around—of wasting time in queues and in traffic (and knowing that it is an electric congestion is of little consolation). It continues to make us think that the car is the main way of getting around in our town and continuing to be dependent on the energy bill for a mere few kilometers.

But, above all, the technological illusion keeps us away from other hypotheses and from other alternatives which, incidentally, would make us less poor. All things considered, walking or cycling to school or work or using a better means of public transport remains a valid alternative, albeit obscured by other interests and by not being able to see this as an alternative due to ignorance.

It is not by chance that the book ends in this manner,

> «Architecture of survival leads to a refusal to believe in promises—those promises that are perhaps benevolent, but definitely deceitful—which the experts have chanted for centuries. We are thus concerned with the application of a long-forgotten uncompromising realism regarding the way we inhabit the earth». (p. 154)

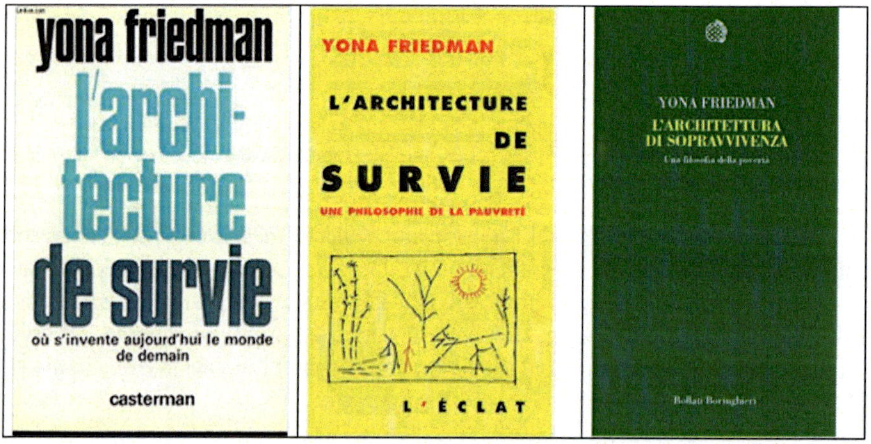

Paolo Pileri is a full professor of urban and ecological planning and design at Politecnico di Milano, his research focuses on deep ecology and sustainability on three axes: i. Soil in urban planning and zero land consumption. ii. Design the Slowness by the way of long tourist walking and cycling routes to regenerate and enhance fragile territories (see VENTO project - www.cicloviavento.it; TWIN project - www.twin.polimi.it/). iii. Qualifying urban public spaces in front of schools (see piazzescolastiche project and www.habitatscuola.polimi.it/). Author of more than 350 publications, columnist for the magazine Altreconomia (altreconomia.it/author/paolo-pileri/), advisor for the Italian Ministry of infrastructures and mobility.

Part II
Planning Pathways

Chapter 7
John Friedmann, *the Good Society* (1979): Panning Pathways for a Just Society

Ute Lehrer

> *Social justice aspect is particularly pronounced in our work here. It is absolutely essential that our theoretical thinking of planning have a value foundation. I would certainly encourage all of us to be very clear about that, and to start thinking specifically about what it means. (…) We come to planning because we want to change the world. We are world changers, that's our ultimate motivation.*
> John Friedmann 2016

Abstract John Friedmann's concept of *The Good Society*, which he developed in the 1970s and published as a book in 1979, shows his dialectical inquiry into the roots of radical practice. While his other writings, such as the *World City Hypothesis; Empowerment: The politics of alternative development*; or *Planning in the public domain: From knowledge to action*, have received countless citations, his book from 1979 is marking its place in the academic world only now. It is a visionary as well as a poetic oeuvre in an unusual writing-style with self-contained paragraphs. It presents a dialectical inquisition of possibilities and challenges of a just society, where hope is at its center. At a time where we see three crises overlapping—climate, health, and racism—this book might demonstrate a path of conquering injustice through dialogue, where every participant is the questioner as well as the responder, where social values, political strategies, and collective action come together with a theory of reality and form the core elements of radical practice.

Keywords Good society · John Friedmann · Radical planning · Dialogue · Just society · Social learning · Democracy · Possibility/challenge/hope

U. Lehrer (✉)
Faculty of Environmental and Urban Change, York University, Toronto, Canada
e-mail: lehrer@yorku.ca

C. Perrone (ed.), *Critical Planning and Design*, The Urban Book Series,
https://doi.org/10.1007/978-3-030-93107-0_7

7.1 A Life of Inquiry

John Friedmann (1926–2017), one of the most authoritative voices in planning theory, a long time university professor at UCLA (followed by engagement at RMIT Melbourne and UBC), with honorary degrees from the Catholic University of Chile, the University of Dortmund in Germany, York University in Toronto, and the University of British Columbia, sometimes called the "pope of planning" (Sandercock 2011, xviii), and in Patsy Healey's words "perhaps the greatest planning scholar of the twentieth century, and undeniably so in the field of planning theory" (Healey 2011, p. xi), has been a strong advocate for how to build better societies. In over two dozen of books and hundreds of articles his mission was to form a place for planning as a discipline that would help to create a better, a more just world. His way of thinking was informed by his own past as well as firmly grounded in social theory (Hannah Arendt's work, for example, had a strong influence on his thinking) and he saw urban planning as a social science-based profession.

What counts first and foremost in his work was the question of justice and value based thinking. In a video interview from 2016, he argued that planning has to be grounded in a clear value: "We have to carefully think about which values we want to further in the world around us. But the oldest one is social justice, and the whole question of (…) how to have a more equalitarian society that is inclusive of all different modes of living" (Friedmann 2016). This search for value based planning has accompanied his entire life as a planner and advisor for regional and international governments in the US, Chile, Brazil, Venezuela, and South Korea as well as a scholar and professor.

Within his prolific publication list, one book stands out: "The Good Society", published in 1979, which is highlighted in this contribution here. It is not his most famous one, in fact it is one that did not get much attention at its time. It stands time-wise between "Retracking America: A Theory of Transactive Planning" (Friedmann 1973) and "Empowerment: The Politics of Alternative Development" (Friedmann 1992). Because of its intellectual relevance of both books, and for showing where Friedmann came from before writing "The Good Society" and where he then eventually went with it, it is important to at least briefly mention the main points of the two books here.

In "Retracking America" (Friedmann 1973), dedicated to his father who had died a year earlier than the completion of the manuscript (the personal history is relevant here), Friedmann suggests a new style of planning, which is in synch with a post-industrial America, responding to, as he calls it, the "crisis of valuing and knowing". The book is a strong plea for mutual learning, for dialogue, and for a learning society. Here he talks specifically about a cellular organizational structure of society, something that he will pick up again in "The Good Society" a few years later, but in a more elaborated form. He proposes transactive planning and with it, foreshadows his analysis of planning history and theory in his seminal book "Planning in the Public Domain: From Knowledge to Action" (Friedmann 1987) where he divides up the field into four traditions: Social Reform, Policy Analysis,

Social Learning, and Social Mobilization. Closest to his own thinking is clearly the Social Learning approach in planning. But he is keenly aware that there are at least two fundamental problems: "the rationalistic bias of the social learning approach and validation of knowledge" (Friedmann 1987, p. 217).

Written almost a generation after "Retracking America", Friedmann argues in "Empowerment" (1992), that it is the mainstream doctrine of economic development and policy that continues to keep people in poverty; in order to lift them out of poverty, he states, we have to have a fundamental shift, a shift that has a solid theoretical foundation. He takes a stand that growth should only happen within an inclusive democracy and with a socially and ecological appropriate approach, which acknowledges the central role that women play within economies not only as a labor force in the organized job market, but also in their productive and reproductive roles within households. Always trying to work within a system, Friedmann suggests that it needs a morally informed framework, which should be implanted into the structures of mainstream development, assuming that the tensions would create a better future for the poor.

7.2 The Good Society

"The Good Society" is a book between those two not only timewise, but also because it shows the thought process of an intellectual, whose life work was (a) to find a way that benefits those that are disenfranchised and (b) to define our profession in a sophisticated way that allows to spread the benefits of planning among all citizens. Never really interested in the built form and urban design, Friedmann's interest was centered in the architecture of a just society.

It is not surprising, however still unique, that "The Good Society" has a format that is very different than any other academic books in the field of planning. Long before poetry and storytelling made it into the planning profession as a method to both analyze and communicate knowledge, Friedmann, who, with his interest in, and active contribution to, music and poetry, was often described by his friends and students as a renaissance man, approached this book in a non-conformist way. Already in the preamble he makes this clear by stating: "The Good Society: A personal account of its struggle with the world of social planning and a dialectical inquiry into the roots of radical practice." (Friedmann 1979, n.p.)

At the time, Friedmann was influenced by Confusion thinking, of a world in balance between Yin and Yang, between receptive and active energy, of dualisms that appear as opposite but in reality are interconnected and complementary, and giving rise to each other. Friedmann also was searching for a different way of knowledge production. And therefore it is not surprising that "The Good Society" stands out among all these writings. It uses an unconventional format, in its structure closer to a musical composition, with a prelude, repetitions for emphasis of points and recapitulations. As Friedmann states:

> The basic elements of the argument are self-contained paragraphs. Of varying length, ranging from a singles sentence to several pages, they are interspersed, at appropriate places, with poems, aphorisms, short selections from philosophical writings, and other "delights" that are intended to illustrate, contradict, confirm, and illuminate the paragraphs in their immediate vicinity, adding depth and concrete imagery to the more abstract portions of the text. (Friedmann 1979, xvi)

Before delving into the core of the book, I would like to situate it in its wider context. The unusual style reminds me of another book that was published a few years earlier: "The Society of the Spectacle" by Debord (1967), where in 221 aphorisms, each a paragraph long, Debord presents his analysis of modern society, deeply grounded in critical Marxist theory, stating that commodification has taken over social life. And then there is the discussion on the "post-industrial society", a term that has been around since the fifties and sixties (Brick 1992), but which was used for the first time in academic writing within the English context through David Bell's book "The Coming of Post-Industrial Society" from 1973 and a few years earlier in French by Touraine (1969). Also Ivan Illich (1973) had referred to the term, but then created a more positive expression as a response that the production process and with it, the entire society had change, kind of along the line as that after the agricultural society had changed, it was not called post-agricultural but industrial. As a response of the dominance of technocrats within industrial society and the institutionalization of specialized knowledge, Illich suggested a framework that is based on conviviality, a concept that goes back to mutual respect and appreciation of each other, and that wants to develop new instruments for the reappreciation of practical knowledge by common citizens. Conviviality also was central in Friedmann's thinking, particularly through the work of his friend and collaborator Lisa Peattie.

7.3 What Are the Core Aspects of the Good Society?

When in the 1970s the shift happened to a newly globalizing, neo-liberal economic world order, with recapitalization of capital, a shirking state and market competition reigned supreme of which we see the results today—increased polarization between rich and poor and a shrinking middle class—Friedmann was interested in a positive change of the existing conditions. The mature and always hopeful planner "saw a way out", as he reflects in his book on *Insurgencies* from 2011. Friedmann understood the local, inevitably temporary associations, or sodalities as he called them, "as 'good societies', small in the numbers of its members, bonding through interpersonal relations of dialogue, non-hierarchical in structure, and oriented towards struggles in the street, the law courts, academia, and democratic elections." (p. 51). And he continues: The good society is a "perennial but temporary social phenomenon in permanent struggle with the hierarchical powers of the state. The struggle (…) [is] for opening up new spaces for dialogical relations and therefore for what [he calls] a communalist social order that could never be totalized." It was assumed that social movements were part of it and would emerge as a result of these struggles.

His reasoning was as following: The quest for good society is not something new, but it has been around since at least Plato in the case of Western Society. Friedmann is keenly aware that this presentation of history leaves out what was going on in the non-Western World at that time. It is important to mention that Friedmann himself has some affinity with Eastern Philosophy. In his approach of dialectics he does not follow the Marxist tradition "in which thesis and antithesis would clash", but rather follows the traditional Chinese understanding of Yin and Yang: "a unity of opposites" which is in a "state of continuing fluctuating tension between the two terms of opposition from which there [is] not release, only a temporal shifting in position and power." He argues that in a Confucius perspective "the idea of inevitable historical progress [is] absent." (Friedmann 2011, p. 52).

The quest might be utopian, but utopia in Friedmann's thinking is a "normative theory: a vision of how social relations ought to be arranged, and how we should proceed to structure them." It is also "a work of fierce discipline and of commitment to the transcendent possibilities of being human." And this is achieved through striving for the Good Society, where the "future rises up in our mind as a transcendent possibility, a challenge and a hope (Friedmann 1979, p. xi)."

Dialogue is at the center of the good society, with each person being a questioner and a responder alike. It is a unity of opposites, of dialogue and dialectical movement, it is a relation within which struggle occurs. The aim is to "open up and expand the spaces for dialogue in a society that is subject to the hegemonic powers of large corporations and the state" (Friedmann 2011, p. 58).

The problem as Friedmann sees is, is that the Good Society has to function within a different system, namely that of the state and societal planning, which have at their core hierarchy and power as their ordering principle. "Good society does not seek power for itself. (…) The good society and the state stand in dialectical opposition to one another, but an opposition that constitutes a 'unity of opposites,' or force field that is inevitably in dynamic tension (Friedmann 2011, p. 58)." This means that two different systems are trying to develop a dialogue with each other. The Good Society model is based on the understanding that every participants is both, a "questioner and a responder alike", while the state and social planning represented a system that believes in hierarchy and power. Not everybody has equal voice at each step of the way. Therefore, there is a permanent struggle for the Good Society to assert itself within the other system, to be in opposition of it, in order to open up and advance into new territories because it refuses hegemonic power. While it is in constant struggle, it does not ask for the abolishment of the state because that would have a totalizing effect. The goal is to transform social planning in a way that it allows dialogue and with it, transforms the world. Because of this, Friedmann calls it "radical practice". There are four elements of radical practice: social values, theory of reality, political strategy, and collective action, which are interconnected but also "constitute a paradigm of social learning." (Friedmann 1979, p. 51).

And he sounds almost apologetic when he states: "To those who fail to be involved, it does not usually disclose itself. You will find it only in practice … now here, now there, wherever you are and whenever you are prepared to join in its work. It makes its appearance in the street, the factory, the neighborhood, the school, … these are its physical settings. But they do not define nor limit the Good Society." (Friedmann 1979, p. xiii). In other words, the Good Society is not bound to physical settings but instead can appear anywhere. It is socially constructed, an ongoing process, never achieved, never finished, always becoming, always changing. The Good Society only exists in time, "and its space is the space of social relations." The Good Society does not need a specific container, it is temporal and created in moments, and it is time sensitive and socially produced through dialogue. In *Insurgencies*, Friedmann describes the Good Society in a way that reminds the reader of Lefebvre's Production of Space (1991): The Good Society "is a temporary social formation which, lacking a territorial base of its own, exists exclusively in its actual practices. Its space is the space of social relations." (Friedmann 2011, p. 58).

While small in size (Friedmann suggests it has between 5 and 12 members), the Good Society should not be mixed up with the family unit. In order to be part of it, the family would need to "break down the barriers between the private and public realms, it must cease to be a place for the accumulation of possessions, for the practice of exclusiveness, for the mere reproduction of social relations. It must instead become a place for the transforming practice of dialogue among its members, each of whom is a dyadic being, free and independent, accepted as an equal and in his difference, or hers, from every other member and by each." (Friedmann 1979, p. xiv).

The Good Society is "a work of fierce discipline and of commitment to the transcendent possibilities of being human." And this is achieved through striving for the good society, where the "future rises up in our mind as a transcendent possibility, a challenge and a hope." (1979, p. xi).

The Good Society is an early attempt of Friedmann to address fundamental questions and practices of planning, for an approach to create a more just work, calling it transactive planning. It is a response to, and an outcome of a seminar series by the name "Good Society" at UCLA, where the aim was to offer courses in "critical studies" that had an explicit value perspective. This concept of the "Good Society" has been at the core of Friedmann's work and has expanded into the concept of the "Good City" (Friedmann 2000).

7.4 Why Does the Book Still Matter?

The content of Friedmann's book from half a century ago with its proposal for a Good Society is still relevant. The fundamental questions of planning for a just society have not been addressed yet, let alone solved. And the year in which I am writing this chapter, 2020, amplified injustices. The three crises we see—climate, health, and racism—makes the message of this book more important than ever.

There is hope, and the hope is that we do not go back to the "old normal", where a large part of the population is disenfranchised, where racism is unrestrained and leads to killing of human beings, where an entire group of people—which we call essential workers without acknowledging that these are the same people that are often underpaid—is exposed on a daily basis to a virus that is rampant (Keil and

Hertel 2020). The pandemic as well as the killings of people of color has brought individuals together in ways that have similarities to the concept of a Good Society as proposed by Friedmann. For example, in the manifesto of BlackSpace, where a group of African American planners, architects, economic developers, activists, and story tellers came together to "protect and create Black spaces" (https://www.blackspace.org/manifesto), challenge themselves and their partners to develop a way of interaction that is based on mutual respect, on "dialogue, inclusion and empowerment", where trust is built and hierarchies are lessened, where the understanding of self and others is deepened, while learning happens without assumptions and predetermined solutions and where criticism is accepted without dispute (BlackSpace 2020).

In her open letter to Canadian urbanist titled "A Call to Courage" the Toronto based place-maker Jay Pitter states: "What we need is collective action and humility from urbanists from all racial backgrounds, professional disciplines (including unsung grassroots leaders) and cities." (Pitter 2020, p. 3). What I would like to add is that when we do that we should follow Friedmann's concept of the Good Society, where in the end the opposite is just as must respected as the own position, all in believe to find mutual respect and understanding of each other, to learn and grow, where Yin and Yang are in balance with each other. And therefore, Friedmann's book on The Good Society, which is a dialectical inquiry into the roots of radical practice, is more relevant than ever.

References

Bell D (1973) The coming of post-industrial society. Harper Colophon Books, New York

BlackSpace (2020) Retrieved from: https://www.blackspace.org/manifesto

Brick H (1992) Optimism of the mind: imagining postindustrial society in the 1960s and 1970s. Am Q 44(3):348–380

Guy D (1967) La société du spectacle. Buchet-Chastel, Paris

Friedmann J (1973) Retracking America: a theory of transactive planning. Anchor Press/Doubleday, New York

Friedmann J (1979) The good society. MIT Press, Cambridge, M.A.

Friedmann, J (1987) Planning in the public domain: from knowledge to action. Princeton University Press

Friedmann, J (1992) Empowerment: the politics of alternative development. Blackwell, Cambridge

Friedmann J (2000) The good city: in defense of utopian thinking. Int J Urban Reg Res 24(2):460–472

Friedmann J (2011) Insurgencies: essays in planning theory. Routledge, London

Friedmann J (2016) A conversation with John Friedmann. Video interview, UCLA Luskin. Retrieved from: https://vimeo.com/165949788

Healey P (2011) Foreword. In: Friedmann J (ed) (2011) Insurgencies: essays in planning theory. Routledge, London

Illich ID (1973) Tools for conviviality. Fontana/Collins, London

Keil R, Hertel S (2020) COVID-19: climate emergency, COVID-19, Black lives matter: urban planning's insurgent moment

Lefebvre H (1991) Production of space. Blackwell, London/New York

Pitter J (2020) A call to courage: open letter to Canadian urbanist. Retrieved from: https://canurb.org/citytalk-news/a-call-to-courage-an-open-letter-to-canadian-urbanists/

Sandercock L (2011) Preface. In: Friedmann J (ed) (2011) Insurgencies: essays in planning theory. Routledge, London, pp xvi–xix

Touraine A (1969) La société post-industrielle. Seuil, Paris

Ute Lehrer is a Full Professor at the Faculty of Environmental and Urban Change, York University (Toronto, Canada). She holds degrees from the University of Zurich, Switzerland, and the University of California, Los Angeles, has previously taught at Brock University and SUNY Buffalo, was a researcher at the Swiss Federal University of Technology in Zurich, and held visiting positions at Université de Montpellier, France, University of the Witwatersrand, Johannesburg, and University of Manchester, UK.

Taking a critical and comparative approach, her research interests include urban design and land use; housing, gentrification and the condominium boom; discourse and mega-projects; as well as the social construction of public space. She has published widely in social science journals and edited books. She was the Principal Investigator and a collaborator in several Social Sciences and Humanities Research Council (SSHRC); funded research projects, including Global Suburbanisms.

Chapter 8
David Harvey's *Urbanization of Capital* (1985): why it helped me so much

Michael Edwards

Abstract This collection of essays set out to show how successive crises in capitalism generate urbanization as capital switches between extracting surplus from production of commodities for sale to land speculation, property development and the production of infrastructure and housing, often with the state as a key intermediary. This approach to analysing the relationships of citizens with their landlords, employers and suppliers at various scales was a huge help to me as a young academic in the mid-1980s. We were trying to make sense of rapidly-changing capitalism as old certainties of the 70s were dissolved and many of us found Harvey's approach illuminating for our students and ourselves.

Keywords Accumulation · Switching · Circuits · Rent · BISS · Teaching · Pedagogy

The Urbanization of Capital (with a z because it's American) (1985b) and *Consciousness and the Urban Experience* (1985a) were collections of previously published papers and some new material, put together by David Harvey in the early 1980s and published in 1985. Neither book is necessarily the best way of starting to read Harvey but I am writing here about *The Urbanization of Capital* because it was so important for me. I describe my intellectual situation in the 1980s in an attempt to explain the value this book had for me and for my subsequent teaching. Rather than summarising the book's content I explore some of the directions in which it has led me and others in the intervening decades.

As a young university teacher in England in the 1970s, I found intellectual and political nourishment mainly in the Conference of Socialist Economists (CSE), a network of hundreds of social scientists and activists which met for an annual conference in Bradford and published a Newsletter which later became the journal *Capital and Class* (and which lives on). It spanned a wide range of fields of study under an implicit aspiration for an integrated social science, more-or-less Marxist. There were some great luminaries in that group including Robin Murray, Diane Elson,

M. Edwards (✉)
The Bartlett School of Planning, Faculty of the Built Environment UCL, London, UK
e-mail: m.edwards@ucl.ac.uk

C. Perrone (ed.), *Critical Planning and Design*, The Urban Book Series,
https://doi.org/10.1007/978-3-030-93107-0_8

Simon Mohun and Cynthia Cockburn and there emerged a sub-group, the CSE Political Economy of Housing Workshop, convened by Steve Merrett and me and with inputs from Doreen Massey, Michael Ball and a few dozen others. Through all this I was beginning to read Marx, as were others in my university faculty who were less concerned with housing, more with construction or architecture and planning.

Out of these friendships in our university (The Bartlett School, UCL) emerged a plan to hold a Bartlett International Summer School on the Production of the Built Environment (BISS), blessed by a start-up grant of £1000 and tacit permission to use the name from the head of the Bartlett, John Musgrove. The first of the Summer Schools was held in 1979, and they continued until 1996, with a printed volume of *Proceedings* emerging in each subsequent year.[1] The notion of a 'school' with distinct teachers and taught had been part of the first meeting but it immediately faded, and all participants had the same status and contributed to the costs thereafter.

The distinctive core of the BISS was the sharing of a broadly Marxist approach to the built environment, insisting on the integral analysis of production (in the narrow sense of construction) with distribution, circulation and the realisation of value (i.e. production in its full sense). For many the labour process, especially on the construction site, was regarded as the driver of change while those of us interested in planning, infrastructure and the markets in buildings were often a bit divergent, discussing land ownership, rent, urban planning and so on.

I think men outnumbered women in the BISS but not heavily and the leader throughout was Linda Clarke, initially a PhD student and now a Professor at the University of Westminster. She and her partner Jörn Janssen from Dortmund remained very productive and are a leading force in the European Institute for Construction Labour Research.[2] Doreen Massey had taken part in some meetings, while Anne Haila of Helsinki joined BISS a few years later.

Out of the group in the Bartlett School there also emerged plans and a syllabus for a Masters programme on Production of the Built Environment which launched in 1986 and led to my first encounter with the actual David Harvey. Harvey's work provided a major part of the reading list for our course, so by December we had all read a lot of his output and I asked whether we should invite him for a seminar. The students were keen, though only on condition that he did not give a talk, that instead they would prepare short provocations or challenges for discussion. We put this idea to him and he accepted—rather intrigued—and he duly came in the spring of 1987. The students did a superb job. I recall a powerful argument from Madeleine Wahlberg that Harvey's minimal treatment of gender issues seriously undermined not just the treatment of reproduction of labour power but the whole analysis of urban structure. Someone else, probably Jonathan Charnley (later Charley), started a discussion on Harvey's rather minimal attention to the state. Alan Spence, former shop steward, Communist Party editor and Covent Garden activist, led a debate on urban social

[1] Long out of print, these volumes have now been scanned by May Day Rooms and the PDFs are free downloads at: https://drive.google.com/drive/folders/1bPR860zXGUcH7sVXvceY9CpDJgRu4hwC.

[2] http://www.clr-news.org.

movements and their potential for the working class. These exchanges, and others which I have forgotten, took all afternoon and we ended up sitting on the grass with bottles of wine. Harvey had been receptive and responsive throughout and ended up saying that, like Marx when he died, he had a lot of unfinished work to do.

I should complete this scene-setting by explaining that our little group of teachers were partly preoccupied with how and what to teach first year undergraduate (bachelors) students in a course (Edwards et al. 2009; Edwards 2014) designed to foster an integrated understanding of urban development, challenging the inherited division of labour between professions and emphasising the social and economic forces at work in settlements. It was in this context that my reading of Harvey's book was especially valuable.

Harvey's quest in this book was to put *space*, and specifically space at the urban scale, into a Marxist analysis, indeed to make it central. Whereas Marx had dealt brilliantly with time throughout his work, space had never seriously figured, save in a few disjointed insights like the famous passage in the *Communist Manifesto* on globalisation and its impacts on established trades:

> They are dislodged by new industries, whose introduction becomes a life and death question for all civilised nations, by industries that no longer work up indigenous raw material, but raw material drawn from the remotest zones; industries whose products are consumed, not only at home, but in every quarter of the globe. In place of the old wants, satisfied by the production of the country, we find new wants, requiring for their satisfaction the products of distant lands and climes. (Marx and Engels 1848)

Formally, this book centres on the proposition that the continuing accumulation of capital will from time to time generate more money capital than can profitably be put to use in the primary process of the economy, the production of commodities. When this happens, one outlet for capital is to switch into a second process, the production of the built environment, i.e. urbanization. Thus we get the physical extension of settlements, transport, urban infrastructure and so on. There are further escape routes too for over-accumulated capital, into education, research and development and into social expenditures, policing and control.

The exposition of this draws on the massive work which Harvey had completed a few years earlier in *The Limits to Capital* (1982), essentially a re-working of Marx for the late 1970s with a strong emphasis on geographical space.

These processes are mapped in the much-reproduced Figure 3 of the later book, Fig. 8.1 here,[3] drawn from Chapter 12 of *Limits*.

The first thing to stress about this diagram is 'the structure of relations...' because the entire book is about relationships: the links between processes, classes of people, institutions. A second key feature is that it can be seen sequentially, with the primary activity along the centre of the chart, secondary (including built environment) above

[3] In this caption and in Harvey's text the term 'circuits' is used though, as Brett Christophers (2011, note 1) rightly points out, there is potential confusion because elsewhere Harvey (following Marx) distinguishes the circuit of productive capital, the circuit of commodity capital and the circuit of money capital, a quite distinct usage. I'm following Christophers here and trying to avoid 'circuits', as indeed Harvey did in the caption of this diagram in *Limits*.

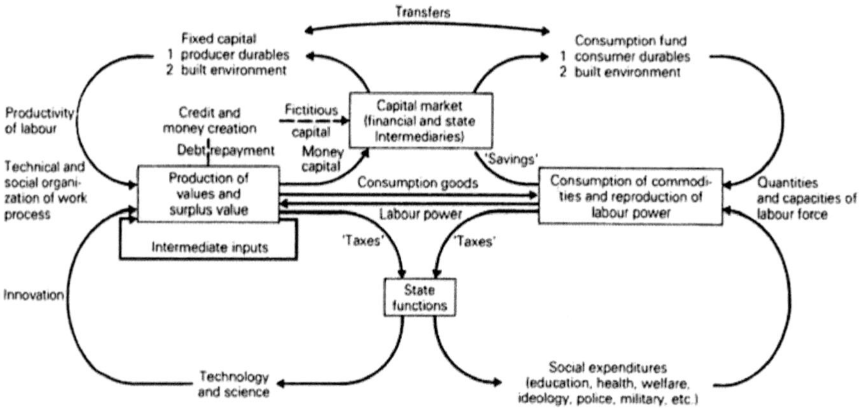

Fig. 8.1 The structure of relations between the primary, secondary and tertiary circuits of capital, Figure 3 of the 1985 book, reproduced by agreement with Blackwells/Wiley

and tertiary below. Indeed this chart, like so many, is the lecturer's friend because it is much more communicative when drawn live by the speaker, or animated as a story. Anyone who has not seen the Royal Society of Arts animation of a Harvey short talk is urged to see it. Search engines know where it is.

The related text in Harvey's foundational Chapters 1 and 2 explores the relationships in some detail, reflecting for example on whether the switching of capital from one function to another tends to be sequential or simultaneous and what may trigger such switches. The key here is crises. That is a term bandied about rather loosely by some Marxists but Harvey is very measured with it. A sectoral crisis arises where a specific part of the economy finds itself stricken, giving capital an incentive to move out of that field if it can, or at least stop reinvesting profits there, and to switch to another sector or perhaps out of commodity production altogether. Another quite precisely defined form is a regional crisis, arising when a geographic area no longer yields good returns, provoking switching to others. Or, of course, a crisis can be global, system-wide, provoking capital to switch everywhere out of the direct production of commodities into other spheres in search of opportunities to accumulate.

I have always found these ideas to be invaluable tools, helping me to make sense of what I see in the world and to help students make sense of their experience and to link spatially scattered phenomena. In our teaching we would always take our students for a week to a foreign city, asking them to describe and make sense of what they could see there: the buildings and urban fabric telling of periods of rapid growth, decline, abandonment of shipbuilding in Venice or of steelmaking in the Ruhr; looking out for international flows of investment in the built fabric of Barcelona or Bilbao and linking those to the supply chains which made them possible. Did you know, for example, that the Guggenheim Museum in Bilbao is clad with titanium after the huge stocks hoarded during the cold war were dumped on the world market in the dissolution of the USSR, making it briefly affordable?

Back in London, it was invaluable in the 1980s and 1990s to have this kind of mental apparatus to hand in making sense of the surges and collapses of property speculation, notably the boom and bust of the office market which surged while capital withdrew from manufacturing in the Margaret Thatcher period and switched to finance and property development—either direct, with the same firm moving from one to the other, or more commonly via the financial sector and capital markets channelling money towards property. Here we had radically new structures in the banking sector—multi-function mega-banks replacing archaic little specialised elite companies in operating the financial markets. They needed a completely new built environment and they got it, very fast, with attendant professions working day and night to produce the buildings, the City of London planning department throwing out their plan for a new one to enable the change and the relatively new London Docklands Development Corporation approving the massive development of Canary Wharf only months after a developer suggested it. Great lessons here about the role of the financial sector and of 'the state'. The English state apparatus was perhaps always very well adapted to the needs of capital (or at least some capitals) and, in the 1980s, was an effective midwife to the new financialised rentier economy which was to dominate the rest of my lifetime.

Harvey's insights were indirectly useful to us in a social movement at King's Cross, opposing developers over a massive, mainly office, development planned behind the railway station. I was working with this campaign and in 1990 we were trying hard to fend off the risk that the local municipality, Camden, would grant permission. I was convinced that the office market bubble would soon burst, others in the campaign agreed and we did all we could to secure delay. I met an ex-student, by then chairman of London's leading property market analysts. He whispered his agreement. It worked. The London office market crashed, taking with it Canary Wharf, the King's Cross developers and many others.

The actual empirical testing of Harvey's 1985 propositions about capital switching received very little attention at the time but Brett Christopers undertook a careful study of the UK data from 2000 to 2007, charting the lead-up to the great financial crisis, thinking this would be a good test. His paper (Christophers 2011) is a great vindication, demonstrating not only that there had been a massive expansion of investment flowing into real estate in the UK in the run-up to the 2007 crash, but also that it really did appear to be investment switched from production processes, much of it via the growing amounts of cash which companies were retaining instead of reinvesting back into production.

Harvey's own accounts of the 2007 crash are now numerous, most interestingly I think in The Urban Roots of Capitalist Crises (Chapter 2 in Harvey 2012), always reworking the original switching analysis from the 1980s emphasising how, by 2007 the urban economy, with its mass of real estate values dominating other assets, WAS the economy. In a sense this way of thinking has become everyone's which is a great tribute. I was having thoughts indirectly inspired by Harvey in 2015 when I compiled this chart, Fig. 8.2, for a government report (Edwards 2015) without even citing him as an inspiration.

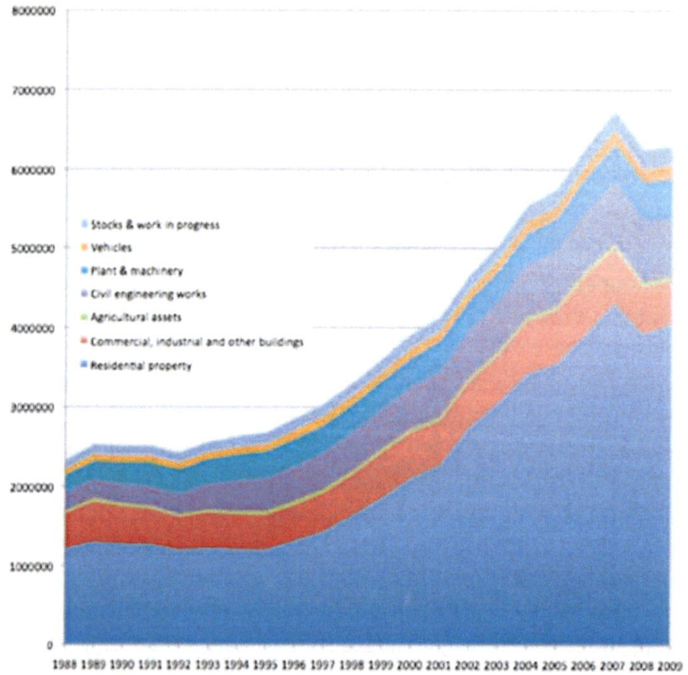

Fig. 8.2 UK Market value of tangible assets, £m, 1988–2009. *Source* Edwards (2015); data from ONS non-financial balance sheets (annual)

I am getting ahead of myself, both in time—leaping into the new millennium—and in scale: leaping to the economy as whole and away from the main learning I gained from Harvey's 1985 book which was about ways of analysing the production of the city and of individual buildings.

One of the valuable adaptabilities of Harvey's circuits diagram (Fig. 8.1) is that almost any part of it can be expanded to explore and explain processes in more detail. There are some brilliant examples in the talks of Christian Zeller in which he elaborates the growth of derivatives markets and the securitisation of property investments by adding successive loops around the housing financing section. I cannot illustrate them here as I have been unable to find an example.

Another expansion is one which I have used a lot in teaching: adding detail to the central horizontal spine of the diagram representing the production process and showing how it could link to processes we are concerned with in the built environment. Figure 8.3 shows a sequence of 6 lecture slides which build from the generic commodity production process, adding a financial sector, then a construction sector, then property investors (landlords) and lastly households. The final slide puts all these together and the talk discusses the flows of materials and money, the extraction of value at various stages and the interdependence of builders, landlords, banks and the working class. It can support, for example, discussion of how workers' living

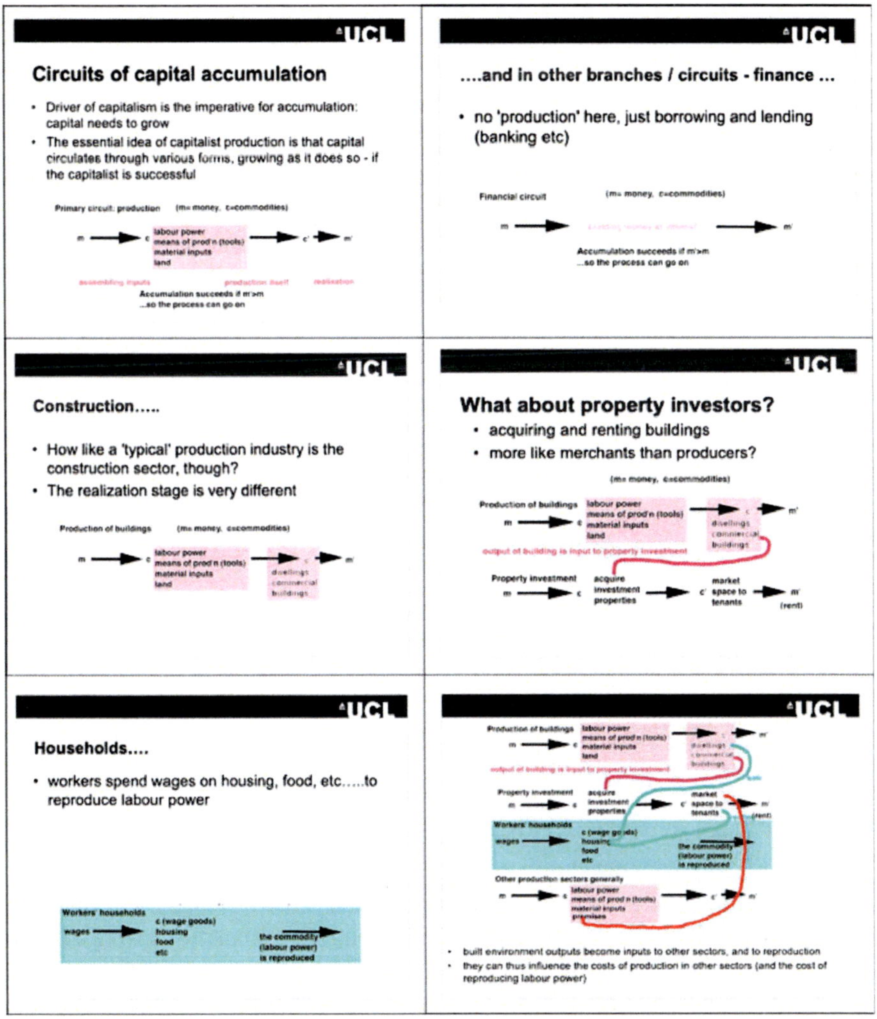

Fig. 8.3 Six slides by the author, unpacking relations between sectors in the production of the built environment

standards depend not only on wages relative to the price of manufactured commodities but on rents and house prices too, a major problem now in so much of the world and dramatically so in London where I work.

The need to explore building or urban development projects in a systematic way, exposing the relationships between the class interests and individuals involved was a strong preoccupation of ours in the 1980s and 90s. It was an approach which enabled us to get beyond the crude application to urban development of the rent relationships which Marx had identified and explained in nineteenth century English agriculture. In the same year that Harvey's Urbanization of Capital appeared, a group of us produced a collected volume of papers on rent theory (Ball et al. 1985). (This

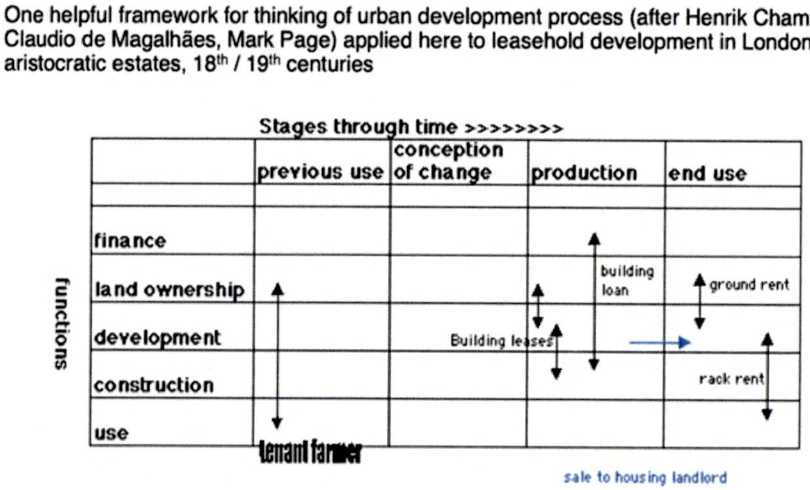

Fig. 8.4 Framework based on the work of Henrik Chambert, from Edwards (1995)

book has just been reprinted and I have mightily enjoyed re-reading it last year in a reading group convened by Elena Besussi and other Bartlett friends). I think it led us to rediscover the potential of rather open-minded examination of urban development as a historical or at least a sequential set of relationships, with Marx and Harvey in the background. (Interestingly, Covid drove us from physical meetings to online sessions which initially felt like a loss but enabled new people to join from Sweden, Finland and Brazil so discussions quickly became much richer.)

I want to end this catalogue of personal spinoffs from Harvey's book by referring to the important, but now almost invisible, work of Henrik Chambert. I am not sure whether he and Harvey ever met, but certainly Chambert (1992) was pursuing a very Harveyish project. He led a team in the Nordic capitals who were exploring processes of capital accumulation in the construction processes of the respective countries, analysing every single social housing project over a number of decades after WW2. They were charting the division of costs and revenues between state agencies, building firms, construction labour force, materials suppliers and so on, working towards a very detailed understanding of the concentration of capital in the house building sector. The remarkably open data environment of Scandinavia enabled this sort of work to be done and demonstrated how one outcome of sustained social democratic and socialist housing production was massive capital accumulation in Skanska and other firms. I am mentioning this because the research team developed a kind of tabulation of agents, functions and relationships which they and others have used to very good effect in the analysis of projects. Figure 8.4 here is an instance, used in this case as illustration in a talk of mine about the development of the Great Estates by aristocratic families in London in the eighteenth and nineteenth centuries. It was further developed and used as a research tool by de Magalhães (1999) and Page (1996).

Notes for that slide: Leasehold development of great estates in London:

Land: huge estates, mainly gifts from King after closure of monasteries.

Building: small firms with little working capital, so contracts small.

Development finance: banks not yet investing much; builders keen to avoid land purchase and borrow from land-owner if possible.

Final users: growing middle classes/bourgeoisie, often enriched by slave trade or colonial trade, buying long leases for their own use or to sub-let. They take on maintenance and insurance costs and pay a ground rent to land owner. [="leasehold"].

State regulation: Land and property law; building regulations against fire risk from 1670.

Planning: purely private urban design by land owners (internalizing the externalities from parks and squares. Bloomsbury; Regent's Park.) Some infrastructure.

Why did it end? I'm not sure. Factors include:

Railways, buses opened up new land for commuter suburbs in nineteenth century.

Ownership of farm land more fragmented further out (and in South London).

Building firms accumulating their own working capital and able to finance developments themselves

My account of Harvey's 1985 book has so far drawn on his Chapters 1 and 2 which set out the main argument. They are followed with two valuable chapters on rent which are classics and extend the analysis of spatial relationships already established. Chapter 5 on residential segregation and 6 on urban politics, however, are very different. Both build a Marxist logic from first principles for handling their topic but were written in a North American context which means they travel less well to European readers. Chapter 6 is a stimulating elaboration of the combinations of labour organisation and kinds of capital which can exist and struggle in urban areas—mainly conceived as labour market areas, though with lots of caveats. He argues that a 'structured coherence' is liable to emerge in an urban area in which coalitions of capital, sometimes with trade unions, coexist tolerably well for a while until disturbed. There is a good discussion of the idea of 'growth coalitions' which was fashionable in 1980s USA and of the scope for radical change to take root in cities, albeit limited in space and time. It is stimulating to read, humane and rather abstract. To me they are like ideas waiting to be used.

The penultimate Chapter, 7, considers how the ideology of planning was, in the late 1970s, beginning to have to be transformed to meet the needs of capital as the harsher confrontations of capital and labour began to bite. He charts a transition from an earlier (almost golden) age of the post-war boom during which benign urbanists and technocrats were more-or-less able to carry through the project of absorbing available capital and labour into the car-based transformation of America with suburbs and freeways, accommodating some redistribution and some environmental awareness when pressed to do so. What the transition was leading to he did not, of course, know but he outlines the increasingly brutal transformations coming in the housing and urban systems as investors intensified their quests for rents and assets in a new class struggle waged by capital. He ends "And if we pursue this possibility, we might come to understand why the planner seems doomed to a life of perpetual frustration, why the high-sounding ideas of planning theory are so frequently translated into grubby practices on the ground, how the shifts in world view and ideological stance are social products rather than freely chosen" (p 184).

The book ends with a chapter bearing the book's name: the Urbanization of Capital. This is both a recapitulation of the analysis and a completely new essay in the history of urbanization as being the history of capitalism (and before that, of other class societies). It looks well into the future too and I would not spoil it for new readers by saying any more. It ends with a clarion call for us to envisage and make the socialist city. "Thinking through the paths to socialist urbanization is to chart the way to the socialist alternative itself." (p 226).

So that is the book which has most influenced my life. Much of its content had appeared before as journal articles or in the massive *Limits to Capital* (1982) and much was to be re-worked later as the decades passed. So it may not now be the key book to buy—indeed that is almost impossible as it is out of print and the internet offers antiquarian copies from across the world for $150, some of which will flow to Mr Bezos. Indeed this chapter nearly did not get written. I have no personal copy of the book since some time in the 1990s when I went on an EU-funded delegation to Cuba led by Paloscia and Paba who asked us each to bring one or two books for the library of the School of Architecture at Havana which was unable to buy what it needed because of the US blockade. Thinking that the Cubans would soon be having to deal with capitalist urbanization again, I decided to take *The Urbanization of Capital*, and I hope that it is being worn out by avid readers. Being cut off from libraries by COVID-19 I was desperate and appealed to Twitter. Within 24 h I heard from a friend from Western Asia, in London to do a PhD, who had a PDF copy. It duly arrived by email and turns out to have been scanned from a library copy at the University of Barcelona. Another friend offered to send a printed copy, formerly in the library of the Centre for Regional Economic and Social Research at Sheffield, but found he had 'lent' it already to someone. All this geography seems appropriate, somehow.

I am a teacher who enjoys being taught, and this book is a great bit of teaching, essentially because every idea is expressed with maximum clarity and with an implied invitation to argue or to elaborate. Where Harvey does not know something he says so and where statements have to be nuanced and qualified they are. But Harvey is capable of holding vast chains of theoretical argument in his head. I am not. I have to hunt about for instances and examples of the abstract relationships he posits. You can probably tell that from what I have written in this chapter. I think many of my students are like me in this respect and find theoretical discourse more informative fleshed out with cases. Harvey just makes all this a pleasure.

References

Ball MJ, Bentivegna V, Edwards M, Folin M (Eds) (1985) Land Rent, Housing and Urban Planning: a European Perspective Reprinted 2018 in Routledge Revivals series

Chambert H (1992) A conceptual model for comparative analysis of housing systems. Royal Institute of Technology, Stockholm

Christophers B (2011) Revisiting the urbanization of capital. Ann Assoc Am Geogr 101(6):1347–1364

de Magalhães C (1999) Social Agents, the provision of buildings and property booms: the case of São Paulo. Int J Urban Reg Res 23(3):445–463

Edwards M (1995) Agents and functions in urban development: frameworks for analysis, published in Spanish as Edwards, M (1995) 'Agentes e funciones en el desarollo urbano' with English text Cartas Urbanas 4: 26–39 and reprinted in English in Edwards, M (2006) Collected Papers 1989–99. https://michaeledwardstemp.files.wordpress.com/2008/05/edwards8999.pdf

Edwards M (2014) Yes, and we have no dentists. In: Borden I, Fraser M, Penner B (eds) Forty ways to think about architecture: architectural history and theory today (ch 28), 280 pages. Wiley, London. http://eu.wiley.com/WileyCDA/WileyTitle/productCd-1118822617.html (e-print Edwards for Forty)

Edwards M (2015) Prospects for land, rent and housing in UK cities. Working Paper 18, Foresight Future of Cities Project, Government Office for Science, download from https://www.gov.uk/government/collections/future-of-cities-working-papers (consulted 1 February 2021)

Edwards M, Campkin B, Arbaci S (2009) Exploring roles and relationships in the production of the built environment. Centre Educ Built Environ (CEBE) Trans 6:1. http://discovery.ucl.ac.uk/15579 (consulted 1 February 2021)

Harvey D (1982) The limits to capital. Blackwell, Oxford

Harvey D (1985a) Consciousness and the urban experience. Blackwell, Oxford

Harvey D (1985b) The urbanisation of capital Baltimore. Blackwell, Johns Hopkins and Oxford

Harvey D (2012) Rebel cities: from the right to the city to the urban revolution. Verso, London

Marx and Engels (1848) The communist manifesto. Marx/Engels selected works (1969), vol one. Progress Publishers, Moscow, pp 98–137

Page MW (1996) Locality, housing production and the local state. Environ Plann D Soc Space 181–202

Michael Edwards studied economics, then planning at UCL 1964–1966. He worked on economic inputs to the Plan for Milton Keynes. He has enjoyed lecturing at the Bartlett School, UCL, since 1969 and has been involved in all the hearings on London Plans since 2000, working with the network of community groups JustSpace.org.uk. His publications are at michaeledwards.org.uk and he tweets as @michaellondonsf. He was one of the founders of INURA.

Chapter 9
A Place in the World? Places, Cultures and Globalization, 1995. Doreen Massey's Lessons: Is the World Really Shrinking or Is the Geography of the World Teaching Us Openness and Diversity?

Cristiana Rossignolo

Abstract The "inspiring book" is a highly interesting work by Doreen Massey, able to trigger points of critical reflection that directly and indirectly make it possible to understand why "geography matters". Despite the fact that the book is not exactly "new", many of the issues addressed still seem highly relevant and unresolved today, such as the increase of inequality in the world, local identities in a shifting world, places in the context of a global world, international migration, Europe and the new boundaries. Doreen's legacy lies in the fact that we can use her ideas as a kind of 'open source', always bearing in mind openness—social and spatial equalities—and a different, better kind of future. She has inspired generations of geographers to see the world differently introducing her "relational" approach to understanding space—the dimension that presents us with the existence of others and place—continuously builts and rebuilts through relations. She was an inspiring and open teacher guaranteeing that the courses dealt with issues that were socially and politically important. The theory was never abstract for her. She had a strong sense of responsibility for the world's complexity, for politics and the injustices of contemporary life.

Keywords Space · Place · Relational approach · Globalization · Social and spatial inequalities

The book I have chosen is not as well-known as many others by Doreen Massey. "Spatial Divisions of Labour" (1984), "Space, Place and Gender" (1994), "For Space" (2005), "World City" (2007), to mention just a few, are pioneering books, well known to scholars of urban studies, not just geographers (e.g. Christophers et al. 2018; Werner et al. 2018). However, while these works are undoubtedly crucial, they have already been widely discussed and much has been written about them.

C. Rossignolo (✉)
Interuniversity Department of Regional and Urban Studies and Planning, Polytechnic of Turin, Turin, Italy
e-mail: cristiana.rossignolo@polito.it

C. Perrone (ed.), *Critical Planning and Design*, The Urban Book Series,
https://doi.org/10.1007/978-3-030-93107-0_9

After Doreen Massey passed away (March 2016), numerous tributes were made to her in the field of human geography, acknowledging the depth of her intellectual influence (e.g. Castree 2016; Clarke 2016; Featherstone 2016; Kitchin 2016; Lee 2016; Meegan 2017; Painter 2016; Rose 2016). However, her thinking and works go well beyond conventional disciplinary boundaries. She introduced conceptual thoughts which have become commonly accepted in sociology, planning, political economics, cultural studies, gender studies, and beyond. She has had an incredible impact on research and debates about how we see and think about space, place and power. She continues to be an inspiring intellectual in geography, for the wider academic community and in political contexts across the world "for the claim that 'geography matters', not as a dry defence of disciplinary turf but as a rallying cry" (Christophers et al. 2018).

When Camilla Perrone and Giancarlo Paba asked me to share an "inspiring book", my choice fell on this highly interesting work, able to trigger—not only in scholars but also in students—points of critical reflection that directly and indirectly make it possible to understand why "geography matters" (Massey and Allen 1984). Her writings have helped to shape the parameters of the subject of geography.

Doreen Massey was committed to student education and through developing popular courses in human geography, such as "The Shape of the World"[1] and "Understanding Cities", she had an influence on teaching and scholarship beyond her university, the Open University.

I would like to highlight three issues that have led to my selection of this book, "A Place in the World? Places, Cultures and Globalization", edited by Doreen Massey and Pat Jess.

Firstly, Doreen Massey was an inspiring and open teacher.

This book was published in 1995 by Oxford University Press as part of the Open University course texts, although it was also presented to the broader market as a potential "stand-alone" text for other courses and is still extensively used, as stated in the common Preface: each book has been designed as "free-standing and can be studied on its own or as part of a wide range of social science courses in universities and colleges" (Massey and Jess 1995: vii).

The five texts for the course "The Shape of the World: Explorations in Human Geography"[2] are "distinguished by a singular style—highly innovative in conceptual and theoretical terms, yet always grounded in material realities and crystal clear in expression" (Lee 2018: 170), excellent examples of textbooks with an integral reader

[1] The "The Shape of the World" course was offered between 1995 and 2005 at the Open University (Undergraduate course, Social Sciences, 30 ECTS credits, 32 weeks, approximately 600 h of study). It was taught "by distance-learning methods, and the books supplement course materials provided to the enrolled students, TV and radio programmes, tutorials and summer schools" (Johnston 1998, 129).

[2] The five books are: "Geographical Worlds" (1995) by Allen and Massey, "A Shrinking World? Global Unevenness and Inequality" (1995) by Allen and Hamnett, "An Overcrowded World? Population, Resources and the Environment" (1995) by Sarre and Blunden, "A Place in the World? Places, Cultures and Globalization" (1995) by Massey and Jess, "A Global World? Re-ordering Political Space" (1995) by Anderson, Brook and Cochrane.

component "extracts drawn from books and articles" (Massey and Jess 1995: vii). Each book develops different aspects of the 'geographical imagination', even if all they follow a standard model to help readers (Johnston 1998): a brief introduction, a series of challenging chapters, each followed by additional readings (printed in two columns to distinguish them), which illustrate or develop a single issue raised by the author of the work, an editorial conclusion. As Johnston observed (2000: 280), "one common feature of almost all of Open University readings is that they are edited. […] Students are given access to only those parts of the whole that the editor considers valuable and relevant. […] This practice is especially important in courses based on distance learning, involving students with no access to either specialised libraries or teachers/tutors who are readily available for consultation. Such students have no alternative but to rely on the provided printed word as their major resource, however much they may be encouraged to approach it critically." In the arguments of this book, the identities of places are frequently contested, "the cases themselves are quite tricky and there are simple 'solutions' in such situations. We are not asking you to take sides, but […] we want you to read, questions and critically assess" (Massey and Jess 1995: 134–135).

Doreen Massey believed in being excellent without being exclusive and discriminatory, and the Open University, rather than traditional campus universities, was (is) open and accessible to all.[3] Despite several offers of professorships from prestigious academic institutions, including Oxford University, she worked for 30 years at the Open University because of its collaborative approach to thinking and teaching: "In a way the Open University has been, for me, the ideal place to teach and to do research because of its social project and because of the way in which it really encourages us as academics to speak beyond the academy as well as within" (Massey 2009).

In this difficult time due to the pandemic, I think this book can also provide us with a valuable pedagogical lesson, as Massey gives us interesting insights from her experience of distance teaching at the Open University. She shows us through this book, as in many of her writings, the importance of the challenge of engaging students who are not physically present. This is an important legacy: communicative writing, the effort to put oneself in another's shoes, to find points of contact and connection, building a basis for dialogue around the issues addressed, outside the rhetoric of 'participation', but dealing in concrete fact. As John Clarke, a colleague of Massey's at the Open University, wrote in the Special Editorial after her death, "her capacity to be critically supportive was much appreciated by younger or emerging scholars—she embodied the principle of taking people and their work seriously. In

[3] The Open University (http://www.open.ac.uk/) was established in the United Kingdom in 1969 to open up higher education to all, regardless of background or circumstances. It was the world's first successful distance teaching university, founded on communications technology and innovation. Its mission is to be open to people, places, methods and ideas—and as such, equality and diversity are at the heart of everything it does. It has promoted educational opportunity and social justice by providing high-quality education to more than 2 million people. It is the leading university for flexible, innovative teaching and world-leading research in the United Kingdom and in 157 countries worldwide.

this, as in so many things, she was an instinctive democrat—not the most common orientation in the academy" (2016: 361).

"For the future to be open, space must be open too" (Massey 2005: 12).

Secondly, Doreen Massey has inspired generations of geographers to see the world differently.

The book's contents are both attractive and inspiring, as it explores the links between place, culture, identity, power and globalization, raising significant cultural issues for student discussion. The six chapters are coherently linked and go into the key themes in greater depth in section summaries followed by student activities.

While space had been conceptualised as a container for a long time, Massey introduced her "relational" approach to understanding space and place. Rather than seeing space dispassionately as a surface on which phenomena were distributed, she imagined space in a contested way as a constellation of different trajectories of activity (Featherstone 2016). In her view, space is "as constituted through interactions, from the immensity of the global to the intimately tiny" (Massey 2005: 9). Moreover, space is always under construction: "it is always in the process of being made. It is never finished; never closed" (Massey 2005: 9).

They use the process of international migration as their starting-point, as a connection to the general issues of globalization. As this conceptual narrative underlines movement and change, King's chapter "Migration, globalization and place" considered migration from a different perspective: migration often involves uncertain engagement with place, particularly one's place of origin and the transformation after the departure, the "places of abandonment". But in the chapter "The conceptualization of place", Massey highlights that even if the "shrinking of the world" had indeed enabled people to be closer, physical proximity is not necessarily a good measure of social and cultural distance. There is cultural diversity within many places: it is here and now in every society, but it is also a question of different others in distant lands. This immense variety of societies and cultures is part of geography. Space is a crucial concept: it is the dimension of things (and people) existing at the same moment. If time is the dimension of change, then space is the dimension of simultaneity. Space is the dimension, therefore, that presents us with the existence of others. Places are continuously built and rebuilt through relations, are the meeting-places at the intersection of a host of connections, of trade, of commerce, and of culture. In a globalised age, re-thinking the "identity of place" means not only recognising it as a claim to exclusive ownership, but also asserting it as a recognition of responsibility—the responsibility of place—within and towards those wider geographies (a politics of place beyond place).

Rose's chapter "Place and identity: a sense of place" analyses the between place and identity from different perspectives: identification with a place (connected to the feeling that people belong to that place); identification against a place (people identify both with and against aspects of place, constructing boundaries to establish insiders and outsiders); and non-identification with places (feelings of displacement or estrangement). Sense of place can work to establish differences between groups of people even if "it is possible to think about interrelations between people and places with tolerance and humility. Identities can be established through positive interrelations. Perhaps the terms of co-existence are being renegotiated, and new ways of thinking about place and identity are developing" (Rose in Massey and Jess 1995: 117).

Jess and Massey in the chapter "The contestation of place" specifically state their concept of sense of place and consider the way in which it is bounded to the relationships between place and identity. Space is the outcome of and ground for social interactions which connect people from different places, "it is where encounters are made and where conflicts arise and can be negotiated" (Benach 2013). They highlight the contestation over place in terms of contending claims to delineate the meaning of place, asking 'whose place is it?' and thereby people's rights to control its use or future. They explore four different cases of conflict over place: "in arguing their positions the different sides in each dispute describe the place differently: they see the place from different points of view, and they emphasize different (even opposing) characteristics. […] each side is laying a claim to how the place should be thought of, how it should be represented—in other words, how it fits into our geographical imagination" (Massey and Jess 1995: 134).

Hall, in the chapter "New cultures for old", suggests that the relationship between place, people and culture may also be assumed in terms of routes, as more people think of their identities and their relationship to culture and to place, in more 'open' ways. The widened examples of sport (cricket), music and religion in the discussion of hybridity also break some welcome new ground.

Places are not static but represent continual processes. 'People make places' as "the influx of new residents is part of the construction of a new uniqueness" (Massey and Jess 1995: 221). Places are not simply delineated by local boundaries within which people meet but are made up of flows and movements and the innumerable interrelations and interdependencies between places: "we cannot understand the character, the uniqueness, of place by looking at that place alone" (Massey and Jess 1995: 222). They are simultaneously local and global, their social, cultural and economic relations stretched out across the globe, shaped by structural processes but preserving local identities. "But to speak, as we just did, of the local and global as being 'opposed' is in fact incorrect, or at least inadequate" (Massey and Jess 1995: 226). The global and the local are inevitably mutually constituted: if space is a product of practices, trajectories, interrelations, if we make space through interactions at all levels, from the local to the global, then places, regions, nations, and the local and the global, must be forged in this relational way, too (where each scale is seen as constitutive of the other). Local/global (and place/space) have been conceptualized in terms of social relations: "it is not just the global which influences the local but also vice versa; or, to put in another way, causality is not all 'top-down'. […] "The local is not merely in passive receipt of global forces. The whole book has been full of examples of people actively *making* their local areas, out of recipes which each involve mixing and moulding both local and wider influences" (Massey and Jess 1995: 227–228).

In this book, the sense of place is constituted by its culture, politics, history and migration from different parts of the world, indigenous and international people (and their diversity). Massey's notion of a progressive sense of place gained much sway in geography as it provided a multiscalar and relational meaning: as Kitchin (2016: 815) argues "to this idea, she added the notion of 'power geometries' noting that places are made through power relations—not simply capital relations—which construct the rules and define boundaries. These boundaries are both social and spatial—they define who belongs to a place and who may be excluded. As such, the socio-spatial processes that help shape and define places do not operate evenly, with different social groups and individuals relatively positioned as a consequence".

Thirdly, Doreen Massey wanted to guarantee that the courses dealt with issues that were socially and politically important.

The theory was never abstract for her. She privileged research into the friction points and contradictions present in reality over abstractions: "She was never prepared to separate theory from the complexity of experience and the possibilities of politics and, in this, Doreen was the epitome of a geographer—wholly aware of the need to understand the multiplicity of relations that interact to shape social and environmental life and of the possibilities for progressive change. And the richness of her conceptions of geography and its relationship to the world also contributed to her many roles—researcher, teacher, political activist, theorist, public intellectual and inclusive communicator" (Lee 2016: 311). In all her writings, the contents are remarkable and challenging: the theoretical and the empirical issues are brilliantly interlinked, the reader comes away feeling both informed and more knowledgeable. For example, in this book, the chapters are simultaneously lively and reflective, as concepts and examples alternate, examining the challenges raised by globalization and the meanings of and relations between place and culture: "Her level of abstraction is always grounded in the most quotidian facts of everyday life, which not only makes her abstract ideas more consistent and understandable but easier to apply to other contexts, thereby serving a variety of objectives" (Benach 2013).

Her geography was at the same time territorial and relational: she wrote much about the endlessly changing and jointly constructive relationship between territoriality and relationality. This notion of the interrelationships between territorial and relational space is both space—and place-making Social and environmental life can be lived only through place-based spatial relationships. "Geography and geographies of life shaped by it are, thereby, heavily and influentially involved in the ways in which social, economic and political practices are conducted and, literally, *take place*. [...] The inherency of political responsibility within the making of geographies was a central trajectory of her work" (Lee 2018: 148).

She was a role model for undertaking socially relevant academic work, and showed that it was possible to combine rigorous scholarship with political activism. She had a strong sense of responsibility for the world's complexity, for politics and the injustices of contemporary life.

Her activity as a "public intellectual" (Hubbard 2008) and political activist did not end with retirement, but she carried on being open to the world, to diversity, to the lives of others, and committed to social justice.

Despite the fact that the book is not exactly "new" and the authors themselves in the Introduction highlight the time when it was written and its context—"This volume has been produced at a particular moment, and a moment in *space*–time than rather just time. That is to say, it reflects concerns, and probably theoretical and conceptual predelictions, which are primarily at issue in, and seen from the perspective of, first world and other industrial countries. Thus the way in which these questions are often couched implies that the relations between place and culture *were* one simple, that local places *were* settled, coherent and bounded, that cultures *were* internally generated and deeply embedded in spatial propinquity, in place. Yet in fact there has since the beginning of human existence always been movement, migration and settlement in new areas; for as long as is known and in most parts of the world, individual places have been open to, and partly constituted by, their contacts with 'outside'. Interconnection is not new, and diasporas are certainly not only a feature of the recent past." (Massey and Jess 1995: 2)—many of the issues addressed still seem highly relevant and unresolved today (the increase of inequality in the world, local identities in a shifting world, places in the context of a global world, international migration, Europe and the new boundaries, just to mention a few). The book as a whole still provides a case for rethinking these issues and recognising their importance to our geographical imagination, to suggesting new pathways for reflection. For example, in Europe, in relation to spatial and social inequalities, the widespread policy of walls, of closed ports, of rejection of migrants (the 'others') at all costs, including the cost of human lives, should make us ashamed to be European citizens and raises questions about why geography matters. Unfortunately, the statement "Europe is thus a place for some, a non-place for others" (Russell King in Massey and Jess 1995: 25) is still valid.

Her legacy lies in the fact that we can use her ideas as a kind of 'open source', always bearing in mind openness—social and spatial equalities—and a different, better kind of future: "The spatial and the social must be considered together. And our geographical imaginations, and how they are acted out in our lives, are a key component in the forging of our social futures" (Massey and Jess 1995: 238).

Acknowledgements The author is grateful to Beppe Dematteis for his important suggestions and Ian Lister for his feedbacks in improving the flow of this chapter.

References

Allen J, Hamnett C (eds) (1995) A shrinking world? Global unevenness and inequality. The Open University and Oxford University Press, Oxford

Allen J, Massey D (eds) (1995) Geographical worlds. The Open University and Oxford University Press, Oxford

Anderson J, Brook C, Cochrane A (eds) (1995) A global world? The Open University and Oxford University Press, Oxford, Re-ordering political space

Benach N (2013) David Featherstone and Joe Painter (eds) Spatial politics: essays for Doreen Massey. Wiley-Blackwell, Oxford, Antipode, http://radicalantipode.files.wordpress.com/2013/09/book-review_benach-on-featherstone-and-painter.pdf

Castree N (2016) A tribute to Doreen Massey (3 January 1944–11 March 2016). Progress in human geography: an international review of geographical work in the social sciences and humanities, vol 40, issue 5, pp 585–592

Christophers B, Lave R, Peck J, Werner M (eds) (2018) The Doreen Massey reader. Agenda Publishing

Clarke J (2016) Doreen Massey (1944–2016): making geography matter. Cult Stud 30(3):357–361

Featherstone D (2016) Doreen Massey obituary. The Guardian, 27 March 2016

Hubbard P (2008) Book review forum: Doreen Massey (2007) world city. Area 40(3):411–412

Johnston R (1998) Book review: replacing geography. Prog Hum Geogr 22(1):129–132

Johnston R (2000) Authors, editors and authority in the postmodern academy. Antipode 32:271–291

Kitchin R (2016) Geographers matter! Doreen Massey (1944–2016). Soc Cult Geogr 17(6):813–817

Lee R (2016) Doreen Massey: 3 January 1944–11 March 2016. Geogr J 182(3):311–312

Lee R (2018) Doreen Barbara Massey (3 January 1944–11 March 2016). Biographical memoirs of fellows of the British academy, XVII, pp 145–178. https://www.thebritishacademy.ac.uk/publishing/memoirs/17/massey-doreen-barbara-1944-2016/

Massey D (1984) Spatial divisions of labour: social structures and the geography of production. Macmillan, Basingstoke

Massey D (1994) Space, place, and gender. University of Minnesota Press, Minneapolis

Massey D (2005) For space. Sage, London

Massey D (2007) World city. Polity, Cambridge

Massey D (2009) Introduction by Doreen Massey. Doreen Massey: space, place and politics. Open Learn, The Open University. https://www.open.edu/openlearn/society-politics-law/geography/doreen-massey-space-place-and-politics

Massey D, Allen J (eds) (1984) Geography matters! A reader. Cambridge University Press, Cambridge

Massey D, Jess P (eds) (1995) A place in the world? Places, cultures and globalization. The Open University and Oxford University Press, Oxford

Meegan R (2017) Doreen Massey (1944–2016): a geographer who really mattered. Reg Stud 51(9):1285–1296

Painter J (2016) Doreen Massey (1944–2016): an appreciation. Soundings 63(Summer):38–44

Rose G (2016) Remembering Doreen. Geopolítica(s). Revista de estudios sobre espacio y poder 7(1):19–20

Sarre P, Blunden J (eds) (1995) An overcrowded world? Population, resources and the environment. The Open University and Oxford University Press, Oxford

Werner M, Peck J, Lave R, Christophers B (eds) (2018) Doreen Massey critical dialogues. Agenda Publishing, Newcastle upon Tyne

Cristiana Rossignolo is an Associate Professor in Urban Geography at Politecnico di Torino, Interuniversity Department of Urban and Regional Studies and Planning. She is the Rector's Advisor for Teaching and Language Lab of the Politecnico di Torino since 2018. She is the Associate Editor of Urban Research and Practice (T&F).

She coordinates AuroraLAB, a project aimed at activating research-action and learning by doing initiatives in a Turin neighbourhood. She is involved in several national and European researches. Her research interests are primarily concerned with neighbourhoods and urban regeneration, European urban policies, networks of cities, post-metropolitan cities, governance and strategic planning and education.

Chapter 10
Patsy Healey and *Collaborative Planning* (2005): Re-thinking Democracy in the 'Reasoning in Public' Arena

Francesco Lo Piccolo

Abstract The interpretation of the notion of citizenship has led some authors to the conclusion of the necessity of overcoming a 'restricted conception' of citizenship, in order to gain a more inclusive recognition of a cosmopolitan, rather than a mere universal, recognition of rights. So, three core issues are tightly intertwined in this paper: the controversial issue of citizenship in its multiple facets: the formal that is institutional, the substantial and the denied one; the role of planning and its ethical implications; and the very idea of democracy in its dichotomy: institutional versus substantial. Patsy Healey's work offers us a wide set of normative perspectives and intellectual stimuli, with all the challenges of their applicability and transferability, in order to nurture (a more substantial) democracy. In all the work of Healey, planning is considered a way to enhance and re-enchant democracy, due to the long history of planning's normative interest in richly participative democratic processes. In line with Patsy's work, this paper considers planning as a way to enhance and re-enchant democracy.

Keywords Citizenship · Collaborative planning · Participation · Patsy Healey · Democracy

10.1 A Normative Ideal: (Questions of) Applicability and Transferability

If I look now at most of my previous as well as present research activities, I recognize that Patsy's work offered to me a wide set of normative perspectives and intellectual stimuli, with all the challenges of their applicability and transferability in my very—so different—research contexts. Given my context-based research work field (and place where I live), I experienced some difficulties in applying normative principles that which are nurtured by promises on behalf of deliberative democracy. They are

F. Lo Piccolo (✉)
Department of Architecture, University of Palermo, Palermo, Italy
e-mail: francesco.lopiccolo@unipa.it

C. Perrone (ed.), *Critical Planning and Design*, The Urban Book Series,
https://doi.org/10.1007/978-3-030-93107-0_10

somehow inappropriate for those contexts that are (in some ways 'extreme'[1] but not irrelevant), and exposed, for example, to violence and to hidden and illegal abuse, as in the case of organized crime.[2] So, these are the questions of applicability and transferability of Patsy's normative principles that I want am trying to explore in the following paragraphs. Applicability and transferability are also a matter of using systematically using reflections deriving from 'extreme', particular and locally determined cases. Moreover, reflections deriving from 'extreme' cases and contexts can be helpful in order to individuate and discuss some (more general) phenomena that are much more evident—and consequently analysable—in 'extreme' conditions.

If we assume the notion of citizenship as the theoretical and political sphere where the inclusive or conflictual relations between 'different' subjects are defined (Bobbio 1990), my research has been focussed on the relation between inclusive and exclusive forms of citizenship, and the recognition of them in urban planning. According to Zolo (1994: 4), the notion of citizenship puts into relation individual subjective rights and pre-juridical reasons and conditions for political inclusion or exclusion. Consequently, the notion of 'substantial' citizenship can be considered as an indicator of the functional level of democratic institutions in providing fair and equal access to public life and activities, including the urban ones (Zolo 1994: 4). In other words, to analyse democratic institutions and their activities, planning included, under the perspective of substantial citizenship allows us to measure and evaluate the level of democracy of political systems *ex parte populi*, as it adopts a double perspective: the entitlement of rights and the endowment of rights (Barbalet 1988).

The instrumental interpretation of the notion of citizenship has led some authors to the conclusion of the necessity of overcoming a 'restricted conception' of citizenship, as explained by Held (1987, 1989), Bobbio (1990) and Kymlicka (1995) in order to gain a more inclusive recognition of a cosmopolitan, rather than a mere universal, recognition of rights. This extensive interpretation of substantial citizenship is the one I am referring to in the following sections of the chapter.

In my perspective, influenced also by the work of Harvey (1996, 2000, 2003), Friedmann (1999, 2011), Holston (1999) and Sandercock (1998, 2003) amongst others, this extensive interpretation of substantial citizenship is strictly connected with and complementary to planning: the juridical recognition of citizenship guarantees the right to forms of representation and participation, related to the processes of transformation of the cities; at the same time, planning initiatives and their results are important in influencing citizenship's boundaries, in order to nurture (a more substantial) democracy (Lo Piccolo 2010). In brief, and in line with Patsy's words (Healey

[1] Here the expression 'extreme contexts' is referred to conditions in which organized crime, corruption and 'disorder' (as a result of 'conflicting orders') are structurally concentrated.

[2] The expression 'organized crime' or 'criminal association' is used in Italy to refer to those illegal groups—Mafia in Sicily, Camorra in Campania and ndrangheta in Calabria—that are locally space-based and with strong territorial identity, links and roots, although their networking activity is on a wider, global scale.

2012), I have been considering planning as a way to enhance and re-enchant democracy,[3] due to the long history of planning's normative interest in richly participative democratic processes (ibid: 20).

So, three core issues are tightly intertwined: the controversial issue of citizenship in its multiple facets: the formal that is institutional, the substantial and the denied one; the role of planning and its ethical implications; and the very idea of democracy in its dichotomy: institutional versus substantial. Behind and at the same time, at the base of this research interest, two other issues and questions arise, which are influenced by the local contexts of my research activity: how to deal with these issues in political and social contexts where the notion of substantial democracy faces several obstacles in its way? And how to set a progressive planning agenda where—also for the same reasons and *milieu*—the role of planning and planners is very weak and with a low level of efficacy, or it is just 'on the side' of power?

From Healey and Underwood (1978) and then Healey (1997) to Healey (2012) there is a normative political and ethical *fil rouge*: the emphasis—when analysing the processes of planning rather than examining their contents—is on individuals and their interactions, without underestimating the importance of the structural contexts and the consequent configurations of power and influence within which these interactions occur. It is also about planning and planners' activity as a complex set of inter-relationships or social networks, where planners interact with people in highly varied positions and roles and at the same time, where planners themselves perform in highly varied positions and roles. It also concerns all the ethical consequential dilemmas of these multiple roles (Lo Piccolo 2009). Also of interest are the potentials of micro-practices in planning activities, and urban dynamics as well; a recognition and appreciation of the multiple identities and ways of life, based on the pluralist multiculturalism approach described by Sen (2006); the importance of discourse, in formal but moreover, in informal contexts through the analysis of normally invisible or unrecorded dialogues with different actors—formal and informal, institutional and not, technical and not, with different levels and nature of knowledge, not to talk about their interests, cultures and values, and 'stretching across boundaries between formal government and people concerned about their daily life living conditions' (Healey 2012: 28). Healey's multi-textured view holds several other elements: the ethical commitment of respecting different arguments, positions and feelings, and a claim for establishing practices of intelligent, multi-sided discussions in order to cultivate 'the intelligence of a polity' according to the principles developed by Dewey (1991), and as deeply discussed in Healey (2009); under the influence of Forester (1989), the normative imperative of developing appropriate and just planning skills, being attentive, asking good questions, exploring meanings and implications, helping to correct the self-deception of others, providing respect, treating the other in a relationship seriously; and finally, the planners' mission of developing skills and sensitivity in

[3] I had not read Healey (2012) before having been asked to write this chapter. And I was surprised at finding in Patsy's article the same intellectual yearning and ethical tensions that were inspiring some of my research, although my fields of observation and analyses have been so different—culturally, politically and geographically—from most of those observed and considered in Healey (2012) as many other of her research works.

order to nurture the capacity of 'reasoning in public' as a key element of a rich and critical democratic life, as developed in Habermas (1981, 1989, 1996).

This is a normative framework that has inspired all my research work. However, what are the conditions and, indeed, the pre-conditions for a fair and substantial public reasoning? The problems, and my own research mission concern:

1. how to practice this normative ideal in those contexts or situations in which the 'reasoning in public' arena is a mere deceitful pretence, or on some occasions, there is no 'reasoning in public' at all, and
2. how to practice this normative ideal when some, either individuals or groups are excluded from the 'reasoning in public' process, due to their citizenship or social or cultural status.

The metaphor of the deceitful 'reasoning in public' in Act I, Scene I of King Lear will be used in order to highlight risks and weaknesses that can occur in some substantially undemocratic contexts. Moreover, the consequences of the opening scene of King Lear will be considered in order to describe the effects of the loss of substantial citizenship rights. This is a crucial point that is not often deeply considered in Healey's reflections. In fact, in Healey's work a recurrent word is 'citizens', which in some contexts does not entirely include all the individuals and groups in the 'world's stage', that is the public planning arena, although Healey (2012: 23) recently highlights the need for innovative problem-solving and delivery capacity for the new agendas of a more demanding citizenry, more conscious of our multiplicities. But what if you do not possess any power? Or others/the institutions do not recognize your right to any power? The 'denied citizenship' still remains an open question.

From this question three problematic issues are:

1. how to gain citizenship status in discursive practices for those who are not recognized as citizens, and the consequent ethical dilemma of imperfect (and consequently not entirely inclusive) systems;
2. how to develop this in contexts where insincerity is the rule, and where the gap and imbalance in role between the public arena and the closed rooms of power and the most powerful elites is so huge; and
3. what might be the role for planners and their technical knowledge/expertise when the nature of conflicts is latent or hidden as a consequence of the structural characteristics of the above mentioned contexts.

In some of my previous research works, I faced the problem of how to assume the inclusive process of communicative ethics (Forester 1989, 1999; Healey 1997) in contexts and in local societies that are unwilling to adopt that model. The experiences I analysed (Lo Piccolo 2008, 2009; Bonafede and Lo Piccolo 2010) show the attempts at facing this issue in a 'substantially un-democratic' context, where the institutional sites provided by the arenas of formal planning rarely become key arenas where the conflicts play out.

Contrary to that which Healey (2012: 34) properly highlights about the benefits of strong local governance institutions in developing interactive, inclusive approaches to rethinking the relations between civil society and the state, the contexts I am referring to differ in two main aspects: (a) the weakness of local institutions; (b) the interference, and in some cases predominance, of 'third parties'—and not just organized crime, but also family and/or religion, or a mix of the all three!—in the relations between civil society and the state. In fact, there are some orders, processes and actions that are beyond or out of the state's control. Most policy and planning analyses consider the state area of intervention, underestimating the role of other institutions, such as family, religion or illegal crime organizations. The Shakespearean metaphors I will use are helpful in order to include those and their role in the theoretical debate.

The effects on planning as well as on urban dynamics are significant: in areas that have a strong presence of organized crime, their powers are capable of bending local governments to support real estate investments encouraging phenomena like unauthorized building, corruption and political patronage. The most insidious causes of these phenomena reside in the capacities to:

- defend, reconcile and promote particular private interests, deliberately jeopardizing the public ones, and
- de-legitimise the public administration, the politicians and professionals, by hook—corruption—or by crook—violence/intimidation—or guaranteeing a general condition of inertia/inaction/inefficiency/mistrust (De Leo 2011, 2013).

From her works prior to 2012, Patsy is fully aware of how imperfect the interactive processes are and how they have to be continuously monitored and revised in a dialogical and communicative approach as well as in an antagonistic dimension, and while not ignoring or refusing the institutional dimension. However, what are these are the conditions for the cultivation of debate and public reasoning in different political and cultural contexts? An analysis of the opening scene of King Lear will provide some insights.

10.2 Cordelia and King Lear: The Disownment of Citizenship in Insincere (Planning) Discourses

The story opens (Act I, Scene I) with the elderly King Lear deciding to give up his power and divide his realm amongst his three daughters, Cordelia, Regan, and Goneril on the basis of which one of them loves him the most. Goneril and Regan, corrupt and deceitful, lie to their father with sappy and excessive declarations of affection, in gushing, insincere speeches; Cordelia, however, refuses to engage in Lear's game, and replies simply that she loves him as a daughter should. Her lacklustre but firm retort, despite its sincerity, enrages Lear and he disowns Cordelia completely. When Lear's dear friend and advisor, the Earl of Kent tries to speak on Cordelia's behalf, Lear banishes him too from the kingdom.

Lear is a planning tragedy, and in my eyes is *the* most outstanding planning tragedy in the entire story of western literature. First, the plot and all the following tragic consequences, starts with a land division (zoning): lands, that is, space, are not just properties (that is, economic values) but power and status. In fact, much more than in other Shakespeare's plays, characters and their identity, status and power are named and identified with the lands they own, administer and rule—Albany, Cornwall, France, Burgundy, Kent, Gloucester. Formally, the lands' assignment (that is, planning decisions) is based on and justified by common principles and shared values, which are argued and expressed in public discourses. As Healey (2012) highlights, decisions are established as consequence of discursive inter-relationships, in formal, and more significantly, informal contexts and unrecorded dialogues, as in the case of the unrecorded dialogues between Goneril and Reagan. Consequently, the interaction among people in Scene I is a practice of intermediation between potentially conflicting interests, using discursive communication as a key tool (Healey and Gilroy 1990: 21). Substantially, most of the public discourses, including those of Lear's daughters Reagan and Goneril, but also those of their husbands Albany and Cornwall, of the Gloucester's bastard and evil son Edmund, and even those of some partially involved stakeholders, as Burgundy and France, are insincere and false, as they are oriented to power and to individual interests. Decisions are formally taken according to equity and rigour in (abstract) priciples and rules, but they are substantially revealed, and considered, as the most unjust and inequitable by all the subjects involved, whether they benefit or are damaged by them. The results of public discourses are not just a land allotment, but a new established system of power, status and citizenship, which impedes some of the involved subjects from participating in the following phases of the institutional decision making process. Cordelia immediately loses her status and rights, and the same happens to the 'technician' Kent. In short, the tragedy of King Lear reveals that planning is not just assigning pieces of land, but recognizing or denying citizenship, as a result of sincere or insincere speeches in public reasoning.

From Lefebvre (1968) to Mitchell (2003), many scholars (see Friedmann 1999; Harvey 2000, 2003; Purcell 2003) have discussed how the right to the city must be defended if we want to live in a diverse, just society, as the very idea of citizenship rights is fundamental to protecting the ideals of liberal democracy. Consequently, citizenship is the theoretical and political sphere where conflictive relations between different subjects are defined. Whatever the explicit or implicit aims of any planning decision—whether favouring functional, economic or aesthetic reasons—every planning initiative contributes to a redefinition of the boundaries of citizenship, consequently shaping spatial or non-spatial forms of social control (Lefebvre 1974; Yiftachel 1998; Hillier 2002). This continuous process of redefinition by means of planning initiatives and decisions may occur directly or indirectly, with intentional or unexpected, witting (often) or unwitting (rarely) effects. The land assignment has immediate and evident effects on the status and the citizenship recognition of Cordelia and Kent, but Lear, too, unexpectedly and unwittingly loses his status as a consequence of his 'planning decisions'. A redefinition of the boundaries of citizenship that are, in other words, the rights of using the city will always occur, as an effect of the redistribution of resources deriving from planning decisions, and whether it is a matter of economies, spaces or rights. As Healey (2012: 20) highlights, "this has encouraged a research focus in the spatial planning and area development field on the micro-practices of democracy-in-action. For it is in these micro-practices that discriminations, exclusions and oppressions become manifest".

The possible 'extensive' or 'restrictive' interpretations of the concept of citizenship—as they are described by Kymlicka (1995) and Held (1989)—have some relevant consequences, because they show, theoretically and practically speaking, how citizenship may include or exclude (Lo Piccolo 2010). The recognition of the differences and the guarantee of an effective equity imply a reconsideration of the idea of citizenship. Nowadays, the characteristic of citizenship status, both legal and political prevails due to a paradox in history and a process of involution.[4] If compared with the political civil struggles and juridical recognitions of the 1970s, democratic rights and justice claims are much more threatened by neo-liberal capitalist globalisation and its effects upon the development and planning of cities (Routledge 2010). In other words, the recognition of citizenship by institutions has recently ignored most of the capacity or, indeed, the potential willingness to encounter and engage with differences.

While a static concept of citizenship, considered as a guarantee of the acquired rights, is widely prevalent, a dynamic concept of citizenship, considered as an activity and a political practice where recognition, defence and plural articulation of the rights are taken into consideration (Kymlicka 1995) is rarely promoted. Citizenship as a status thus creates new geographies, distinguishing between centres and peripheries of the right holders (Roche 1992); and in this way juridical inclusion/exclusion has repercussions for spatial inclusion/exclusion. Coming back to King Lear, the three layers of time overlap, and with implications for citizenship and status. Cordelia's rise and fall present a clear example of the disownment of citizenship in planning processes. Based on her speech, Cordelia loses her status and consequently, she has no right; she is not a citizen anymore, and presents as a metaphor of the marginalization of some groups into unacceptable 'others' within the political arena and due to the 'reasoning in public' effect. The same destiny (from presence to absence) is reserved also to the technician Kent. In the first case, of Cordelia, I am referring to the loss of status of all the minority groups or their advocates that are ignored in every political decision-making process, and the consequent disownment of their citizenship rights. In the second case, of Kent I am referring to the loss of the role, that is, the substantial disclaimer of the institutional validity of technical expertise when technical expertise faces power, reclaiming at the same time critical thinking and justice: Kent appears,

[4] Bobbio (1990), describing the increase in the number of the spheres of rights, highlights how now more than ever rights cannot be ascribed to an abstract category—the generic man—but rather to different and peculiar categories—the specific man—according to the characteristics and the privileges of the various social statuses. As status, e.g. family status, the status of the free citizen, the status of the legitimate child etc., was in the past. Today citizenship is an instrument for the differentiation—and consequently the separation and the discrimination—of some subjects within the social body from others; hence, the status of citizenship is a privilege. This phenomenon is more evident in the urban dimension, especially if we consider the inability of the contemporary welfare-state to cope with 'a more demanding citizenry, more conscious of our multiplicities' (Healey 2012: 23).

in other words, as he who is punished for speaking truth to power, as an aggressively radical planner.

Here is the problem of truth and insincerity in public planning discourses. According to Young (2000), weaker parties can achieve sometimes their political goals when the democratic process is open and fair, and when there is sustained public discussion in which they have a chance to persuade fellow citizens of the justice and wisdom of their cause. Under the influence of Forester (1989), Healey and Gilroy (1990) highlight the importance of a people-sensitive planning, which is based on an epistemological, ethical, political, socio-psychological, structural and self-reflective consciousness; most of Patsy's subsequent works (Healey 1992, 1993, 1996, 1997) are devoted to developing such an approach, method and sensitivity. If, according to Healey and Gilroy (1990: 27) "we must be able to 'ground' not merely our knowledge claims but the ethics of our choices about conversational strategies in the politics of the interactive situation", the failure of the reasoning in public in the first scene of King Lear shows the absence of a theory of truth or of some truth-criteria on which planning discursive inter-relationships can rely on.

Healey and Gilroy (1990) highlight as one of the critical ingredient of a people-sensitive planning the importance of an ethical consciousness, but how to nurture and defend such consciousness still remains an open question. In the Habermasian theoretical framework they refer to, rationality and ethics rely on accuracy, integrity, accountability and sincerity of what participants say in communicatively rational discussions, and on a planner's duty of responsibly constructing dialogues and knowledge according to truth and rigour. This requires planners to be aware of the nature of the knowledge they use, how it relates to the knowledge used by others, and the kind of 'validity claims' brought forward (Healey and Gilroy 1990: 26). Discourses have to be 'fair' and 'right', and based on an ethical dimension of justice as well as honesty. However, as often happens, Goneril and Reagan's discourses are not true, and they are based on rhetorical devices that are instrumental and entirely false; Cordelia is honest but her speech is not persuasive: Lear's choice consequently lacks justice. It is profoundly unjust and unethical.

In fact, if we analyse the 'reasoning in public' process in Lear, some unsolved problems still remain. Cordelia refuses to make a big public fuss about her love for Lear; her honesty and integrity contrast with her sisters' selfish insincerity, which sets and perverts the context for the conversation. As highlighted by Healey and Gilroy (1990: 27), "every interaction is a power relationship, partly 'structured' by the roles and rules which set the context for the conversation, and partly by the 'power of personality' which everyone possesses". Cordelia's character is not one-dimensional; she is more than just 'the good daughter', as her father's power would like. Cordelia refuses to go along with Lear's little love test because she is simply too principled, indeed, too authentic to take part in something so socially low and fake; or, she appears to be completely alien to strategic action. A lot of Cordelia's motivation comes not from frustration with her father but from anger at her sisters' insincerity, that is, from speaking truth in the face of power.

The ethical dilemma posed by the effects of insincere speeches in public reasoning requires further insights; it is still a highly debated issue. Although Healey (1992, 1996, 1997) considers Habermasian communicative ethics a valuable conceptual resource from which follows the ethical challenge of mediating conflicts in inclusionary processes, Hillier (2003) analysed the Habermasian concept of rational consensus formation and its fallibility, describing the possibilities of the permanence of conflict, non-reciprocity, domination and the related power games that are inherent in planning decision-making. In my experience, the argumentative turn fails if the discursive approach is not based on a truthful and open level of confrontation: it is not just a matter of imbalance of powers; it is a matter of making conflictive positions clear and visible, according to agonistic theory (Hillier 2003; Gunder 2003). In other words, in Lear as in many of my research experiences, public planning discourses are neither honest nor authentic and sincere; but this happens in many other situations! There is a further problem, and one peculiar to some undemocratic contexts: as a consequence of Cordelia's public refusal to take part to an insincere discourse that hides conflicts of interests as well as of values, she is abruptly and violently interrupted by the undemocratic power of Lear. The result is that Cordelia loses her status and consequently her rights and, so, there is no longer room for representation, as in the metaphorical figure of homo sacer used by Agamben in order to describe those circumstances when hegemonic power suspends the law with the intention of excluding, depriving of rights and marginalizing within the juridical order. This is exactly what happened, for example, in the case analysed by Bonafede and Lo Piccolo (2010), which explored the ambiguous role that participative practices tend to assume in fragile and substantially undemocratic contexts.

In my metaphor, we could say that Cordelia is not the planner, as she could be considered a disempowered citizen, like asylum seekers or refugees, but also, the technical expertise that is, of planners in our case is subjected to a similar treatment: in fact, Lear's advisors are silent, becoming, sooner or later, victim of the conflicts of power, as in the case of Gloucester,[5] or they are banished after reclaiming their intellectual independency, losing their status and rights of citizenship, as in the case of Kent.[6]

In such circumstances, conflicts can arise or just remain hidden, when, as in most of my experiences, there is no moral shock, there is no room for representation, 'local voices' are disappearing, powerlessness is not contested by the powerless, and protests are latent. It is not just a matter of imbalance of powers (Hillier 2002; Hoch 1994), or of progressive and radical urban movements involved in resisting

[5] When the loyal Gloucester realizes that Lear's daughters have turned against their father and are planning to murder the King, he decides to help Lear in spite of the danger. Regan and Goneril discover Gloucester has warned Lear of their plot, and Cornwall, Regan's husband, accuses him of treason and gouges out Gloucester's eyes.

[6] Kent is notable for speaking honestly, which is a rather rare position in King Lear. Kent is not afraid to be blunt when he tells Lear he is a fool for believing that Goneril and Regan care about him or, when he tells Cornwall that he doesn't like the looks of his face. Kent, although banished by Lear, remains to try to protect the unwitting King from the evils of his two remaining daughters. He disguises himself and takes a job as Lear's servant, named 'Poor Tom'.

the assaults on their life worlds and working for better living conditions (Routledge 2010); rather, it is a matter of making conflictive positions clear and visible in order to have the opportunity to deal with them. Whether on the side of collaborative planning or of the agonist (Hillier 2003; Gunder 2003; Pløger 2004), what about those contexts where conflicts are eluded, implicit or drained, that is, devoid of their oppositional power? The stories that I dealt with (Lo Piccolo 2008, 2009; Bonafede and Lo Piccolo 2010) show such a context where deprived and minority groups such as immigrants and local poor, children and young people have passively accepted policies that confirm their inferior status or the absence of public policies in favour of de-regulating and privatizing the public realm, and without expressing any particular form of protest and/or conflict.

10.3 Re-thinking Democracy

In her recent work, Patsy warns us that a people-centred polity, "requires continual attention to resisting subversion back into the well-established pathways of elite technocracy and corporatism" (Healey 2012: 34). She also suggests (ibid: 30) that nowadays there are different strands for re-thinking democracy in the face of a corroded public realm. Amongst them, the strands of 'deliberative democracy' and 'network governance' take an ontology of mutuality more seriously than others, and have two main qualities: (1) the significance of public argumentation, that is, 'reasoning in public' in ways that recognize the diversity of perspectives, the complexity of many issues faced and the uncertainty of knowledge; and (2) the importance of micro-practices and networks of relations that connect people, drawing back into-correlation the spheres of civil society and the state.

My reflections are concerned with how to transfer and apply such a normative ethos. In my 'extreme' territories, blatant as well as latent conflicts place themselves and reproduce themselves within the gap between public objectives and the distortion of private interests, in the forms of corruption, patronage and un-democratic (hidden) control.

When considering the issues of citizenship and status, we should highlight the political role that planning may have for minority groups and individuals, representing a potential political arena for them through the micro-practices of democracy-in-action: "small struggles which hardly seem to make a difference may sometimes, if built on in the flow of time, lead to major changes in political cultures" (Healey 2012: 20). The appropriation of spaces and the construction of new uses and practices in them, in 'institutional' or 'insurgent' forms, provide a guarantee of the claims of those rights of citizenship that often are denied at the political and/or juridical levels. Local actions of participation, formal or informal practices of coexistence in urban space, as well as inclusionary planning initiatives become all-significant for a redefinition of the category of citizenship with respect to the substantial changes (and plural articulation) of the social corpus in the contemporary city (Lo Piccolo 2010); according to Healey (2012: 28), "this, however, demands a changing ethos and practice from those working in the public sector".

The Habermasian (1981, 1989) substantive norm, which is at the base of Patsy's work, sometimes fails not just because of the gap between its theory and practice (Gunder 2003) or because it represents a restrictive model that does not apply to most decision-making processes (Hillier 2003); in some cases, such as those I refer to (Lo Piccolo 2009; Bonafede and Lo Piccolo 2010), it also fails due to the absence of a basic level of access to citizenship rights that are the indispensable ingredients—as I would say, pre-conditions—for Arendt's space of democracy, which Cordelia lacks and which is the physical and metaphorical arena where 'reasoning in public' practices can really take place and flourish. According to Arendt (2005), the public space of democracy can be defined as the ambit where all the discursive issues can show up their many-sidedness and people can freely show up their own plurality by acting on and expressing their plural opinions. Thus, the public space of democracy coincides with the political space of freedom. The sense, but not the aim, of politics is the freedom of plurality; therefore, understanding a political situation means acknowledging a large framework of different viewpoints and positions from which the situation can be considered and judged.

I highlighted previously (Bonafede and Lo Piccolo 2010) the need for the existence of political spaces where conflicts can fully take place and agreements emerge through discursive and dialogic approaches, using some concepts as developed by Arendt (1958, 2005). In this approach, a substantial precondition for any form of 'public reasoning' is the existence of political spaces such as Arendt's idea of a discursive and active 'infra-space': the tangible space of the *agora* and the metaphorical space of democracy. This exists among equal individuals freely debating; wherever this active and conversational freedom is lacking, a proper political space does not exist (Arendt 2005).

Accordingly, citizenship rights do not only represent a framework of reference that is predominantly theoretical, or an abstract body of principles to which to rhetorically appeal, but are a litmus of daily analysis assessment (Zolo 1994), debates, choices and deliberations that in our ambit of thought and intervention are all specific, minute, practical and tangible. This commitment, however, assumes recognition of an ethical dimension to the planning discipline that is neither obvious nor automatic. Healey (2012: 28–29) suggests a changing ethos and practice from those working within the public sectors, in the face of the current weaknesses of elite democratic practices and the protests and/or challenges against the 'technocrats' as the source of authority. In my experience, planners cannot act as Gloucester, otherwise they will be blinded. They can learn lessons from time to time, from Kent, the disavowed, democratically committed planner who does not renounce to speak truth to power; Edgar in his role of 'Poor Tom', the insurgent planner who acts outside the formal framework of institutionally established procedures.

If Healey and Gilroy (1990: 26) properly highlight that, 'planners interact with people in highly varied positions and roles', I would add that planners themselves should learn to act in different roles and positions at the same time, with all the ethical dilemmas arising from playing multiple roles in multiple dramas (Lo Piccolo 2009) and developing that "epistemological consciousness" (Healey and Gilroy 1990: 26), which is necessary in order to cope with the diversity of interactions and multi-dependent, ethical and contextual issues. In potentially emergent polities and methodological approaches such as enhance the plurality and multiplicity of roles, we may assume at the same time that the formal institutions of representative democracy are but one dimension, and not the dominant one (Healey 2012, p. 30). The role of informal networks is consequently a major issue, but in which way the formal and the informal levels cope or clash, even frequently, with the other, especially in very latent conflictive and substantially un-democratic contexts?

According to Arendt (1958), without the existence of a common space—both physical and metaphorical—initiatives fail or do not even take place. In planning terms, without a common and sincere place of debate, there is no space that is also efficacious for informal networks, because of the patronizing attitude of the political actors and the subjugated condition of all the other subjects. It is not just the old, classical contrast between elite and participative forms of democracy that I am suggesting, but a search for making conflictive positions clear and possibly, sincere: although Kent belongs to the elite, he is forced to set up informal networks out of the formal arena and of the 'reasoning in public' process; the same strategy is adopted by Hamlet in its 'use' of the players, in order to defeat insincerity. Otherwise, without a public space of freely truthful confrontation, Cordelia will never regain her status—which was lost as a consequence of the disregard of her distinctive and different character as well as her authenticity in the 'reasoning in public' game.

References

Arendt H (1958) The human condition. The University of Chicago Press, Chicago, IL

Arendt H (2005) The promise of politics. Schocken Books, New York

Barbalet JM (1988) Citizenship. Rights, struggle and class inequality. Open University Press, Milton Keynes

Bobbio N (1990) L'età dei diritti. Einaudi, Turin

Bonafede G, Lo Piccolo F (2010) Participative planning processes in the absence of the (public) space of democracy. Plann Pract Res 25(3):353–375

De Leo D (2011) Public sphere and illegal settlements: a case from the Naples metro-region. In: Cremaschi M, Ekardt F (eds) Changing places, urbanity, citizenship, and ideology in the new European neighbourhoods. Teknè, Amsterdam, pp 199–220

De Leo D (2013) Trading with enemies? The trading zone approach in successful planning processes in Sicily. In: Balducci A, Mäntysalo R (eds) Urban planning as a trading zone. Springer, Dordrecht, pp 125–142
Dewey J (1991) (1927) The public and its problems. Swallow Press/Ohio University Press, Athens, OH
Forester J (1989) Planning in the face of power. University of California Press, Berkeley, CA
Forester J (1999) The deliberative practitioner: encouraging participatory planning processes. MIT Press, Cambridge, MA
Friedmann J (1999) Claiming rights: citizenship and the spaces of democracy. Plurimondi 1(2):287–303
Friedmann J (2011) Insurgencies: essays in planning theory. Routledge, London
Gunder M (2003) Passionate planning for the others' desire: an agonistic response to the dark side of planning. Prog Plann 60(3):235–319
Habermas J (1981) The theory of communicative action: reason and the rationalization of society. Beacon Press, Boston
Habermas J (1989) The theory of communicative action. Lifeworld and system: a critique of functionalist reason. Beacon Press, Boston
Habermas J (1996) Between facts and norms. Polity Press, Oxford
Harvey D (1996) Justice, nature, and the geography of difference. Blackwell, Cambridge, Mass
Harvey D (2000) Spaces of hope. University of California Press, Berkeley
Harvey D (2003) The right to the city. Int J Urban Reg Res 27(4):939–941
Healey P (1992) Planning through debate: the communicative turn in planning theory. Town Plann Rev 63(2):143–162
Healey P (1993) The communicative work of development plans. Environ Plann B Plann Des 20(1):83–104
Healey P (1996) The communicative turn in planning theory and its implications for spatial strategy formations. Environ Plann B Plann Des 23(2):217–234
Healey P (1997) Collaborative planning: shaping places in fragmented societies. Macmillan, London
Healey P (2009) The pragmatic tradition in planning thought. J Plann Educ Res 28(3):277–292
Healey P (2012) Re-enchanting democracy as a mode of governance. Crit Policy Stud 6(1):19–39
Healey P, Gilroy R (1990) Towards a people-sensitive planning. Plann Pract Res 52(2):21–29
Healey P, Underwood J (1978) Professional ideals and planning practice. Prog Plann 9(2):73–127
Held D (1987) Models of democracy. Polity Press, Cambridge
Held D (1989) Political theory and the modern state. Polity Press, Cambridge
Hillier J (2002) Shadows of power. An allegory of prudence in land-use planning. Routledge, London
Hillier J (2003) Agon'izing over consensus: why Habermasian ideals cannot be 'real.' Plann Theor 2(1):37–59
Hoch C (1994) What planners do: power, politics, and persuasion. Planners Press, Chicago
Holston J (ed) (1999) Cities and citizenship. Duke University Press, Durham, NC
Kymlicka W (1995) Multicultural citizenship. Oxford University Press, Oxford
Lefebvre H (1968) Le droit à la ville. Anthropos, Paris
Lefebvre H (1974) La production de l'espace. Anthropos, Paris
Lo Piccolo F (2008) Planning research 'with' minorities in Palermo: negotiating ethics and commitments in a participatory process. Plann Pract Res 23(2):187–209
Lo Piccolo F (2009) Multiple roles in multiple dramas: ethical challenges in undertaking participatory planning research. In: Lo Piccolo F, Thomas H (eds) Ethics and planning research. Ashgate, Farnham, pp 233–254
Lo Piccolo F (2010) The planning research agenda: plural cities, equity and rights of citizenship. Town Plann Rev 81(6):i–vi
Mitchell D (2003) The right to the city. Social justice and the fight for public space. The Guilford Press, New York, London
Pløger J (2004) Strife: urban planning and agonism. Plann Theor 3(1):71–92

Purcell M (2003) Citizenship and the right to the global city: reimagining the capitalist world order. Int J Urban Reg Res 27(3):564–590
Roche M (1992) Rethinking citizenship. Welfare, ideology and change in modern society. Polity Press, Cambridge
Routledge P (2010) Introduction: cities, justice and conflict. Urban Stud 47(6):1165–1177
Sandercock L (1998) Towards Cosmopolis: planning for multicultural cities. Wiley, Chichester
Sandercock L (2003) Cosmopolis II: Mongrel cities of the 21st century. Continuum, London, New York
Sen A (2006) Identity and violence: the illusion of destiny. W.W. Norton and Co., New York
Yiftachel O (1998) Planning and social control: exploring the 'dark side.' J Plann Lit 12(2):395–406
Young IM (2000) Inclusion and democracy. Oxford University Press, Oxford, New York
Zolo D (1994) La strategia della cittadinanza. In: Zolo D (ed) La cittadinanza. Appartenenza, identità, diritti. Laterza, Roma-Bari, pp 3–46

Francesco Lo Piccolo, Dr., is a Professor of Urban Planning and Head of the Department of Architecture at the University of Palermo, Italy. Fulbright Fellow (The University of New Orleans and the University of California, Berkeley) and Marie Curie TMR Fellow (Cardiff University), he has carried out extensive research on the multi-ethnic city and forms of participation and empowerment of local communities. He was the AESOP President for the years 2014–2016. His current research interests are mostly directed towards social exclusion, ethnicity and diversity, public and civic participation, ethics and justice with a specific focus on Southern European contexts. Prof. Lo Piccolo has led or co-led over 38 research projects in the field of urban planning over the past 20 years, participating in research partnerships and scholarship exchanges with universities and research institutes across Europe and the US.

Chapter 11
Utopian Tension: Sandercock's Inspiring Journey "*Towards Cosmopolis*" (1998)

Giovanni Attili

Abstract In the connection of postmodern theories and radical/transformative practices, Towards Cosmopolis: Planning for Multi-Cultural Cities represents a foundational pillar in the attempt of finding out ways to manage our coexistence in increasingly diverse urban landscapes. It is a book that is able to nurture and inspire new paths that are potentially able to challenge mainstream planning. From this perspective Towards Cosmopolis is able to talk about the theory that difference makes, destabilizing the rational, detached, and a-political approach that is embedded in the modernist planning.

Keywords Planning · Difference · Epistemology · Theory · Empowerment · Co-existence

11.1 Strange Multiplicity

In 2005, during a talk entitled "Transforming Cities: the Power of Imagination" (given as part of a UBC Lecture Series), Leonie Sandercock imaginatively focuses her reflections around a huge bronze sculpture called *The Spirit of Haida Gwaii* that was realized by Bill Reid, a member of the Haida Gwaii Nation from the Pacific Northwest. This sculpture is a boatload of strange, mythical creatures and spirits. This is how she describes it: "The canoe has thirteen passengers, spirits or myth creatures from Haida mythology. The bear mother, who is part human, and the bear father sit facing each other at the bow with their two cubs between them. The beaver is paddling menacingly amidships, and behind him is the mysterious intercultural dogfish woman. Shy mouse woman is tucked in the stem. A ferociously playful wolf sinks his fangs into the eagle's wing, and the eagle is attacking the bear's paw. A frog (who symbolizes the ability to cross boundaries between worlds) is partially in, partially out of the canoe. An ancient reluctant conscript paddles stoically. In the centre, holding a speaker's staff in his right hand, stands the chief, whose identity

G. Attili (✉)
Urbanism Division, Department of Civil, Building and Environmental Engineering, Sapienza University of Rome, Rome, Italy

C. Perrone (ed.), *Critical Planning and Design*, The Urban Book Series,
https://doi.org/10.1007/978-3-030-93107-0_11

(according to the sculptor) is deliberately uncertain. The legendary raven (master of tricks, transformations, and multiple identities), steers the motley crew" (Sandercock 2005).

According to Sandercock's powerful capability of using evocative languages, this complex sculpture represents a playful symbol of the *strange multiplicity* of life. A contemporary human condition in which people hitherto unused to living side by side are thrust together in what she calls the mongrel cities of the twenty-first century. In this respect, Bill Reid's sculpture not only recalls the intermingling of diverse existences within the same fragile space but also speaks of the needed mutual recognition, unending dialogue and affirmation of all cultures that respect other cultures and the earth. In other terms this sculpture is not only a descriptive statement but also a highly imaginative and utopian metaphor. It is the future that needs to be built. It is a powerful and transformative image of the artistry of living together.

Sandercock's work has always been deeply shaped by this utopian tension. She has always tried to fill the gap between the world as it is and the world as it should be. But filling this gap is not an easy thing. It cannot be merely evoked in a naïve statement. It requires political and sociological imagination and the capability of dealing with complexities. It requires deconstructing old ways of thinking as the foundational step to reconstruct and envision other possible futures.

In this respect, *Towards Cosmopolis: Planning for Multi-Cultural Cities* represents a key work that is able to powerfully encompass the needed de/re-constructive route. It embodies Sandercock's persistent commitment and sensitive imagination towards the exploration of Cosmopolis, intended not as an end-state but as transformative process aimed at improving the quality of life of the cities we live in.

Her passionate and rigorous narrative is consciously rooted in the field of planning. But planning is not a natural fact. As every discipline, it is the emerging result of an historical process located in space and time. Bourdieu would argue that this historical process inevitably leads to an institutional consolidation that makes other ways of framing knowledge unexplored. In this respect, Sandercock's work aims at reopening other ways of framing planning.

In the book, this objective is preliminarily addressed by the production of a severe critique of the modernist planning and its roots: the obsession for more rational decision making processes; the propensity for a positivistic approach based on quantitative modelling and analysis; the value-free framework and gender/race neutrality in the interest of "the public"; the project of state-oriented futures. In this deconstructive stance, Sandercock refers to a "heroic model of planning" that has been showing its discrepancies: "planner's historic role has been above all to control the production and use of space. In their state-designed role, they have acted as spatial police, regulators of bodies in space, deciding who can do what and be where, and even, when. They have closed public parks at night so that the homeless cannot sleep in them, created ordinances to prevent street vendors, blocked permits for mosques, determined what kinds of housing renovations are permissible, and invented no end of both blunt and subtle ways of keeping certain bodies (marked bodies, marked by colour, by race, by gender, by sexual preference, and by physical ability) out of the sight and out of the way and out of the neighborhoods of certain other bodies. These discriminations and repressions are the noir history of planning. They mark the profession as reactionary, as complicit with the dominant culture" (Sandercock 1998: 16).

In her book, Sandercock outlines the dark side of planning by making the invisible visible: she unveils histories of sexism and racism, of silenced women and neglected minorities, of repressed cultures and forgotten communities. If the struggle of people against power is a struggle of memory against forgetting, Sandercock succeeds in giving space to what has been hidden by an arrogant and self-referential modernist account: a white, masculine, Eurocentric narrative. In this respect, the book advocates for the recovery of those stories that have been too often ignored, disregarded and erased from the record. But these absences are not innocent. They are part of a systematic exclusion that Sandercock wants to focus on, outlining the urgency to move "from planning history to planning's histories" (Sandercock 1998: 54).

Articulating the history of planning in plural terms, means paying attention to differences. This positionality is powerfully performed all through the book, even when Sandercock addresses a core issue regarding the de/re-construction of planning epistemologies. In outlining what kind of knowledge planners consider valid, the book destabilizes the rational, detached, and a-political approach that is embedded in the modernist planning: an approach dominated by a scientific rationalism rooted on quantitative analysis and technical jargons. What Sandercock claims is that the Enlightenment legacy and its universalistic principles need to be overcome in favour of an epistemology of multiplicity. This epistemological openness necessarily embraces plural and contextualized ways of knowing: "experiential, intuitive, local knowledges; knowledges based on practices of talking, listening, seeing, contemplating, sharing; knowledges expressed in visual and other symbolic, ritual, and artistic ways" (Sandercock 1998: 5). Through the recognition of these alternative epistemologies, planners have to learn to access, use and, therefore, legitimize multiple ways of knowing rather than adhering to one single predominant, and manifestly ineffective, modernist epistemology.

Multiple are also the theories (marxist, feminist, post-colonial, post-structuralism and queer studies) and approaches (advocacy, radical, communicative and collaborative planning) that have progressively contested, implicitly or explicitly, the rational comprehensive planning paradigm since the 1960s. Sandercock is able to reconstruct this variety of contributions leading to one of the most groundbreaking compendium of postmodern and critical planning theory. A courageous capability of voicing the need to radically deconstruct the implicit and untouched theories of the modernist project, challenging its deep-rooted "representation of planning as the voice of reason in modern society [and] the carrier of the Enlightenment mission of material progress through scientific rationality" (Sandercock 2003: 55).

But theories work if they are grounded. Sandercock is very well aware of this. Her effort to deconstruct old ways of thinking and rebuild new lens is consciously rooted in the attempt of discovering new forms of planning that can effectively change our world in a better place. Through a conscious reconsidering of the concepts of social justice, citizenship, community and multiple publics she advocates for examples of bottom-up mobilization around the world. These are inclusive, diverse and insurgent practices that can be interpreted as the foundations of new planning paradigms: transformative actions in which new claims are spatially translated. Sandercock refers to "a thousand of tiny empowerments" that are able to reshape a new role for civil society (and its different articulations) in fostering radical, democratic, multicultural planning practices.

In the connection of postmodern theories and radical/transformative practices, *Towards Cosmopolis: Planning for Multi-Cultural Cities* represents a foundational pillar in the attempt of finding out ways to manage our coexistence in increasingly diverse urban landscapes. It is a book that is able to nurture and inspire new paths that are potentially able to address a very difficult task. These difficulties are partly connected to our crystallized way of thinking, partly due to the major societal changes that are reshaping our cities. The impetuous growing of consistent flows of immigrants, in fact, oblige us to rethink who we are and to reframe the very same idea of the world in which we live (Attili 2008). In Bourdieu-ian language, we are talking about the destruction of the *habitus*, the destruction of our comfort zone, of all that is familiar and homely (Sandercock 2003). Both Zygmunt Bauman and Ulrich Beck have observed that "strangers bring the outside in" and, in so doing, they appear to come as a threat to the known order in which we are able to feel 'at home' with ourselves (Bauman 1998; Beck 1999). The distinction between the inside and the outside collapses. "Where 'we' are today globally is a situation in which every 'we' discovers that it is in part a 'they': that the lines between 'us' and 'them' are continuously redefined through the global realities of immigration, travel, communication, the world economy, and ecological disasters" (Benhabib 1995: 244).

In this highly complex reality, up-rooting and destabilization occur. Fear, aversion, loathing, anxiety end up nourishing the darker side of difference (Sandercock 2003). I am talking about the fear of the Other in its broadest sense, which includes a wide range of groups (indigenous peoples, migrants and refugees, women, gays and lesbians, poor and homeless folks, those with disabilities, …) and intersections among different dimensions (race, ethnicity, gender, sexual orientation, culture, class, age, …). These fear often lead to conflicts, wounds, distresses and pains. These fears are increasingly becoming constitutive in cities of difference and planning and planners have to find proper and effective ways to respond to such fears. If planning can be interpreted as the attempt to manage our co-existence in shared spaces (Healey 1997: 3), then the capability of dealing with difference remains a crucial challenge.

11.2 Looking Back and Moving Forward

This is the challenge *Towards Cosmopolis: Planning for Multi-Cultural Cities* is able to powerfully address. After 20 years since its publication I really think it remains one of the most important contribution in making difference an analytical and transformative category. I had the chance to interview Leonie Sandercock inviting her to look retrospectively at what she wrote. I asked her what kind of role her book played within the planning theory debate. She precisely named a few key issues: "The single most important contribution was to talk about 'the theory that difference makes'. Before *Towards Cosmopolis*, planning theory was not only gender blind but also ignored race and culture as analytical categories (because its authors were almost exclusively white males). Hence the book had a huge impact in places I never expected, like South Africa, Hawaii, and among minority communities. Many readers assumed I was a woman of colour and were surprised when they actually met me! I also think the attempt at challenging Enlightenment epistemology and outlining an epistemology of multiplicity was pretty path breaking, as was the attempt at challenging mainstream planning history as a story of the great white men and their ideas and talking about 'insurgent planning histories'" (Sandercock 2019).

Despite the fact the book was written in a specifically framed historical period, I really think these main achievements make the book, still today, an incredible and influential source of inspiration. Sandercock agrees on that. "I don't think it's outdated in any way. Intellectually, it's a pretty coherent challenge to what was then mainstream planning theory, and since its publication planning education can no longer ignore the challenge of 'difference'. And as a political project, I think it is still inspirational because of certain concepts such as 'insurgent planning practices', and 'a thousand tiny empowerments', and focusing on planning's role in community building rather than or as well as city and nation building" (Sandercock 2019).

I was lucky enough to work with Leonie Sandercock on several research projects (Sandercock and Attili 2009, 2010, 2012, 2013, 2014) and one of the things I instantly appreciated was her self-reflective positionality. The same kind of positionality that she performed when she retrospectively outlined some of the weakness of her book. "Just as my early work (pre-1989) was gender blind, *Towards Cosmopolis* foregrounded 'difference' as an analytical category but was blind to/silent about the systemic injustices directed at Indigenous peoples, globally. I began to be aware of that in *Cosmopolis 2*, but still had not re-framed this as a 'rights' issue rather than one of 'inclusion'. Obviously too, I was not aware of the work being done on climate change/global warming. There was not much environmental awareness in *Towards Cosmopolis*. When Susan Fainstein published The Just City in 2010 I was very critical of that book for ignoring those issues. But then I'm a social or community planner, not an environmental planner, if I may revert to silos! And even as I say that, I realize I am currently trying to articulate a convergence between Indigenous worldviews and what some of recent western science calls the Gaia hypothesis or chaos/complexity theory, in other words seeing everything as necessarily interconnected, and therefore the danger in perpetuating those familiar silos that excuse us from having to know about climate change, etc." (Sandercock 2019).

The recognition of some of these weaknesses made Sandercock move forward along new research paths, paying a renewed attention to the resurgence of Indigenous peoples and their rights claims. This is where her work is now concentrated. *Towards Cosmopolis* is still the pulsing hearth of these new research experimentations, but in an expanded way. The claim for difference now is more inclusive of what her book left out, namely indigenous people. The need for other ways of knowing is now consciously performed trough film-making as a way of expanding the language of planning and, more importantly, as a planning intervention.

In this renewed framework, 8 years after the publishing of *Towards Cosmopolis*, we worked together on a 5-years action-research project called *Finding our Way*. The history of segregation of indigenous peoples and settlers in Canada as a result of four centuries of colonization, led us to imagine a decolonizing role for planning, using film as a way of approaching collaborative and transformational planning: as a mode of inquiry, a form of meaning making, an expression of resistance to municipal colonialism, and as a planning intervention. In this project the use of film in a deeply divided community, played a major role in opening up a new space for dialogue about the past, present, and future. And, beyond dialogue, it leads to action, to different ways of doing things, to alternative imaginings that helped re-shaping the fragile co-existence of two peoples, Native and non-Native Canadians, towards reconciliation and partnership.

The attempt at addressing the history of injustices by the Canadian state towards Indigenous peoples, through community planning work is still the core motive that led, in 2014, Sandercock to build a new research project, *Edge of the Knife*, in partnership with the Haida Nation: a dramatic feature film conceived as a way to address language and cultural revitalization as well as economic development and environmental conservation for the Haida people. The same people whose spirit Bill Reid carved in the sculpture described at the beginning of this paper. A spirit of the *strange multiplicity* of life that Sandercock constantly recalls in all of her writings/projects. Her capability of looking and working through differences is what defines her in the most enchanting way. This has been her work since the publication of her ground-breaking book. This is her walking towards Cosmopolis.

References

Attili G (2008) Rappresentare la città dei migranti. Jaca Book, Milan
Bauman Z (1998) Globalization: the human consequences. Blackwell, Cambridge-Oxford
Beck U (1999) Che cos'è la globalizzazione. Rischi e prospettive della società planetaria, ed. italiana: Carocci. Roma
Benhabib S (1995) Cultural complexity, oral interdependence, and the global dialogical community. In: Nussbaum M, Glover J (eds) Women, culture and development. Clarendon Press, Oxford
Healey P (1997) Collaborative planning. Macmillan, London
Sandercock L (1998) Towards cosmopolis: planning for multi-cultural cities. John Wiley, London
Sandercock L (2003) Cosmopolis 2: Mongrel cities of the 21st century. Continuum, London
Sandercock L (2005) Transforming cities: the power of imagination. Presentation at "Living out the metropolis" lectures series. Vancouver, University of British Columbia, 14th March 2005
Sandercock L (2019) Personal interview
Sandercock L, Attili G (2009) Where strangers become neighbours: integrating immigrants in Vancouver, Canada. Springer, Dordrecht
Sandercock L, Attili G (eds) (2010) Multimedia explorations in urban policy and planning: beyond the flatlands. Springer, Dordrecht

Sandercock L, Attili G (2012) Multimedia and urban narratives: film as policy inquiry and dialogue catalyst. In: Fischer F, Gottweis H (eds) The argumentative turn revisited: public policy as communicative practice. Duke University Press, Durham, NC
Sandercock L, Attili G (2013) The past as present: film as community planning intervention in native/non-native relations in British Columbia, Canada. In: Jojola T, Natcher D, Walzer R (eds) Reclaiming indigenous planning. Queens-McGill University Press, Montreal
Sandercock L, Attili G (2014) Changing the lens: film as action research and therapeutic planning practice. J Plann Educ Res (JPER) 34(1):19–29

Giovanni Attili works as an Urban Planning Associate Professor at the University La Sapienza (Rome, Italy) where he teaches "Analysis of Urban and Regional Systems" and "Sustainable Development". In the last years, he has been working in the field of urban analysis and in the construction of planning processes aimed at building social bonding through learning and knowledge exchange. He is the recipient of many national and international awards. Among his publications: Rappresentare la città dei Migranti (Jaca Book), Il Pianeta degli Urbanisti (DeriveApprodi, with E. Scandurra), Multimedia Explorations in Urban Policy and Planning (Springer, with L. Sandercock), Civita. Senza aggettivi e senza altre specificazioni (Quodlibet).

Chapter 12
Zwischenstadt | Inbetween City. Thomas Sieverts, *Cities Without Cities*: An Interpretation of the Zwischenstadt, 2004

Roger Keil

Abstract This chapter discusses the life's work of German planning theorist Tom Sieverts focusing on arguably his most important work, *Cities Without Cities: An Interpretation of the Zwischenstadt*, first published in German in 1997 and subsequently in English in 2003. Sieverts's salient intervention must, in the first instance, be seen in the context of the debate around the European city. Urban planning scholarship and practice in Europe that had been largely critical of the urban prospect in the 1950s and 1960s, when cities were seen as centers of economic and social crisis, had changed their tone in the final decades of the twentieth century to normatively re-appreciate the—dense, compact, central—"European city" as a cultural, ecological, and social example for future urban development. Sieverts's *Zwischenstadt* questioned the assumptions at the basis of this re-evaluation and demonstrated that the European city centre, if it even still existed in the way it was portrayed, was in fact only part an emerging urban landscape that stretched over a regional scale and beyond. This chapter revisits *Zwischenstadt* and its contribution in light of current debates on postsuburbia and extended urbanization through the themes of infrastructure, nature, everyday suburbanism and design and planning.

Keywords Zwischenstadt · Tom Sieverts · European city · Postsuburbia · Extended urbanization · Infrastructure · nature · everyday suburbanism · Design · Planning

It is an impossible task to approach Thomas Sieverts's life's work in a brief sketch like the one intended here. Yet his masterpiece *Cities Without Cities* offers a lucid glimpse into the complexity of the German planning theorist's urban thinking. The book is the hub of a creative mind and practice that is condensed into a slim volume of high impact and significance. The book is equally synthetic, critical, and generative of an entire domain of research in Europe and elsewhere. Originally published in German in 1997 under its iconic title *Zwischenstadt*, the book has subsequently been translated into English and French.

R. Keil (✉)
Faculty of Environmental and Urban Change, York University, Toronto, Canada
e-mail: rkeil@yorku.ca

C. Perrone (ed.), *Critical Planning and Design*, The Urban Book Series,
https://doi.org/10.1007/978-3-030-93107-0_12

I might just admit it at the outset that *Zwischenstadt* has been one of a handful of books over four decades that have influenced my intellectual life and perhaps career most. I had met Tom Sieverts and been familiar with his work in the late 1980s and early 1990s, especially in the context of my work with the Frankfurt GreenBelt office in 1990–1991.

I read and wrote about *Zwischenstadt* for the first time just after it was published in German (Keil 1999). Traveling virtually and physically through a recently united Germany, I made some observations for the journal *Environment and Planning D: Society and Space* in a special symposium on German language urban themed publications. It is useful in this context to recall what I thought the significance of Sieverts' groundbreaking publication was.

> Sieverts's acute view of Germany's postdivision landscape has generated a lively debate here on the values and substance of planning. The transition from a divided to a unified Germany has come with a reconsideration of the country's development paradigm. The 'old city' is discarded as a developmental mode for German cityscapes' newer, more new-world ways of looking at human settlement are considered, yet the European city's promise of social mix and ecological sustainability is meant to be preserved. *Zwischenstadt* opts for attending to the spaces in between that have been undefined traditionally and newly created as the crevices of globalization and modernization. The book is a passionate plea to get a handle on the unknown and the unfolding rather than extrapolating the known and the existing. It argues for a novel kind of social ecology which has the capacity to integrate the unintegratable.
>
> Sieverts is fascinated with the possibilities of posturbanized landscapes. Specific attention is given to the societal relationships with nature. Urban planning's singular trust in the creative potential of the built environment is replaced by a free-falling entry into the world of urban—nature dialectics: '*The open space of the landscape becomes the actual creative field which must preserve and restore the identity, the unique character of the* Zwischenstadt' (Sieverts, 2003, page 121–2; emphasis in the original). This inversion—to think of the urban from the perspective of 'nature' and of the city from the perspective of its periphery—is certainly new to German thinking (if one excludes from view, for a moment, some of the experiments of social planners and urban ecologists with greenbelts in the 1920s and 1990s)."

While it is important to remind ourselves of the specific historical significance of Sieverts's work, *Zwischenstadt* has had an intellectual and practical afterlife that made it a classic milestone of planning literature.

The value of *Zwischenstadt* lies in its generative capacity to combine a synthetic account of the actual state of urban landscapes (mostly in Europe but also elsewhere) with a progressive realism in redirecting urban planning towards confronting the challenges posed by these landscapes. What appeared new and surprising in the 1990s—"the dissolution of the compact historical European city and … the treatment of a new completely different and new form of the city, which is spreading across the world: the urbanized landscape or the landscaped city" (Sieverts 2003: xi)—has now become a universally recognizable phenomenon of extended urbanization (Keil 2018a, b). This phenomenon has occupied many scholars and practitioners over the last generation. In one of its most recent incarnations, the *Zwischenstadt* has come under scrutiny from Swedish-American architect Lars Lerup who had this to say about what he calls "the continuous city" (Lerup 2017). Declaring the "*idea* of [the] old European city (Sienna, Arezzo, Pompeii) had passed, supplanted by a new city still searching for a name," Lerup concludes that he is speaking about "[a] city that in its shocking spherical reach seems to have erased itself in favor of total urbanization—*the continuous city*" (2017: 11).

Similarly, the debate on postsuburbanization has captured some of the same dynamics of urban expansion and contraction. As Nick Phelps has pointed out in an important and comprehensive new book on "interplaces", a "mid-urban realm" is now taking shape which is not only "morphologically complex" but "also a complex landscape in functional or economic terms" (Phelps 2017: 126). He adds that the city in its traditional understanding of being economically successful "has been joined and perhaps eclipsed in this urban economic triumph by a variety of settlements that make up a mid- or inter-urban realm—an economy increasingly housed in between the cities we take for granted as the primary containers of economic activity. This mid-urban realm has an economic dynamic that is often quite independent of historic city economies" (Phelps 2017: 127).

Sieverts rather immodestly claimed that "in the future about half of the population of the world will live in *Zwischenstädten*" (2003: 3). More importantly still, "([t]he fate of the *Zwischenstadt* is shared by all humanity" (Ibid.). The ethical and political consequences of this statement are far-reaching as it is anti-colonial in its core and invites instead an intercultural dialogue and learning process (Ibid.). The global social justice potential of the generalized *Zwischenstadt* is established accordingly. But it also carries with it a possible commitment to a global sharing of environmental responsibility: "The urban areas in the *Zwischenstadt* must also meet all the requirements of an ecological balance, since there will be no other areas to offset any imbalance" (2003: 11). In these elegant passages at the beginning of the book, Sieverts postulates the idea that the *Zwischenstadt* is not just a neological invention for academic purposes but the true and practical introduction of a concept that grasps real developments in a critical manner. The passages here and elsewhere in the book are reminiscent of the ways in which Henri Lefebvre has theorized urban society, especially the dialectic of concentrated and extended urbanization (2003). The first avenue through which Sieverts approaches the subject is the political economy of contemporary global capitalism which he sees inscribed into the fabric of the in-between landscape. Yet he is not interested in ceding the space of the *Zwischenstadt* entirely to what he calls the "increasingly comprehensive economic instrumentalisation of time and place" (2003: 73). Instead, he proposes to pay particular attention to the "organization of everyday living space" (2003; Chap. 3). The peculiar problem constellations encountered by the inhabitants of the "flexspace" (Lehrer 1994) of the *Zwischenstadt* interest Sieverts most. Here is where planning solutions need to focus.

The problem constellations are general in terms of their scale and pervasiveness but always specific in terms of who is concerned: "Individuals and groups of people will experience only a tiny fraction of the *Zwischenstadt* as a place of proximity, slowness and personal life" (2003: 74). This experience occurs in a space that is polarized into "a global system of production, supply and disposal as part of the worldwide division of labour and into a world of immediate experience of perception, personal encounter with family and friends and self-determination" (2003: 75).

While *Zwischenstadt* is relevant to planning theory and practice, I want to isolate four areas where it has spoken to me specifically: infrastructure, nature, everyday suburbanism in the habitat and planning/design.

12.1 Infrastructure

The *Zwischenstadt*, like Phelps's interplaces, is rich in interconnectivity. Phelps notes the relevance of its spaces to the connection, especially in terms of providing an intersection of virtual and physical mobilities. Examining iconic inbetween mobility spaces such as airports (and aerotropolises), rail corridors, and the exit ramp economy, Phelps concludes: "the infrastructures and technologies associated with progressively enhanced personal physical and virtual mobility should produce some of the best examples of fixed places that are inbetween" (2017: 149). Phelps emphasizes specifically the association of "the largely 'new' places associated with fixed infrastructures for trains, planes, and automobiles, on the one hand, and the liminal spaces of transit, on the other" (Ibid.). This astute observation corresponds to Lerup's treatment of the infrastructural spaces of the continuous city where the map that has "thickened into a veritable meshwork" (2017: 187) built around the "suburban machine," once considered a parasite but now providing the integrative mesh of the "distributed cities" (2017: 39–42). Technology is central to this imaginary and it is changing—to return to Phelps—as physical interaction is increasingly interlaced with and programmed by digital code. The machinic character of the suburban is the groundwork on which such interlacing now takes place. Riffs Lerup over the cacophony of Houston's oil-producing landscape: "The seemingly rigid mathematical distribution of all these objects across the land further recalls the suburbs, with the assemblage of metal bodies and machines again appearing to expose the real workings of the city, in this case suburbia's housie-houses and garish green lawns" (2017: 186). This type of thinking is, of course, reverberant of Keller Easterling's idea of suburbs as disposition where "[a] field of mass-produced suburban houses is a common phenomenon in infrastructure space" (2014: 73).

Sieverts himself precisely observes the roles of infrastructure networks for the functioning of the *Zwischenstadt*. Keenly aware of the automobile dependency of the networked *Zwischenstadt*, which enables "inhabitants with a car to compose their personal metropolis *a la carte* in the form of highly specialized islands of activity" (2003: 78), Sieverts also notes that inhabitants "can reach and connect with a large number of diversely specialized uses and places in a short time" (2003: 71). Like Phelps, Sieverts trusts in the "connecting channels" and sparsely used inbetween spaces of the *Zwischenstadt* when creating the new technology enhanced life of the continuous city. From this results a triple problematique: "Read and used as a 'system', the *Zwischenstadt* is thus problematic from several perspectives. It exerts stress on the environment, it does not serve those sectors of the population which do not have access to a car, and it fragments living space and living time" (2003: 71). Sieverts's complex and layered view of infrastructure in, of and for the *Zwischenstadt* foreshadows, then, Jean-Paul Addie's sophisticated attempt at theorizing suburban infrastructure in an intertwined matrix of conceptual triads that show both the constraints and potential of servicing the inbetween city (Addie 2016).

12.2 Nature

Sieverts never goes as far as Henri Lefebvre, who stated that "capitalism destroys nature and ruins its own conditions, preparing and announcing its revolutionary disappearance" while he also admits, via Marx, that post-capitalist possibilities might rectify this situation as "the exchanges (organic as well as economic) between the social and the natural, the acquired and the spontaneous, [will] be able to reestablish themselves" (Lefebvre 2016: 122). That is one hope that one can pin on the emergence of a completely urbanized society (Keil 2018c).

Sieverts shares with Lefebvre a rootedness of his thinking on the city in what is not the city: the countryside. Remembering his childhood in the 1930s and 1940s, he recalls "the old rural world… the old, almost autarchic farms and the rustic cultural landscape that kept the rural circular economy stable over centuries, where also the energy as oats for the horses and draft animals came from the soil (and not from the gas station)" (Sieverts 2017: 1). Like others before him, Sieverts notes that mechanization and industrialization, along with urbanization, sounded the death knell for this rural mode of life. In a double dialectic, he significantly adds, though, that the *Zwischenstadt* may be the historico-geographic terrain on which new forms of "rurbanity" might help sustain life on a planet of 10 billion. This would mean the "merging of urban and rural, of cultural and natural characteristics in this urbanization process" (2017: 3) including an increase in food production, heightened contradictions of industrial agriculture with more diverse forms of cultures in and around cities, and the spread of "horizontal metropolises" that will have to develop "their wildnesses, their areas of adventure and recreation, in themselves, as fractal urban landscapes" (2017: 4). Sieverts ends with a (rhetorical) question: "Why should, under the constraint of inclusion into natural metabolisms, the greatest urban transformation in human history that we have sketched here not lead to fascinating forms of an urban–rural continuum, fascinating new urban landscapes" (2017: 4).

12.3 Everyday Suburbanism in the Habitat

Tom Sieverts and Henri Lefebvre have some uncanny similarities in the way they are both observers and critics of the "habitat", the state sponsored modernist environment that provides a core feature of the urban periphery and a marker of the *Zwischenstadt* landscape. Interestingly, the large housing estates that for Lefebvre (2003), so monumentally symbolized the "bureaucratic society of controlled consumption" also turned out to be the core elements of the new inbetween cities across Europe and Canada, and not least across the (developing) world since the second half of the twentieth century (Güney et al., 2019).

Sieverts points to the central and irresolvable contradiction that these housing forms represent: "the problem of the temporal, spatial and mental conclusiveness of these urban concepts. For the "new towns" were mostly designed not just as spatially bounded but also as temporally limited settlement structures—for a finite image of a 'timeless future'. In other words, they were '*ready-mades'* for a state of completion and wholeness, without conceptual openness for historical change" (Sieverts 2006: 164). This type of finality contradicts the dynamic character of the real "exploding" (Lefebvre 2003) landscapes of which these housing estates have become a distinctive part.

These and other segments of the inbetween city have become synonymous with a sense of loss, deficiency, deprivation, incompleteness, and unsightliness. They have also become, if we believe Sieverts and others, the home and workplace of the majority of urban dwellers worldwide. They are the anaesthetic everyday world of the suburban century.

The interest in the "day-to-day world of the *Zwischenstadt*" leads Sieverts to questions of design and planning, as well as aesthetics. It culminates in opening up a view "on the chaotic richness of forms of the *Zwischenstadt* which, when measured against standardized ideals of beauty, are regarded as ugly but have long ago been discovered by contemporary art" (2003: 92–3).

12.4 Design and Planning

In some ways, *Zwischenstadt* revolves around the question of design and planning. Sieverts fundamentally adheres to the professional conventions and traditions of architecture and city building while challenging the conservatism still characterizing the profession as the old city of houses and buildings diffuses into a landscape of intertwined structures or as Easterling (2014) would have it, "dispositions."

In *Zwischenstadt*, Sieverts builds his argument from the core of the city and the core of the disciplines of architecture and planning and challenges the assumptions that emanate from both. He specifically references a debate among Italian architects and urbanists on the periphery which some consider "the sick part of the city" (2003: 90; for a revision of this view see Perrone et al. 2017). But quoting Pierre Luigi Nicolin, Sieverts welcomes that the debate ends up accepting that "in our modern society a comprehensive and complex fabric of differences is being made" (2003: 91). Hard to comprehend through "sensory perception," the "citta diffusa" as Italian scholars have called it, "is constituted of more or less dense areas of activity, characteristics, appeals, signs, messages and recollections, of stable and elusive elements, such as an old village on the one hand and a mobile home settlement for new city nomads on the other" (2003: 91). Its diffused reality cannot easily be understood along the conventional aesthetic registers underlying the urbanist professions. In the

Zwischenstadt, "the non-aesthetic, what is normally not consciously perceived, has an excessive weight" (2003: 93–4).

The "unsightliness" and "anaesthetics" of the *Zwischenstadt* are both a challenge and opportunity for the planning profession (Sieverts 2015). Addressing the needs of the *Zwischenstadt* "must be recognized as an important task of suburban governance" (Sieverts 2015). This involves steering the designer's eyes away from the individual object in the urban landscape to the regional scale of the post-indstrial "rurbanity" we have already discussed above. Importantly, as Sieverts has explicated in a manifesto style text on suburban governance "the production of meaningful and life- enhancing space, both in the city and in suburbia, always involves material, technical and functional manipulations together with cultural practices" (Sieverts 2015).

12.5 Conclusion

Tom Sieverts's *Zwischenstadt* remains a relevant text in urban studies and urban planning. It predicted and speaks to the tremendous transformations our urban world continues to undergo in this present century. In this short sketch, I outlined the book's significance in the context of a small sample of newer literature on the "distributed city". Much that was found on the pages of *Zwischenstadt* when it was first published in 1997 was first experienced as an intellectual shock in continental European urban research and planning circles. In the more than 20 years of its reception, and especially after the publication of its English language edition in 2003, has now become standard in professional and academic debates about the future of the city (Carr and McDonough 2016; Hesse and Siedentop 2018; Keil 2018a). In all this, Sieverts remains an original voice. His permanent update on his own thoughts on the development of the *Zwischenstadt* are fresh and original still. His latest incursions into debates on the anthropocene (Sieverts 2017) are just one powerful example of the continued relevance of the ideas Sieverts presented first under the heading of *Cities Without Cities: An Interpretation of the Zwischenstadt* (2003).

References

Addie J-P (2016) Theorising suburban infrastructure: a framework for critical and comparative analysis. Trans Inst Br Geogr 41:273–285. https://doi.org/10.1111/tran.12121

Carr C, McDonough E (2016) Integrative planning of post-suburban growth in the Glatt valley (Switzerland). Raumforschung Und Raumordnung: Spatial Res Plann 76:109–122

Easterling K (2014) Extrastatecraft: the power of infrastructure space. Verso, London

Güney KM, Keil R, Üçoğlu M (2019) Massive suburbanization: (re) building the global periphery. University of Toronto Press, Toronto

Hesse M, Siedentop S (2018) Suburbanisation and suburbanisms—making sense of continental European developments. Raumforschung und Raumordnung: Spatial Res Plann 76:97–108

Keil R (1999) Germany: a middle landscape in winter. Environ Plann d: Soc Space 17:32–39
Keil R (2018a) Suburban planet. Polity, Cambridge
Keil R (2018b) Extended urbanization, "disjunct fragments" and global suburbanisms. Environ Plann D: Soc Space 36(3 June):494–511
Keil R (2018c) The empty shell of the planetary: re-rooting the urban in the experience of the urbanites. Urban Geogr. https://doi.org/10.1080/02723638.2018.1451018
Lefebvre H (2003) The urban revolution. University of Minnesota Press, Minneapolis
Lefebvre H (2016) Marxist thought and the city. University of Minnesota Press, Minneapolis
Lehrer U (1994) Images of the periphery: the architecture of Flexspace in Switzerland. Environ Plann d: Soc Space 12(2):187–205
Lerup L (2017) The continuous city. Park Books, Zurich
Perrone C, Giancarlo P, Paolo P (2017) Post-metropolis—between equipment and flows, places and corridors, fixity and motion. In: Alessandro B, Valeria F, Francesco C (eds) Rethinking the urban question. Regionalization of the urban in Italy and scenarios of innovation. Guerini e Associati, Milan, pp 23–52
Phelps NA (2017) Interplaces: an economic geography of the inter-urban and international economies. OUP, Oxford
Sieverts T (1997) Zwischenstadt. Zwischen Ort und Welt, Raum und Zeit, Stadt und Land. Vieweg, Braunschweig/Wiesbaden
Sieverts T (2003) Cities without cities: an interpretation of the Zwischenstadt. Spon Press, London and New York
Sieverts T (2006) Die Geschichtlichkeit der Großsiedlungen. Informationen zur Raumentwicklung 3(4):163–167
Sieverts T (2015) On the relations of culture and suburbia: how to give meaning to the suburban landscape? In: Pierre H, Roger K (eds) Suburban governance: a global view. University of Toronto Press, Toronto
Sieverts T (2017) Rurbane Landschaften: Vom Aufheben des Ländlichen in der Stadt auf dem Weg ins Anthropozän. Personal communication from author

Roger Keil is a Professor in the Faculty of Environmental and Urban Change, York University. He researches global suburbanization, urban political ecology, cities and infectious disease, infrastructure, and regional governance. Among his publications are *Suburban Planet* (Polity), the *Globalizing Cities Reader* (Routledge, with Xuefei Ren) as well as the forthcoming *After Suburbia* (UTP with Fulong Wu).

Part III
Conceptual Frames

Chapter 13
"*Inquiry and Change: The Troubled Attempt to Understand and Shape Society*", 1990. The Radical Contribution of Charles E. Lindblom's Self-Guiding Society and Probing Volition

Alessandro Balducci

Abstract Inquiry and Change has been the last and most radical book of Charles Lindblom, the political scientist that with his thoughtful reflections changed the way in which we interpret and try to shape public decisions. This chapter offers a reading of the book putting it in relation with the previous work of the great scholar. Following his work on incrementalism and the relation between scientific expert knowledge and ordinary knowledge, Lindblom dedicates this book to explore the consequences of an approach to social and political change that is based upon a critical understanding of how people gather and analyze information to cope with social problems. The concept of volition is central. Volitions are not only defined in interaction and are not given, but they are formed through a process of probing. But probing is not free, it is frequently impaired by institutions in order to keep and maintain the power structure of a centrally guided society. In this sense, scientific knowledge is unveiled as an intensified exercise of lay knowledge but at the same time frequently used as an instrument of impairment. The perspective assumed by Lindblom is that of the self-guided society, that requires a commitment towards the removal of barriers to the free probing volitions of citizens. The chapter offers in conclusion a reflection about the implication of Lindblom thought on planning theory and practice.

Keywords Volitions · Probing · Inquiry · Change · Expert knowledge · Lay knowledge · Self-guiding society

Charles Lindblom was born in 1917 and is one of America's leading political scientists. He passed away in 2018, at more than 100 years old. In his long life, he made fundamental contributions to understanding the structure of social and political processes and, in particular, decision-making. In one of his most recent articles, he wrote:

A. Balducci (✉)
Department of Architecture and Urban Studies, Politecnico di Milano, Milan, Italy
e-mail: sandro.balducci@polimi.it

C. Perrone (ed.), *Critical Planning and Design*, The Urban Book Series,
https://doi.org/10.1007/978-3-030-93107-0_13

> I have been working a long time, still with inadequate success, to try to think clearly about the market system and about democracy. One difficulty may be that we—meaning people all over the world—have actually tried the market in many of its possible forms, learning greatly from both its flaws and its merits; however, we have not yet tried democracy, only distant approaches to it (Lindblom 1995).

I find in this statement all of his wisdom, and at the same time, an explanation of his research program.

Lindblom was a born economist. During the immediate post-war period, he worked at the Rand Corporation and later at the American Defense Research Centre, which assembled hundreds of scholars in a cutting edge research effort ranging from the study of decision-making in situations of uncertainty to game theory.

During that period, he performed in-depth research on the role of trade unions in the American system and published his work *Unions and Capitalism* (Lindblom 1949). This text is virtually impossible to find, but it is not by chance preserved in the Library of the Olivetti Foundation.

Soon, Lindblom began his collaboration with Robert Dahl—a teaching colleague at Yale—and in 1953, they published *Politics Economics and Welfare: Planning and Politico-economic Systems Resolved into Basic Social Processes* (Dahl and Lindblom 1953).

In the following years, he expanded the theme of the structuring of decision-making processes in a series of articles, defining his "incremental model" of policy-making. *The Science of Muddling Through*, published in the Public Administration Review of 1959, is considered as one of the most widely read articles in the social sciences; it has been translated into practically all languages and reported in more than 40 anthologies.

With his distinctive flat and clear style, he accompanies the reader in looking at problems and society from a new perspective, rediscovering the roots of a different and more complete rationality in the complex game of actors.

The central argument is that incrementalism is a better model to describe both the way decisions are made and how they should be made. It is worth recalling the underlying argument, because to some extent, it is the necessary premise for *Inquiry and Change.*

In contradiction with the advocates of the rational model, Lindblom argues that decision-making processes are not coordinated by anyone, but that they are mutual partisan adjustments between a set of actors made up of anyone expressing interest in the problem addressed, regardless of their formal role.

Everyone depends on the other actors with whom they are obliged to interact to achieve their objectives, and everyone has a partisan position. All actors adjust their positions in relation to other actors in a direct form—depending on interaction, confrontation, explicit conflict—or in an indirect adaptive form. When they do not conflict with the position of actors who could harm the achievement of their goals, they adjust their position even without interaction.

Each actor applies an incremental attitude; that is, they evaluate a few alternatives that differ little from each other and from the current situation. What appears to be an incomplete approach is instead, in Lindblom's opinion, the one that allows the progress of society, as each actor is a watchdog of specific aspects of the issue and everyone is forced to interact with the other positions. This induces a much better process of rationalization of choices and decisions than would happen by giving the power to analyze the situation and to decide only to an expert or to a central coordinator.

Lindblom also points out that because of the incremental approach, it is very difficult for actors to fully define their objectives before taking action; however, the opposite is often the case: objectives are defined by actions, and not before it.

These themes were then developed further in two subsequent books: *A Strategy of Decision*, written in 1963 with the philosopher David Braybooke and the book that reorders all his thinking—with the evocative title *The Intelligence of Democracy*—written in 1965, a work of extraordinary clarity.

From this moment on, his works deepen and develop the insights that are already contained in the inaugural texts: the first edition of *The Policy-Making Process* was published in 1968. This allowed him to establish himself fully in the scientific community of political science as the founder of the policy orientation.

In 1975, he published his only well-structured text on planning, *The Sociology of Planning: Thought and Social Interaction*, which appears as the first essay in a book edited by Morris Bornstein, entitled *Economic Planning, East and West*. This text is thought by many to be fundamental; unfortunately, little known of planners in the scientific community.

In 1977, when he was already internationally known, he published *Politics and Market*. It was the only book translated into Italian by Etas Kompass in the series edited by Alessandro Pizzorno. The book received a great response.

In 1979, he published a short but important book along with David Cohen from Harvard, which, to some extent, opened a new direction in a field that has not previously been explored. This book, *Usable Knowledge,* addresses the relationship between ordinary knowledge and scientific knowledge in the social processes; the two authors explore what knowledge is really used in policy processes. They posit that problem-solving involves not only scientific and professional knowledge, but also ordinary knowledge that arises from interactions that frequently prove to be more relevant than scientific knowledge. In other words, we often do not act on professionally produced information, but rather because we obtain information by acting directly.

In 1979—20 years after *The Science of Muddling Through*—Lindblom published a new article in the Public Administration Review that clarifies a series of nodes related to the original text and reflects on the previous work, two decades later. In this article—"Still Muddling, Not yet Through"—Lindblom responds to the many criticisms on his 1959 article by making some limited admissions but essentially confirming his theses.

Finally, in his most recent work published in 1990, *Inquiry and Change, the Troubled Attempt to Understand and Shape Society,* Lindblom retraces part of the path taken years earlier with David Cohen in *Usable Knowledge*. Here, he goes much further into fully understanding the implications of the distinction between professional and ordinary knowledge and how to manage social problems. He also discusses being more open to the definition of a new general philosophy of social reform.

He writes it when he is 73 years old; one can therefore assume that this is a book borne of maturity, which he presents as follows:

> This book is a study of how people in contemporary industrialized societies, competently or not, go about gathering and analyzing information in grappling with social problems. By people, I mean all kinds: ordinary citizens, politicians and other public officials, heads of private organizations, opinion leaders, and experts of various kinds, including for special attention, social scientists.

13.1 Part One of Inquiry and Change: "Probing Volitions"

Volition is a central concept that is immediately contrasted with demand, preference, need, or interest. None of these definitions adequately explain what people really pursue when trying to address a problem. All problem solving is based on the concept that needs, demands, and interests—fixed or variable—are somehow foreseeable and that the foundation of social problem-solving lies in discovering people's preferences, wants, and needs, or what is functional to their interests. Lindblom maintains that this is not the case; that demands, preferences, and interests are not discoverable and that it is useless to try to do so. The consequences are relevant because they recall the need to describe problem-solving in terms that make sense, despite the impossibility of discovering people's preferences, demands, and needs.

While discussing the impossibility of discovering needs and interests, Lindblom highlights that people often act against their own interests, for example, not quitting smoking, voting for a party candidate who will penalize their interests, or buying the wrong house.

The concept of volition helps us to clarify the rationality of behaviors that are difficult to fully understand. Defining actors' positions as their volitions means admitting that people create their interests, needs, and demands, rather than discovering them: the decision does not respond to demand, but decides what the demand should be. The difference between demands, interests, or needs that are considered to be facts and probed volitions, as commitment or creative acts, is not only terminological. It means seeing actors as the writers, not the readers, of their own stories.

According to Crosta (2006) "Lindblom (1990, p. 21–24) is skeptical even about the possibility of the actor knowing his own preferences and needs, that the actor would not 'discover' but 'create' during the action and through the action, while interacting with others and 'adjusting' to the circumstances put in place by the interaction. In this sense, the adjustment does not derive from decisions based on preferences (uncertain to be determined), but from the idea that one decides on what one may want based upon the circumstances of the interaction (and from a judgment on the opportunities—positive or negative—to do something, which is built by the determination to activate oneself to 'do'). Lindblom's volition recalls here, in particular, Dewey's investigation in 1954, as a judgment that marks the passage from an 'indeterminate' problematic situation to a 'determined' one. "

Probing activity is the activity of testing our way of facing a problem that occurs continuously, understanding what others do, adapting our first impulse to other possible comparisons, and being forced to look at the problem from different points of view. However, we all measure ourselves by testing our own volitions, consciously and unconsciously, at different times of the day. We add investigations when we discuss with friends the merits of candidates for a political position, reflecting on the way we live and the changes we might try, and look for role models among our acquaintances. We seek help from people we believe are more experienced than we are in a particular field: doctors, mechanics, priests, and astrologers. In some societies, people even gut chickens to read the entrails.

These investigations mix and advance through action; we do not necessarily test first and act then. However, we continue to test and learn with every action we take (p. 30). There may be many examples. Problem-solving, therefore, is associated with conflicts of volitions and thus with the problem of which volitions matter most.

From this stems one of the central theoretical junctions of the Inquiry and Change argument: if needs, preferences, and demands are a given, the crucial subjects for problem-solving are those capable of discovering them: experts or social scientists; if needs, demands, and preferences are volitions built through interaction, the crucial subjects are also all ordinary citizens.

If the probing activity is weak, we can argue to allow a political or technocratic elite to carry out a volition-defining activity by compressing public participation to the minimum level required in democracy.

For this reason and others, the way ordinary people and their representatives behave while investigating possible volitions becomes a significant part of how society achieves problem-solving.

No one defines their volitions in a linear form. Each individual rather examines many interactions between volitions in all directions and, at most, reaches a very imperfect combination that they call "coherence," ex post.

As an example, when we are deciding on buying a home in the city or in the suburbs, we must consider means of transport, cost, the location of schools, the neighborhood characteristics, and possible leisure activities. All of these variables must be balanced with the family budget and all other expenses that compete for the same resources. It is this combination or "coherence" that is not a linear relationship at all, but rather a series of tests and relationships, instead of a straight line.

A common aspiration is to increase the level of information, insight, and persuasion in policymaking and to reduce the weight of other factors. It makes sense to a point, but beyond a certain level, it becomes senseless: a solution obtained only through scientific investigation and persuasion implies that everyone gradually becomes convinced of the goodness of the solution, and therefore adheres to it. However, there is always a need to use other paths to move toward the solution. This often leads to disagreement because the research is costly, takes long, and often does not achieve results. Therefore, there is always a need to operate through interaction.

Identifying the solution to a problem that concerns a group can take the form of an investigation that disregards the different individual positions and addresses the problem from a perspective of "what is best for the group" as a community.

Alternatively, interaction can be used by comparing the positions of all members of the group and asking each to express their preferences and defend them. This follows the theory of the "Intelligence of Democracy" and is generally considered to be more efficient by Lindblom.

13.2 Part Two: "Impaired Inquiry"

This theme displays a surprising radicalism, which is uncharacteristic of Lindblom, who has always been accused by United States liberal culture of being an inspirer of the pluralist and neoliberal approach.

The argument for impaired inquiry is strong, presenting statements such as "Society teaches people to read, then circulates misinformation to them. Alternatively, adults provide children with schooling, then use the classroom to indoctrinate them" (p. 62).

Understanding all the mechanisms of impairment and the constraint exerted on the forms of inquiry in which each of us is engaged is crucial to understanding power relationships.

"The whole history of humankind reads, in some large part, as a history of impairment of inquiry: ignorance, superstition, barriers to inquiry, exile and execution of dissenters, the many intimidations of tyranny, illiteracy, the steady impositions of peer pressure, and the use of the media for propaganda, among many manifestations" (p. 69).

Only the free competition of ideas offers an escape route from impairment and by careful examination of all forms of limitation to probing, Lindblom affirms that scientific and professional knowledge is only a form of intensified lay probing.

In the third part of the book, entitled *The Social Science Contribution,* Lindblom addresses the use of scientific knowledge as expert probing as well as a contribution to impairment.

In fact, professional knowledge is largely based on lay probing; it does not originate from anything else. Although there is a close interconnection between lay and professional knowledge, professional knowledge has always struggled to detach itself from lay knowledge to defend its advantages and positions of power that allow to keep citizens away from the possibility of deciding on the basis of a more open probing possibility.

13.3 Part Three: "Towards Prescription"

This section of the book opens with a comparison between a scientific society and a self-guiding society.

The former puts science—including social science—at the center. In this model, social problem-solving and social progress is primarily based on scientific observation of human behavior. This means that ideally, mankind can discover the requirements of being a "good citizen" and a "good society" and the results of scientific observation can be used to move in the right direction.

Conversely, the model of a self-guiding society focuses on the probing activity of citizens and their organized forms, with the support of science and social science adapted to the lay role of probing different volitions.

The self-guiding society model is almost a philosophical program in which Lindblom outlines its main characters. It is based on the elimination of barriers to improve the impaired probing of citizens.

It rejects the role of science as a substitute for ordinary investigation but sees scientific knowledge as support and a deepening and extension of lay knowledge.

It also foresees learning processes. It does not assume that there are absolutely correct solutions. It foresees that one learns from mistakes and that learning takes place in the course of action. It tries to compress the role of elites and limit social control. It seeks to promote democracy and the search for solutions that can be found through confrontation or as indirect effects of actions aimed at other ends. Solutions are not valid in general, but time-bound and space-bound. It starts from a realistic and limited confidence in the human intellect's ability to effectively address and solve complex problems.

Addressing Lindblom's contribution to urban planning and planning in general can be said to be unnecessary, as it is largely self-evident.

The rational model against which he opposes his incremental model is the basis of the theoretical conflict between the rational comprehensive paradigm of planning and the political approach proposed by John Dyckman, Melvin Webber, John Friedman, as well as several European planners, such as Patsy Healey, Louis Albrechts, and Pier Luigi Crosta. The incremental model is the theoretical framework that leads to the discovery of actors in the planning process. This is the discovery that there is not a planner, a politician, or a world out there, just the object of the planning activity, which is open to participation of all interested parties, either as a way to describe the planning process, or as an indication to improve the results. At the same time, it considers all the different actors' positions, which can lead to better probed solutions.

However, *Inquiry and Change* suggests much more, in three different directions: digging into the very process of the formation of these actors' positions, interpreted as probed volitions; analyzing the continuity between lay and professional knowledge; and arriving at the idea of a self-guiding society, for which professional and scientific knowledge should operate as a support, an extension, and a refinement of lay probing.

It is a radical contribution to the understanding of the scope of a planning process that recalls the position of Jane Jacobs. Thirty years before, in her battle against obtuse and rigid urban planning that cannot explain how the better urban streets and neighborhoods are not the ones designed by urban planners, but those created through the continuous process of adjustment of a myriad of actors in their activities—Lindblom would say—of probing volitions.

The sense of the profound message of Lindblom for planning is well represented by John Dewey's quote, with which I would like to conclude this review:

> Until secrecy, prejudice, bias, misrepresentation, and propaganda, as well as sheer ignorance, are replaced by inquiry and publicity, we have no way of telling how apt for judgment of social policies, the existing intelligence of the masses may be. Dewey, *The Public and its Problems*, 1927.

References

Crosta PL (2006) Interazione: pratiche, politiche e costruzione di pubblico.un percorso attraverso la letteratura, con attenzione al conflitto. In: Cru, Critica della Razionalità Urbanistica, n 19

Dahl RA, Lindblom CE (1953) Politics, economics, and welfare. The University of Chicago Press, Chicago

Lindblom CE (1949) Unions and capitalism. Yale University Press, New Haven

Lindblom CE (1959) The science of "muddling through. Public Adm Rev 19:79–88

Lindblom CE (1965) The intelligence of democracy. Free Press

Lindblom CE (1975) The sociology of planning: thought and social interaction. In: Bornstein M (ed) Economic planning east and west. Ballinger, Cambridge MA

Lindblom CE (1977) Politics and Markets: the World's Political-economic Systems. Basic, New York

Lindblom CE (1979) Still muddling, not yet through. Public Adm Rev 39:517–526

Lindblom CE (1990) Inquiry and change: the troubled attempt to understand and shape society. Yale University Press, New Haven

Lindblom CE (1995) Market and democracy. Obliquely. Polit Sci Polit 28(4):684–688

Lindblom CE, Braybrooke D (1963) A strategy of decision: policy evaluation as a social process. Free Press

Lindblom CE, Cohen DK (1979) Usable knowledge: social science and social problem solving. Yale University Press

Alessandro Balducci, PhD, is a Professor of Planning and Urban Policies at Politecnico di Milano. He has been the President of AESOP, the Italian Society of Urban Planners, and the National Centre for the Studies of Urban Policies. Working on the new forms of governance and planning in the conditions of regional urbanisation. His recent books are: *Post-Metropolitan Territories. Looking for a New Urbanity*, (Routledge 2017) co-edited with V. Fedeli and F. Curci, *Situated Practices of Strategic Planning. An International Perspective* (Routledge 2017) co-edited with L. Albrechts and J. Hillier, and *Risk and Resilience* co-edited with D. Chiffi, and F. Curci, (Springer 2020).

Chapter 14
Rediscussing Robert Nozick's *Anarchy, State and Utopia*, 1974: Property Titles to Land and Issues of Distributive Justice

Stefano Moroni

Abstract The American philosopher Robert Nozick's fame rests mainly on one book: *Anarchy, State and Utopia.* Discussion of the classic issue of distributive justice is one of the book's central parts, and it has had the greatest impact and sparked the most animated debates. The present chapter critically revisits this issue.

Keywords Distributive justice · Entitlement theory · Property rights

14.1 Introduction

The American philosopher Robert Nozick's fame rests mainly on one book, *Anarchy, State and Utopia* (Nozick 1974). Discussion of the classic issue of *distributive justice*[1] is one of the book's central parts, and it has had the greatest impact and sparked the most animated debates.[2] In this part of the book, Nozick famously criticises the usual way of dealing with the issue of distributive justice. As an alternative, he proposes his own specific theory of justice, which he terms the "entitlement theory of justice". Nozick's theory is mainly a response to John Rawls' theory of justice (1971), which had enjoyed enormous success a few years previously.

[1] As known, Aristotle already spoke of distributive justice. He defined it as that type of justice "exercised in the distribution of honour, wealth, and the other divisible assets of the community" (*Ethica Nicomachea*, 1130b 30–4). For a history of the concept of distributive justice, see, for example, Fleischacker (2004).

[2] Nozick would go on to write several other books without ever systematically returning to the issue of justice (see Nozick 1981, 1989, 1993, 1997 and 2001). On Nozick's life, career, reception and influence see, for example, Bader (2010). For an overview of his philosophical work see, for example, Lacey (2001).

S. Moroni (✉)
Urban Planning and Design Division, Department of Architecture and Urban Studies, Politecnico di Milano, Milan, Italy
e-mail: stefano.moroni@unifi.it

C. Perrone (ed.), *Critical Planning and Design*, The Urban Book Series,
https://doi.org/10.1007/978-3-030-93107-0_14

14.2 The Entitlement Theory of Justice: Legitimate Holdings, Minimal State and No Redistribution

Nozick propounds a rights-based theory,[3] typically deontological and radically anti-consequentialist, in which rights are *absolute*, as opposed to being *prima facie* (Francis and Francis 1976).

Although Nozick is not completely clear on this point, it may be assumed that while he recognises the formal *right to own private property* as one of the basic individual rights,[4] when discussing distributive justice he focuses only on what legitimate ways individuals (who enjoy such a right) have of acquiring *specific property titles* on something.[5] In this latter case, for Nozick the acquisition and transfer of property titles on *land* become the paradigmatic case.

The entitlement theory of justice is compatible only with the idea of a minimal state in the strict sense. From a radically libertarian perspective, it implies that redistribution is unacceptable (wherever it is assigned a moral value in itself on the basis of any given distributive model that is deemed preferable wholly a priori) (Nozick 1974: 168).

14.2.1 *Three Principles: Acquisition, Transfer, Rectification*

Nozick begins his discussion of justice by pointing out that those who concern themselves with distributive justice usually consider a certain distribution at a given moment and ask whether or not it is fair *in and of itself*. However, Nozick observes, the real problem is whether the individuals concerned have acquired what they have at a certain moment in time legitimately or otherwise: for example, a plot of land in the city centre or an apartment in a building.

In this regard, Nozick proposes his entitlement theory. As known, this theory states that (i) an individual who acquires ownership of a good in accordance with the terms established by the *principle of justice in acquisition* has a legitimate right to that ownership; (ii) an individual who acquires ownership of a good in accordance with the terms established by *the principle of justice in transfer* (from someone else

[3] For an overview of rights-based approaches, see, for instance, Waldron (1984), Donnelly (1985), White (1985) and Freeden (1991).

[4] Nozick simply assumes (i.e. presupposes) that certain basic rights exist, without ever providing a thoroughgoing list of what they are, and without furnishing specific justification for their importance (Nagel 1975; Barnett 1977; Pettit 1980: 94–96; Ryan 1982; Campbell 1988: 45; Kukathas and Pettit 1990: 87; Machan 2006: 165–166 and 2009: 78). Nozick himself acknowledges this shortcoming in the introduction to his book—"This book does not present a precise theory of the moral basis of individual rights" (Nozick 1974: XIV)—and in subsequent works: "In *Anarchy, State and Utopia* I presented a political philosophy based upon a certain view of the content of rights but did not (as I said there) present any moral foundations for that view" (Nozick 1981: 498).

[5] For the difference between (i) the formal and general *right* to own private property and (ii) specific property *titles*, see Moroni (2018).

who had the right to that ownership) has a legitimate right to it; (iii) an individual who acquires ownership of a good through *the principle of rectification of injustice* has a legitimate right to that ownership. In general, therefore, no one has the right to a good, a resource or anything else except through repeated applications of (i), (ii) and (iii).

The underlying idea is that anything that derives from *just* initial conditions, through *just* processes, is itself, automatically, *just*. In short: "Whatever arises from a just situation by just steps is itself just" (Nozick 1974: 151). In this sense, the entitlement theory of justice is a purely and radically *process-oriented*—rather than *outcome-oriented*—theory (Davis 1977; Hamlin 1983; Bojer 2003: 14, 53; Konow 2003). In this case, justice depends not on how much individuals own but on how each of them came to own it (Smith 2008: 202).

Nozick never actually specifies in what the specific substantive content of such principles consists (Goldsmith 1979; Replogle 1984; Waldron 1988; Wolff 1991; Knowles 2001; Bader 2010), and actually admits it in the introductory part of his book: "This book does not present [...] a precise statement of the principles of the tripartite theory of distributive justice it presents" (Nozick 1974: XIV).[6] Nevertheless, we shall examine some additional insights that can be deduced from Nozick's book.

The *principle of justice in acquisition* establishes that someone may come into possession of something that does not belong to anyone provided that, in appropriating it, that person respects a weak interpretation of the so-called "Lockean proviso".[7] The Lockean proviso impedes anyone from taking possession of a scarce resource in its entirety in such a way that it is impossible for others to gain access to it. It rules out, for instance, the legitimacy of someone appropriating all the oases that contain all the water available to drink in a desert, preventing others from drinking what they need for their survival. This restrictive clause seems to be understood by Nozick as not only *necessary* but also *sufficient* (the same interpretation is made by Lyons 1977; Kymlicka 1990; Christman 2002: 68 and Duff 2005). In other words,

[6] After providing a general overview of his entitlement theory of justice, Nozick (1974: 153) again notes: "To turn these general outlines into a specific theory we would have to specify the details of each of the three principles of justice in holdings: the principle of acquisition of holdings, the principle of transfer of holdings, and the principle of rectification of violations of the first two principles. I shall not attempt that task here".

[7] Locke (1690/1960) argued that it is possible to appropriate goods that are not held by anyone provided that "enough and as good" remain for others. Nozick notes that two interpretations of this clause are possible: one "stringent", the other "weak". Indeed, there are two ways in which appropriation by one person may worsen the conditions of another. This may happen (i) when the second person does not have the opportunity to improve his/her condition through the appropriation of something, or (ii) when it is no longerQuery possible for him/her—without appropriation—to make use of what he/she could use previously. A stringent interpretation of the restrictive clause would preclude both situations, while a weak interpretation—which Nozick embraces—considers only the second one to be undesirable. Nozick's interpretation is actually one of the weakest versions of the (Lockean) proviso yet to be proposed (Werner 2015).

once satisfied, nothing else would be required to justify an original acquisition by the firstcomer.[8]

The principle of justice in transfer determines that it is possible to become the legitimate owner of something as the result of a voluntary transaction—market-based or otherwise—conducted with the legitimate previous owner; for instance, with someone who in turn came into possession of something by means of an exchange with someone else who had title to it, and so on and so forth. In the same way as valid rules of inference are *truth-preserving*, any transition from one situation to another in accordance with the principle of justice in transfer is *justice-preserving* (Nozick 1974: 151). This is the case provided that such transactions do not themselves entail a violation of the Lockean proviso. Indeed, the Lockean proviso acts *indirectly* also with respect to the principle of justice in transfer. If it is not legitimate completely to exclude others from the enjoyment of certain resources by means of an original acquisition, likewise it will not be legitimate to achieve the same situation through subsequent transactions: for example, by acquiring the entire stock of a vital resource at a certain point in the series of exchanges. As Nozick (1974: 180) puts it, each claim to ownership of a resource contains the "historical shadow" of the Lockean proviso.

The *principle of rectification* states that, in cases where acquisitions or transfers of ownership have occurred in violation of the two principles just mentioned (in other words, through theft, fraud, etc.), it is necessary to rectify matters. In other words, it is a principle that rights any wrongs that have accumulated historically during the acquisition and transfer of ownership. It compensates for past injustices.

In short, and as a consequence of these three principles: "A distribution is just if everyone is entitled to the holdings they possess under the distribution. A distribution is just if it arises from another just distribution by legitimate means" (Nozick 1974: 151).

From the Nozickian perspective, the argument concerning justice in the distribution of goods is thus inseparable from the argument concerning justice in the production of goods. "Things come into the world already attached to people having entitlements over them" (Nozick 1974: 160). Different approaches constitute manna-from-heaven models.

In conclusion, the fundamental maxim of justice becomes, for Nozick (1974: 160), "from each as they choose, to each as they are chosen".

---

[8] Another interpretation of Nozick's thinking is possible. It might be held that the Lockean proviso is merely necessary. Other conditions would therefore need to be met for an original acquisition to be just. The point is that Nozick says nothing about these possible additional conditions (Wolff 1991: 107). The interpretation of the Lockean proviso as both a necessary and sufficient condition is also based on the conviction that Nozick's severe criticism of Locke's theory of "added value" would seem to imply his rejection of that theory (as Varian 1975; Davis 1976; Lyons 1977; Wolff 1977; Ryan 1982; Waldron 1988 and Roark 2012: 690, also argue). As known, according to Locke, even before a restrictive proviso of any kind comes into play, "improvement work" on a resource is required in order to justify its (original) acquisition. Here is a first criticism of such a position by Nozick (1974: 175): "If I own a can of tomato juice and spill it in the sea so that its molecules [...] mingle evenly throughout the sea, do I thereby come to own the sea, or have I foolishly dissipated my tomato juice?". And a second objection: "Why should one's entitlement extend to the whole object rather than just to the *added value* one's labor has produced?" (Nozick 1974: 175).

14.2.2 Four Dichotomies: Historical Principles Versus Non-historical, Patterned Versus Non-patterned, Organic Versus Non-organic, Aggregative Versus Non-aggregative

Entitlement theory, Nozick states, is thus a theory based on *historical principles*. The justice of the current distribution of goods depends on how it has been achieved historically.

Most other theories of justice, by contrast, apply *end-state principles* (*end-result principles*), that is to say, unhistorical principles (Nozick 1974: 153–155). Nozick also refers to such principles (or, at least, to some specific cases of them) as *current time-slice principles of justice*.

These (non-historical) theories focus on a final state of things, and they evaluate "who owns what" at the end of the process. In comparing different distributions, we would therefore need to verify solely the matrix representing (only) current information about distributions. Views of this type include utilitarianism and welfare economics.

At this point, Nozick (1974: 155–156) provokes a certain confusion by stating that historical principles are divided between *patterned* and *non-patterned* (as in the case of his entitlement theory). He thus seems to overlook those which he has just called end-state principles, and he does not make clear whether or not these are patterned. In general, the Nozickian classifications presented in the chapter devoted to distributive justice are unclear on an initial reading (as pointed out by Chiassoni 1985 and Schmidtz 2006, for example). One possible way to make sense of Nozick's argument is to claim that virtually *all* end-state principles are patterned in some way and that *some* non-end-state (that is, historical) principles are patterned as well.[9]

Nozick (1974: 156) claims that a principle is *patterned* if it requires a distribution to vary along with (i) some natural dimension, (ii) weighted sum of natural dimensions or (iii) lexicographic ordering of natural dimensions (i.e. hierarchical order of them). The term "natural" appears to serve no purpose here (do distributions that concern "artificial" dimensions also exist? Van Parijs 1983).[10] It thus suffices to refer to a dimension of any type that is taken as a single criterion to evaluate a distribution fully and comprehensively. Indeed, Nozick considers as patterned those theories which state that the distribution must vary, for instance, according to need, according to moral merit, according to effort (i.e. how hard one tries), etc. or according to some weighted sum thereof. In short, contending that the main task of a theory of distributive justice is to fill in the blank in the sentence "to each according to his/her ..." is to look for a pattern (Nozick 1974: 159–160).

[9] For a partially different interpretation, see Hevia and Spector (2008).

[10] On this, Nozick (1974: 156) writes (somewhat abstrusely) thus: "I speak of natural dimensions, admittedly without a general criterion for them, because for any set of holdings some artificial dimensions can be gimmicked up to vary along with the distribution of the set".

If we agree that people freely participate in transactions of goods, any distributive pattern according to Nozick would be undermined. A patterned distribution could be maintained only by constant interference in the lives of citizens in order to re-establish the preordained configuration. Nozick illustrates the issue with a well-known story (to remain faithful to the text, I cite the original figures). Let us assume an initial distribution, D1, which we deem just, for example, an egalitarian distribution. Let us now suppose that highly-regarded basketball player Wilt Chamberlain's contract entitles him also to 25 cents on every ticket purchased to attend one of his team's matches. Let us assume that by the end of the season around one million fans have attended the matches. Chamberlain will have earned 250,000 dollars from this single item. We thus have a new distribution, D2, in which considerable inequality exists between Chamberlain and all of the others. The question at this point is this: if we considered D1 to be a just distribution, and if the individuals concerned passed voluntarily from D1 to D2, should we not then consider D2 to be just as well? Nozick concludes that freedom inevitably upsets any patterned distribution.

Nozick also accuses patterned theories of completely ignoring the act of giving. Any distributive model would in fact be inevitably disrupted by the practice of voluntary donation (Nozick 1974: 168 ff.). Patterned theories of justice "are theories of recipient justice; they completely ignore any right a person might have to give something to someone" (Nozick 1974: 168).

Before concluding, it should be noted that Nozick (1974: 209) introduces two further classifications: *organic* versus *non-organic* principles, and *aggregative* versus *non-aggregative* principles.

A principle is non-organic if a distribution which a principle deems just remains so also by (hypothetically) removing some people and their distributive shares. The opposite applies in the case of an organic principle.

A principle is aggregative if, given two just distributions (in regard to two different sets of individuals), the sum of the two is also just. The opposite will be true in the case of a non-aggregative principle.

In conclusion, Nozick's entitlement theory is based on (i) historical, (ii) non-patterned, (iii) non-organic, (iv) aggregative principles.

14.3 Discussion: A Weak Point, an Unexpected Point, a Strong Point

Nozick's entitlement theory is certainly challenging. This section will discuss three key matters.

14.3.1 Issues Related to Acquisition

We begin with the principle of acquisition. A weak point in Nozick's argument is the idea that one may take ownership of things that initially were nobody's property—including goods available in finite quantities, such as land—simply by being the first to secure them and limiting oneself to not violating an extremely weak clause.

Even if we overlook the assumption that certain things are initially nobody's property,[11] and the excessively weak version of the Lockean proviso that Nozick accepts,[12] a fundamental problem remains. How exactly can things owned by no one, such as land, be acquired in an original state of nature? (Becker 1977: 26–28).

Does it suffice, for instance, for A to express his/her intention to appropriate certain plots of land? Is it enough for him/her to announce that those plots are his/hers, or to write to that effect from somewhere (even a place far away from them)? If this were possible, A could appropriate all the lands of the Earth with a mere declaration. It would seem, therefore, that A must take possession of the land in some "more concrete" way. But how? Is his/her presence enough? What share of land does a person acquire through his presence? Might he/she only claim ownership of the portion of land that he/she is able to "cover" with his/her body? Then again, it is not enough to be in a place in order to take possession of it (even in an imaginary state of nature). Perhaps A would have to add something to the land that he/she appropriates, for example through his/her labour (which, as we have seen, was Locke's position). Yet as we have already noted, Nozick criticises—and appears to reject—this idea. Some might think that it is enough for A to build a fence or something similar. Yet, how could we prevent A from claiming that the land of which he/she has taken ownership by building the fence is not the land "inside" it but all of the land "outside" it?

We shall now discontinue the sequence of imaginary questions of this type. However, they should suffice to show that the issue of original acquisition is extremely awkward and complex[13]; and it is so—it must be conceded—for *any* position.[14]

---

[11] Why should we suppose, as Nozick assumes, that land is initially "unowned" (that is, free)? Why not imagine instead that land is initially collective property, so that each individual is entitled to an *equal* share of it or at least has a *veto* on certain unilateral appropriations? (see e.g. Roemer 1996: 205–235). Nozick and libertarians in general might counter-argue that *any* conception—and not just theirs—needs some theory of original acquisition (even those who propose a collective or semi-collective notion of property must for example explain *their* kind of original acquisition). Such a rejoinder is pertinent; however, the fact remains that Nozick should justify *his position* on the matter.

[12] On this, see Cohen (1986), Kymlicka (1990: 112–117), Munzer (1990: 269–272), Christman (1994: 61–63) and Duff (2005). Obviously, the problem here arises from the fact that land is available in a finite quantity and the supply of land cannot be expanded indefinitely (Ryan 1987: 69).

[13] For an attempt to find a solution to the *impasse* from a perspective that embraces the Nozickian theoretical framework, see, for example, Kirzner (1978). For a different attempt, yet still in keeping with the Nozickian framework, see Van der Vossen (2009).

[14] It should be noted that it is the conviction that the original acquisition of finite resources—such as land—cannot be wholly justified that gave rise to so-called left-libertarianism (Vallentyne and

14.3.2 Issues Related to Rectification

We now focus on the principle of rectification. The concrete application of this principle is anything but simple, for both logical and empirical reasons (Davis 1976; Goldsmith 1979; Kavka 1982; Graham 1988; Waldron 1992; Valls 1999; Tebble 2001). Suffice it to consider the impossibility of having precise information about *all* of the lawful and unlawful transfers of property that have taken place in history (and the impossibility of being able to manage such information adequately). Consider for example the extremely complex case of land.

In this regard, in a particularly interesting (but often overlooked) passage, Nozick (1974: 231) maintains that a *pragmatic* rectification of procedural injustices might be necessary. His premise is that the material positions of individuals today may depend on past injustices; that is, single or repeated violations of basic liberties that have occurred in previous times. For example, the fact that A today has 3000 and B 30 may depend on the fact that yesterday A illegitimately appropriated 2000 from C (B's father).[15]

However, since we are certainly not able to identify and rectify all these possible injustices *directly*, we might proceed differently (Nozick 1974): for instance, we might opt for an institutional system that continues to recognise only negative rights as fundamental rights but which allows everyone to access a minimum subsistence grant/income to protect against the possibility that those who suffer poverty today do so as a result of injustices (for example theft or fraud) perpetrated by others in earlier times.

This is a significant admission on Nozick's part (Lund 1996; Valls 1999). In fact, such an approach might allow—even from a Nozickian libertarian perspective—forms of redistribution in favour of the most vulnerable (albeit for pragmatic reasons, and still within a purely historical framework).[16] The aim in this case would be to counter some form of *absolute* poverty.

14.3.3 Issues Related to Transfer

The foregoing criticisms notwithstanding, it is undeniable—and this is a convincing component of Nozick's argument—that no theory of distributive justice can simultaneously tackle *all individual* social positions and expect to impose upon them, and maintain over time, an *overall* pattern. This point is connected with the principle of justice in transfer in a situation in which certain basic liberties are granted to actors.

Steiner 2000). According to this theory, the acquisition of finite resources (i.e. resources available in a fixed quantity) must be at least heavily taxed in order to compensate others.

[15] That the assets owned by some people may depend on injustices perpetrated by others in earlier periods is a recurrent theme in Balzac (for example, in "L'auberge rouge" and "L'interdiction"), a problem that creates doubts and remorse in some of his characters.

[16] This possibility had already been discussed by Spencer (1891/2010), although he rejected it.

This argument was anticipated by, for example, Hume (1751),[17] Pareto (1896) and von Hayek (1944 and 1960).[18] A particularly interesting reformulation of the Nozickian idea that liberty upsets patterns can be found in Mack (2002).

In this regard, John Rawls was forced to clarify his position in a direction not so distant from that of Nozick's (and, it should be recalled, also that of Hayek 1944 and 1960).[19] In pointing out that his two famous *principles of justice* apply to the basic structure of society (that is to say, to background institutions), Rawls writes (1993: 282): "The two principles of justice [...] incorporate an important element of pure procedural justice in the actual determination of distributive shares. They apply to the basic structure [...]; within appropriate limits, whatever distributive shares result are just". In short, "the two principles of justice do not insist that the actual distribution conform at any given time (or over time) to any observable pattern" (Rawls 1993: 283). Even the famous *difference principle* therefore "applies to the announced system of public law and statutes and not to particular transactions or distributions" (Rawls 1993: 282).

The same point is stressed in Rawls (2001: 54): "Taking the basic structure as the primary subject enables us to regard distributive justice as a case of pure background procedural justice: when everyone follows the publicly recognized rules of cooperation, the particular distribution that results is acceptable as just whatever

---

[17] Hume (1751: Sect. 3, part II) stressed the difficulties and risks of authoritarianism connected with the pursuit of patterned models of distribution, for instance those based on equality of goods or resources: "Historians, and even common sense, may inform us that, however specious these ideas of perfect equality may seem, they are really, at bottom, impracticable; and were they not so, would be extremely pernicious to human society. Render possessions ever so equal, men's different degrees of art, care, and industry will immediately break that equality. [...] The most rigorous inquisition too is requisite to watch every inequality on its first appearance; and the most severe jurisdiction, to punish and redress it. But besides, that so much authority must soon degenerate into tyranny, and be exerted with great partialities".

[18] Hayek (1944: 102) observes that it is necessary to impose a dirigiste system "if we want to secure a distribution of wealth which conforms to some predetermined standard". He makes a similar point (Hayek 1960: 232) when he notes that "those who pursue distributive justice will in practice find themselves obstructed at every move by the rule of law. They must, from the very nature of their aim, favor discriminatory and discretionary action". Furthermore: "Distributive justice is irreconcilable with freedom in the choice of one's activities: it is the justice of a hierarchic organization, not of a free society" (Hayek 1960: 440). Unfortunately, the "conflict between the ideal of freedom and the desire to 'correct' the distribution of incomes so as to make it more 'just', is usually not clearly recognized" (Hayek 1960: 232). In short, Hayek (1960: 87) writes "Our objection is against all attempts to impress upon society a deliberately chosen pattern of distribution".

[19] As Nelson (2002: 201) observes "Insofar as Rawls's theory addresses the question of the justice of particular holdings of particular persons, his theory is actually similar to Nozick's. Both hold that there ought to be rules of property and contract, that these ought to be respected, and that any particular person's entitlements are determined by the free operation of that system. The question is just which rules, which system of rights, we ought to adhere to". Compare with Biesenthal (1978: 164): "If one lays particular stress on Rawls's claim [...] that justice as fairness attempts to use pure procedural justice to settle questions of particular distributive shares, then the differences between his and Nozick's accounts of distributive justice are not as pronounced as they first might appear. The rights defined by the principles of justice as fairness provide the background against which the dealings of community members occur, and ensure that the resulting distribution will be just".

that distribution turns out to be [...]. This allows us to abstract from the enormous complexities of the innumerable transactions of daily life and frees us from having to keep track of the changing relative positions of particular individuals".

The more general point in this regard seems to be the following: a theory of distributive justice can only be a theory that focuses on the basic framework (i.e. a structural system of liberties, prerogatives, permissions, obligations, duties, formally set and granted by public institutions) within which actions and interactions take place. Hence, it cannot regard (all of the) single contingent distributions *in themselves* because they will always be at least partially determined by free choices and interactions over time. In other words, the distributions of individual instances of property titles to specific assets will be inevitably and legitimately determined in the main by historical factors. Depending on the substantive theory of justice adopted, the framework within which this may occur will be different. And it will be more or less legitimate to correct *some* aspects of the outcomes (for ethical reasons, as Rawls would argue, or for pragmatic reasons, as Nozick would acknowledge). However, a large proportion of individual holdings will remain legitimate, simply, if acquired in accordance with the background framework.

References

Bader RM (2010) Robert Nozick. The Continuum International Publishing Group, New York
Barnett R (1977) Whither anarchy? Has Robert Nozick justified the state? J Libert Stud 1(1):15–21
Becker LC (1977) Property rights. Routledge, London
Biesenthal L (1978) Natural rights and natural assets. Philos Soc Sci 8(2):153–171
Bojer H (2003) Distributional justice. Routledge, London
Campbell T (1988) Justice. Macmillan, London
Chiassoni P (1985) Natura e artificio nella dottrina dello stato di Robert Nozick. Materiali per Una Storia Della Cultura Giuridica 1:233–254
Christman J (1994) The myth of property. Oxford University Press, Oxford
Christman J (2002) Social and political philosophy. Routledge, London
Cohen GA (1986) Self-ownership, world-ownership, and equality. In: Lucash F (ed) Justice and equality here and now. Cornell University Press, Ithaca, pp 108–135
Davis L (1976) Nozick's entitlement theory. J Philos 73(21):836–844
Davis M (1977) Necessity and Nozick's theory of entitlement. Polit Theory 5(2):219–232
Donnelly J (1985) The concept of human rights. Croom Helm, London
Duff DG (2005) Private property and tax policy in a libertarian world: a critical review. Can J Law Jurisprud 18(1):23–45
Fleischacker S (2004) A short history of distributive justice. Harvard University Press, Cambridge, Massachusetts
Francis LP, Francis JG (1976) Nozick's theory of rights: a critical assessment. Western Polit Q 29(4):634–644
Freeden M (1991) Rights. Open University Press, Milton Keynes
Goldsmith MM (1979) The entitlement theory of justice considered. Polit Stud 27(4):578–593
Graham G (1988) Contemporary social philosophy. Basil Blackwell, Oxford
Hamlin AP (1983) Procedural individualism and outcome liberalism. Scottish J Polit Econ 30(3):251–263
Hayek FA (1944) The road to serfdom. Routledge, London

Hayek FA (1960) The constitution of liberty. Chicago University Press, Chicago
Hevia M, Spector E (2008) The bizarre world of historical theories of justice: revisiting Nozick's argument. Soc Theory Pract 34(4):533–549
Hume D (1751/1998) An enquiry concerning the principles of morals. Oxford University Press, Oxford
Kavka GS (1982) An internal critique of Nozick's entitlement theory. Pac Philos Q 63(4):371–380
Kirzner IM (1978) Entrepreneurship, entitlement, and economic justice. East Econ J 4(1):9–25
Knowles D (2001) Political philosophy. Routledge, London
Konow J (2003) Which is the fairest one of all? A positive analysis of justice theories. J Econ Lit 41(4):1188–1239
Kukathas C, Pettit P (1990) Rawls: a theory of justice and its critics. Polity Press, Cambridge
Kymlicka W (1990) Contemporary political philosophy. Clarendon Press, Oxford
Lacey AR (2001) Robert Nozick. Acumen, Chesham
Locke J (1690/1960) Two treatises of Government. Cambridge University Press, Cambridge
Lund B (1996) Robert Nozick and the politics of social welfare. Political Studies 44(1):115–122
Lyons D (1977) The new Indian claims and original rights to land. Soc Theory Pract 4(3):249–272
Machan T (2006) Libertarianism defended. Asghate, Aldershot
Machan T (2009) The promise of liberty. Lexington Books, New York
Mack E (2002) Self-ownership, Marxism, and egalitarianism: Part I: challenges to historical entitlement. Politics, Philos Econ 1(1):75–108
Moroni S (2018) Property as a human right and property as a special title: rediscussing private ownership of land. Land Use Policy 70:273–280
Munzer SR (1990) A theory of property. Cambridge University Press, Cambridge
Nagel T (1975) Libertarianism without foundations. Yale Law J 85:136–149
Nelson W (2002) Liberal theories and their critics. In: Simon RL (ed) Social and political philosophy. Blackwell, Oxford, pp 197–217
Nozick R (1974) Anarchy, state and Utopia. Basic Books, New York
Nozick R (1981) Philosophical explanations. Harvard University Press, Cambridge, Massachusetts
Nozick R (1989) The examined life. Simon and Schuster, New York
Nozick R (1993) The nature of rationality. Princeton University Press, Princeton (N.J.)
Nozick R (1997) Socratic puzzles. Harvard University Press, Cambridge, Massachusetts
Nozick R (2001) Invariances. Harvard University Press, Cambridge, Massachusetts
Pareto V (1896) Cours d'économie politique. F. Rouge, Lausanne
Pettit P (1980) Judging Justice. Routledge, London
Rawls J (1971) A theory of justice. Harvard University Press, Cambridge, Massachusetts
Rawls J (1993) Political liberalism. Columbia University Press, New York
Rawls J (2001) Justice as fairness: a restatement. Harvard University Press, Cambridge, Massachusetts
Replogle R (1984) Natural rights and distributive justice: Nozick and the classical contractarians. Can J Polit Sci 17(1):65–86
Roark E (2012) Applying Locke's proviso to unappropriated natural resources. Political Studies 60(3):687–702
Roemer JE (1996) Theories of distributive justice. Harvard University Press, Cambridge, Massachusetts
Ryan CC (1982) Yours, mine and ours: property rights and individual liberty. In: Paul J (ed) Reading Nozick. Basil Blackwell, Oxford, pp 323–343
Ryan CC (1987) Property. Open University Press, Milton Keynes
Schmidtz D (2006) Elements of justice. Cambridge University Press, Cambridge
Smith P (2008) Moral and political philosophy. Palgrave, New York
Spencer H (1891/2010) Justice. General Books, Memphis
Tebble AJ (2001) The tables turned: Wilt Chamberlain versus Robert Nozick on rectification. Econ Philos 17(1):89–108
Valls A (1999) The libertarian case for affirmative action. Soc Theory Pract 25(2):299–323

Vallentyne, P., Steiner, H. (eds.) (2000), Left-Libertarianism and Its Critics. The Contemporary Debate. London: Palgrave Macmillan.
Van der Vossen B (2009) What counts as original appropriation? Politics, Philos Econ 8(4):355–373
Van Parijs P (1983) Nozick and Marxism: socialist responses to the libertarian challenge. Rev Int Philos 37(146):337–362
Varian HR (1975) Distributive justice, welfare economics, and the theory of fairness. Philos Public Aff 4(3):223–247
Waldron J (ed) (1984) Theories of rights. Oxford University Press, Oxford
Waldron J (1988) The right to private property. Clarendon Press, Oxford
Waldron J (1992) Superseding historic injustice. Ethics 103(1):4–28
Werner PJ (2015) Self-ownership and non-culpable proviso violations. Politics, Philos Econ 14(1):67–83
White AR (1985) Rights. Clarendon Press, Oxford
Wolff R (1977) Robert Nozick's derivation of the minimal state. Ariz Law Rev 19:7–30
Wolff J (1991) Robert Nozick: property, justice and the minimal state. Polity Press, Cambridge

Stefano Moroni is a Professor of planning at the Department of Architecture and Urban Studies (Polytechnic University of Milan, Italy). His main research interests are in planning theory, applied ethics, theory and philosophy of regulation.

Chapter 15
Blade Runner or: Do Androids Dream of Electric Sheep? Philip K Dick's Science Fiction and Maschinenmenschen in Metropolis

Peter Ache

Abstract By presenting Philip K Dick's novel in the current collection, an appeal is formulated. Planners need to provide space and time for utopian exercises (Bloch) and, as society, we need to learn again how to develop a capacity of utopian thinking. Phil K Dick's novels help us to re-learn this; by exploring the future as 'inexhaustible reserve of difference' (Anderson); triggering our 'Möglichkeitssinn' (sense of possibilities, Musil); formulating a critique of the current (Bloch); and, making us step ahead, in the sense of 'opening horizons' (Pinder). Novels like 'Do Androids dream of electric Sheep?' and movies like 'Blade Runner' provide to us a cultural surplus (Bloch) that is irritating but also inviting, which shatters the order of things but which also educates our desire (Abensour in Revolutionary Romanticism. City Lights Books, San Francisco, 1999), a desire to imagine life otherwise (Bloch in Das Prinzip Hoffnung. Berlin, Suhrkamp, 1985/1954).

Keywords Utopia · Futuring · Scenario · Urban society · Bloch · Trans-humanism

Why writing about a science fiction novel in a book compiling classic works in the field of urban planning and scholarly reflection? And why not use classic fiction, as with Thomas Mores? The answer to that question will take a few thousand words and comprises the following six elements: (1) Philip K Dick will be introduced, which will be followed (2) by an outline of the content and relevance of Philip K Dick's novel and its (3) reception. (4) Next, the issue of futuring in urban planning will be addressed, in theoretical and conceptual terms. (5) Then, a scenario will be outlined, men machines in metropolis, which will be combined with the formulation of an (6) appeal: there is a 'right to utopianism' and we need to re-establish utopian thought, not least by re-learning utopian thinking, to create futures that would otherwise not be.

P. Ache (✉)
Nijmegen School of Management, Radboud University, Nijmegen, Netherlands
e-mail: peter.ache@ru.nl

C. Perrone (ed.), *Critical Planning and Design*, The Urban Book Series,
https://doi.org/10.1007/978-3-030-93107-0_15

15.1 Shakespeare of sci-fi

To begin with, the author Philip K Dick requires a certain level of introduction, as readers of this compendium will not necessarily be sci-fi fans or close observers of the scene. Philip K Dick (or PKD, born 1928, deceased 1982) is considered to be one of the most productive authors in the field of science fiction. His oeuvre comprises more than 40 novels and more than 100 short stories. His writing is challenging and at times psychedelic, not least because of his own experiments (some would say excesses) with mind enhancing substances. In the culture sections of mayor newspapers or journals, PKD is sometimes even labelled 'Shakespeare of sci-fi', addressing the richness but also the immense quality of his writing.

The un-initiated non PKD-reader will nevertheless have encountered PKD, most prominently through movies and film. PKD has a huge media presence and is actually very 'visible' in our times. Releases in movie theatres include amongst others the Matrix, eXistenZ, Total Recall, Minority Report. In TV streaming, more recently, the Man in the high castle has been created from a PKD novel with the same title. With respect to the novel discussed in this chapter, Do Androids dream of Electric Sheep?, we finally need of course to mention the Blade Runner (1982, 2017) directed and/or produced by Ridley Scott, probably the most prominent and influential movie produced from PKD's ouevre.

15.2 Do Androids Dream of Electric Sheep?

The novel, published in 1968, is considered to be one of Phil K Dick's more accessible pieces. In the book, PKD sketches out a dystopian future of our society in the year 1992 (later editions speak of 2021), using San Francisco as actual location. It is a post-apocalyptic world, as the environment has been largely damaged by a nuclear global war, called World War Terminus, leaving mankind with an irreversibly damaged environment. Most species are endangered or extinct from extreme radiation poisoning. Society is clearly divided. There are those humans who are deteriorating, literally, in abandoned sections of San Francisco, where the radiation is turning them into 'chicken-heads', a slow but unstoppable process to the brain at the end of which individuals fall into the abyss of mental darkness.

Down there, on street level but also within houses, a strange phenomenon exists: Kipple. "Kipple is useless objects, like junk mail or match folders after you use the last match or gum wrappers or yesterday's paper. … When nobody's around, kipple reproduces itself. For instance, if you go to bed leaving any kipple around your apartment, when you wake up the next morning there's twice as much of it. … No one can win against kipple. … It's a universal principle operating throughout the universe; the entire universe is moving toward a final state of total, absolute kipple-ization." This is how Philip K Dick foresaw the accumulation law in our wasteful society, in 1968.

Not only through kipple-isation life on earth is effectively killing you, which is why the other, more affluent part of society pays huge amounts of money to escape to colonies on Mars. The United Nations encourage emigration to off-world colonies, mainly to preserve humanity's genetic pool. If you emigrate, you will be granted a free personal android to serve you; at least, that's what the marketing brochure says. Those androids, programmable cyborgs made from organic materials, are manufactured by the Rosen company with its headquarter being located on Mars. The androids are to such an extend human like, that a bone and marrow analysis does not help and instead, a polygraph style of test needs to be devised, the notorious Voight-Kampff test, which tries to probe the emotional reactions of any test person, memories and especially empathy, which are considered to be completely absent in androids. With the androids or replicants, another segment of society is established. The future society is based on slave labour provided by androids; an interesting parallel to Thomas Mores Utopia (Morus 1516), which is also based on convicts as slaves!

Both, in the novel and the movies, people live in hyper artificial environments, in a literal sense an anthropocene; in Blade Runner 2049 a dead tree represents nature, the sight of which, even when only printed on paper, triggers intense amazement in viewers. In the novel, owning a real live animal, in other words owning a piece of nature, is a status symbol for the rich, which is why also robot animals are available for the less affluent sections of society. Empathy, emotions, memories, or dreams can be shaped and evoked. The permanent question is, whether what I feel, enjoy, touch, or my emotional relations with other humans is fake or real, cuts as theme across the entire novel. Deckard, the main figure of both, novel and movie, uses a 'mood organ' to control his emotional situation, in a literal sense 'dialing' him happy, sad or angry. In terms of a quick reality check, this resembles a current mobile phone user listening with his headphones to the mood play list of a music streaming service. Society has seen the arrival of a religion called Mercerism, which uses empathy boxes to link users in a virtual reality of simultaneous and collective suffering. They follow Mercer, a martyr character in his eternal climb up a hill, while being hit with crashing stones, something felt by all persons connected to the VR! Again, applying a reality check, one can think in our times about scripted reality TV broadcasting on any sort of 'camp life' allowing spectators to be or not to be empathic or enjoy vicarious embarrassment (in German Fremdschämen).

Robot sheep, mood organ, android humans, virtual reality. The individual finds itself immersed in a kind of simulation which brings to the front the question, what does it mean to be human, to have an identity, and to exist in relation to other humans? (Sleight 2009). The aspect of simulation is central to the PKD universe. In his introduction to PKD's novel Ubik, Michael M Smith, refers actually to Baudrillard and his America book: "Everything is destined to reappear as simulation. Landscapes as photography, …, thought as writing, … events as television. … You wonder whether the world itself isn't just here to serve as advertising copy in some other world" (Smith 2006). In Blade Runner zeppelin airships hang in the skies and screen promised land advertisements for the outer colonies on Mars or elsewhere.

As becomes hopefully clear from that short summary, one can see that Dick's work, repeat 1968, is a dark but also at times very absurd and ironic comment on central questions of the (post)modern society. And with San Francisco as the backcloth of that society, it is definitely also an urban setting. However, PKD is less explicit about the exact conditions; actually, predicting technological advances is not the main skill of any science-fiction author, including Dick. Dick comes close to some sort of prediction and reliable scenario, at the same time however, the human condition is what he is more interested in, as was made clear. The world around that human is depicted in measured doses, revealing at times surprisingly realistic technical dimensions. Focusing on this world in his novels, humans also continue to fail, dramatically; the movies follow a different emotional economy compared with that.

15.3 Blade Runner

To fully grasp that stimulus and to discuss the human condition in an urban setting, we have to leave the paper universe and step into the movie reception. Focusing on the movies, we obviously bring other actors into the plot, which are the director(s), producers, writers, but also directors of photography, who have congenially adapted, or better, trans mutated the novel (Shanahan 2014). By now, we have seen two such transmutations of Do Androids Dream of Electric Sheep? as the Blade Runner sequel, the original Blade Runner (1982, directed by Ridley Scott) and placed in 2019, and the continuation as Blade Runner 2049 (2017, directed by Denis Villeneuve, produced by Scott) with a scenario for 2049 and stepping out of the novel's narrative. Philip K Dick's relation to the Blade Runner was mixed in the beginning; in particular the way androids are depicted was seen critically by him. In his novel, they are clearly less human, mainly due to the absence of empathy; Ridley Scott depicts them more than human, supermen who cannot fly (Shanahan 2014, p6). At later stages, Philip K Dick became more acceptable to the movie, which he actually never saw completed because he died a couple of months before the official premiere.

Both movies can be considered master pieces in terms of the visualization of the dystopian urban society outlined in the novel. The movies are actually located in Los Angeles, leaving San Francisco behind. The opening scene of Blade Runner (1982) is by now a classic image, with a wide-open eye reflecting on its eyeball a dark urban scenery, unfolding endlessly to the observer. The urban fabric is a combination of skyscrapers and industry buildings, fire bursting from chimneys, one does not know whether it comes from industrial or from energy production. Continuing the fly over, the final destination emerges on the horizon, a kind of pyramidal building, towering high over all others, sturdy and defensible, a city within the city, a power icon; the headquarter of the Tyrell company (or Rosen association in the novel), which produces the androids or replicants. In the 2017 version of Blade Runner 2049, the opening does not start with such an overview but focuses immediately on the main actor who is searching for replicants. However, in a later scene, we finally see the city, Los Angeles 2049, which has been designed even more compact, with housing blocks almost cut into a continuous layer of building stock, and which is surrounded by a huge wall, that keeps out the ocean water. That wall is symbolic also for the main divide in that society, the humans and the replicants, a transgression of that wall is not allowed. The environmental catastrophe, the terminal war, results in 2049 in an even more containerized built environment, hermetically closing off the unhealthy living conditions. Citizens are first invisible, they only appear when zooming into the urban landscape. In the original Blade Runner (1982), people come together in the low street level of LA. A brimming and busy multi-ethnical setting, Asia meets America, chaotic and dangerous, covered in endless rain showers. Compared with that, the 2049 version foresees a more sterile environment, controlled and regulated, also in its clearly more exotic elements. Virtual Reality looms everywhere, interacting with passers-by offering proposals what to consume next. And always looms the issue over every actant, who is human, who is replicant, what is real and what is constructed as a simulation.

The mega-city or endless city of Los Angeles is the iconic reference point throughout the movies, only in a small side scene, one sees in Blade Runner 2049 the hinterland of Los Angeles. San Diego is used as the dumping side for LA; a very ironic comment on the state of, so the actual marketing slogan, America's Finest City in 2049. Another scene of the movie plays obviously in Las Vegas, a dead grandezza filmed against a poisonous yellow-orange sky. The epitome of 'simulacrum' (Baudrillard) is not more than an urban desert and the urban environment is certainly alien to humans, and replicants, too. Nature is the escape, as symbolised in the director's cut of Blade Runner (1982), where Deckard finally hopes to find peace and freedom.

The dialogue between PKD and the movie makers created a world visible to us, which deserves our attention and can guide our discussion on the future of the urban, but obviously only in addition to other scholarly work. The main argument being, that, when thinking about futures for specific cities or the urban in general, images play an important role. Many planning and design competitions work essentially with a kind of visual scenario depicting how the future state of the intended development looks like. Virtual reality apps of today provide a direct visual control of any design in a real-life and real-time setting; you can walk through a city and compare the actual with the planned situation by looking at the city through a mobile device. This visual element has been strongly influenced by movies, giving quite some stimulus to our discipline. Think of Fritz Lang's Metropolis (1927), which not only provided an enhanced version of New York city, which Lang visited before shooting the movie, but which also screened new modes of transport, like cars and busses flying, or railway high lines, and which introduced also the first android, ever. Those pictures were important for future generations and designers to come, with clear references that can be found, for instance, in the Fifth Element (1997, director Luc Besson) or in Blade Runner. The latter provides an experience that, in a way, combines a bird's eye (planning) perspective a la Corbusier with street level perspective following Jane Jacobs; the first element certainly paired with control, the other element combined with (potential) chaos and resistance. In any case, again following from Jane Jacobs, the look of things and how they work are inextricably bound together, and are of course an essential element in and for planning.

15.4 Futuring the City

Now, both the novel and the movies have obviously not been made with the intention to have a direct impact on urban planning and development. But, as reflections of the actual and possible modes of development of our postmodern societies, they provide food for thought, and to state this clearly, in addition to academic reflections of the current and probably coming state of urban society.

With that, we enter the second line of argument in this paper, related to the novel as a modern outline of a scenario of a specific urban future. This corresponds with a classic task within planning, the conceiving of future states of the city or the urban. Connell (2009, p 86–87) provides a handy list of quotes that speak about the general orientation and engagement of planning with the future:

- 'the exercise of deliberate forethought by people' (Alexander 1992: 13);
- 'all conscious attempts to organize action in order to affect future outcomes' (Paris 1982: 6);
- 'the specification of a proposed future coupled with systematic intervention and/or regulation in order to achieve that future' (Byrne 2003: 174);
- 'an explicit exercise in imagining the future' (Healey 1996: 242);
- 'ability to prescribe the future urban life' (Davidoff 2003: 210);
- 'to prepare for future activity' (Myers and Kitsuse 2000: 221);
- 'attaining a preferred future physical environment (Hodge 2003: 5);
- 'persuasive storytelling about the future' (Throgmorton 1992: 17);
- 'the guidance of future action' (Forester 1989: 3); and,
- 'it identifies a (future) order, together with the steps that must be taken to bring it about based on knowledge concerning the present order of things' (Faludi 1973: 51).

What is visible from this select set of quotes is, that futuring takes many different approaches in planning: forethought, imaginations, storytelling as quite broad and generic approaches on the one hand, prescription, future order, preferred state as concrete definitions on the other. In planning as practice, many activities are preparations for some sort of forecasted development, be it housing demand, schools, energy demand, water supply and more. Predict and provide, as the (in)famous notion goes, is certainly at the core of planning activities and connects the field seamlessly to a kind of futuring. Methods like trend extrapolations are paired with scenario exercises or future workshops. Strategic or master plans reach out 15 years into the future; revisions on a regular basis try to keep the development paths clear.

True, a scenario as developed in the novel by Dick does not easily sit with those techniques, but it is neither totally incompatible. It is a story told persuasively (Throgmorton) about a possible future development and it has many elements in it, the social and the physical, which are central to our professional and academic concerns. We touch with that the ontological and epistemological field, inviting in particular for the ongoing debate of positivist versus constructivist positions—a perspective which becomes difficult of course, when being actually caught in a total and complete simulation.

The related aspect, the capacity of anticipation is essential in our profession but also restricted. We cannot predict the future in its entirety, but we can create the capacity to anticipate and to create resilience and allow for adaptation. A lesson which we better learn quickly, given the challenges which we are currently faced with for instance with respect to climate change; see Dick's version of world war terminus. An aspect which will be addressed further down.

15.5 Maschinenmenschen in Metropolis?

Reflecting on Philipp K Dick's work and reception, in this section we finally can discuss and reflect on scenarios regarding urban development. A possible outline could be labelled *Maschinenmenschen in Metropolis*, men machines in the metropolis, with men machines which could be androids or allude to transhumanism.

To explore this further, we start with the metropolis. The triumphant city (Glaeser 2011), the 'megapolis' (Gottmann 1961), or further still, the "Promethean city of ideas" (Gottmann 1961), in short the metropolis seems to be the most attractive and challenging hypothesis in our field. In a variant, the metropolis really becomes a machine, the smart city (Herrschel 2013; Townsend 2013), the motoric and propulsive steam machine of the modern society. The smartness of the smart city relates to information and big data and to the technical excess, literally speaking, which industry prepares for our cities. Sensors everywhere, observing our conduct inside built environments, providing in real time all the necessary services. The urban environment augmented with technical devices and paired with artificial intelligence works as an extended machine, a life support system that optimises the physical reproduction of labour. The smart city will certainly not stop outside our bodies and soon include our emotional life and needs; the super-accumulation in modern capitalism does not stop at the outer shell of our body, it intrudes and explores our deeper layers, like emotions and feelings, in search for opportunities for extended forms of consumption. The 'likes' in social media are already today exploited as markets, with a tendency to push the final frontier of consumption further.

The Internet of Things, all those sensors and other technical devices that have as a task to observe and analyse real time the consumption of services provided for by the built and un-built environment, creates an augmented space around us which is not only seen as the salvator regarding the climate challenge but also as the source of new growth of the economy, creating businesses and jobs (United Nations 2014). The strategies of industry consortia in the smart city field departed already on the construction of this new big machine, reducing our idea of a city to a mere urban structure that provides us with the conditions to sustain our life in alien environments. It is an extension of our bodies and maybe also our personalities that exceeds the old fashion 'urban living'; it directly connects 'us' to the environment. The question will be, whether this contains, using Foucault's governmentality frame, a momentum of 'disciplining' cities and, actually, disciplining citizens through, political-technological assemblages'? Or, to the contrary, whether the fuller potential of the smart city can be developed as a utopian project, based on the aspect of 'insurgence' in the sense of experiments and conflict (Vanolo 2014)?

In any case, the smart city and 'us', we become an extreme form of trans-human entity (Bo 2012). Transhumanism is already five decades old and still young in terms of its expectations. Discussing transhumanism merges with sci-fi now: what is the difference between the movement of 'transhumanism', and the androids or cyborgs presented in Blade Runner or Metropolis? We step out of the box now, leaving the usual ground of our common understanding of what sort of city we are planning, why, how and with or for whom? The planning project is certainly one that has the 'human' dimension in central view; we plan for people and citizens. Planning literature is filled with relevant statements, and to name a few, scholars like Friedmann (2002), but also others like Swyngedouw (2009) or Novy et al. (2012) see either the individual or the societal human as the main subject of the planning enterprise. What if that subject changes and takes features, especially in relation to the built environment, which open up entirely new perspectives for what we demand from and the ways we use that environment? The transhumanism movement talks about sensory and physical extensions of our body; cameras translating pictures into sound, colour or other stimuli; prosthetics allowing 'us' to run like an ostrich; chemicals extending our brain capacity; modifications of our human genome introducing new sensorial capacity for the individual. On the basis of that, could we conceive new ways of forming and using the urban environment, transcending our biological restrictions? The normative, regulative, societal, ethical, moral or other consequences of such ideas will be huge. New divides will be introduced to our societies; based for instance on commercial products we will have those who can afford and those who cannot; or those fully embracing such developments and those deliberately staying away from it, very similar to the digital divide of today.

This will confront our societies with considerably sized challenges. How is the 'right to the city', Lefebvre (1968), to be understood from a trans-human perspective. In the augmented city, are trans-humans dreaming of digital democracy, twisting the title of Dick's novel? And are we talking about fifty years or one hundred years and can we simply wait and see this happen? The average speed of the technical part of this scenario has accelerated massively during the last century, based on innovation cycles that become shorter and shorter. Planning, as seen in above quote, addresses a future of usually ten to fifteen years ahead, of investment cycles (infrastructure) spanning 30 and more years, of generation successions within 25–30 years; obviously we are not speaking of the very short termism usually preferred by politicians. In addition, we actually might reach already today into those sci-fi scenarios, without clearly recognizing them (the issue of weak signals, Miller et al. 2012).

Despite the actual material or other content of such scenarios, the issue of time horizons and our actions today is the important element to be considered in a discussion of planning: there be three times, a present of things past, a present of things present, and a present of things to come (Augustinus, Confessiones). This trinity of times being present, the past, present, future, is what planning operates with. And, what is especially needed, is a view towards the future which is present today. This is typically the field of foresight and exploration, trying to capture 'weak signals' and to extrapolate from those important development trends shaping our future living (Ache 2015). Philip K Dick and his novels, as well as the Blade Runner movies can help us develop such explorations of weak, or also stronger, signals. It is their imaginative capacity, which we can help us envision and, literally, imagine possible alleyways towards the future.

15.6 Pursuit of Unknowable Novelty

Phil K Dick and his novels do not replace other work on urban future, specifically not theoretically and also empirically grounded academic work; they come in addition, as was hopefully made clear so far, and they can help us to re-learn utopian thinking. In a way, they help us explore the future as 'inexhaustible reserve of difference' (Anderson) and appealing to our *Möglichkeitssinn* (Musil). In short, by presenting Philip K Dick's novel in the current collection, an appeal is formulated: we need to provide space and time for utopian exercises (Bloch) and we need to learn again how to develop that capacity. That is of course not saying, we are in need of more dystopian stories, like 'Do Androids Dream of Electric Sheep?' On the contrary, we should use those to start with and attach a critique of the current (Bloch), we should step ahead in the sense of 'opening horizons' (Pinder 2013 also with reference to Lefebvre).

Based on an analysis of vision processes and strategic documents (Ache 2017), one can state that visions expressed by participating actors in published documents are rather a repetition of the 'real' (Pinder 2013) and are definitely no experiments in utopian thinking. What is in particular absent, is an element of 'strife' (Pløger 2004, referring to Mouffe (2000) and her theory of agonism). As can also be read from vision documents, again using Pløger's (2004) ideas for analysis, the processes are at the moment more geared to a consensus style of operation. Participation is aligned towards the elimination of conflict; consensus is the norm set for the process. Actors learn to be good visionaries and vision making is not more than a normalizing discourse (Pløger 2004, p 80 refers here to Huxley 2002). However, what is needed instead is a process where we move from a minimalist consensual solution of antagonistic behaviour, to a co-creative attitude of adversaries, emerging in form of a dialectical utopianism (Lefebvre).

We start from a negative position: we tend to see exercises which try to formulate visions for future development frequently as consolidating and comforting, and also utopistic, in the sense of being non-consequential. It is about time to take these exercises serious. In a way reformulating another claim of Lefebvre, there is a right to utopianism, and vision making creates 'moments of experiments in dialectical utopianism'—"to create an alternative urban life that is less alienated, more meaningful and playful but, as always with Lefebvre, conflictual and dialectical, open to becoming, to encounters (both fearful and pleasurable), and to the perpetual pursuit of the unknowable novelty" (Harvey 2012, paraphrasing Lefebvre).

Reflecting on a planners' role, instead of knowing the exact outcomes in advance to implement a future according to 'plan', we might rather be looking into trajectories and possibilities (Ganis quoting Hillier and Healey 2008). We do not own the future as experts with capacity to control, we instead are stewards of a "place [that] becomes a 'participant' in the 'flow of action' and entrusts the design professional to deliver a relevant place. [...] midwifery of the future implies that there is no fixed expectation for a place; rather, the place becomes what it needs to be." (Ganis 2015).

What the place 'needs to be' will be established, in line with the utopianism formulated by Lefebvre, by challenging closed political horizons. Pinder (2013) develops for this some propositions: uncover the desires and dreams of urbanism today; transduce from given real's to possible's; focus on everyday life and its critique (with a reference to Bloch (1985, 1954)) and his 'concrete utopia'); experiment and invent; and finally, 'demanding the impossible is as realistic as necessary'.

In conclusion, it is high time to develop our utopian thinking again and to teach this also to students and practitioners alike. Following Levitas (2013), novels like Do Androids Dream of Electric Sheep? and movies like Blade Runner provide to us a cultural surplus (Bloch) that is irritating but also inviting, which shatters the order of things but which also educates our desire (Abensour 1999), a desire to imagine life otherwise (Bloch 1985, 1954)).

References

Abensour M (1999) William Morris: the politics of romance. In: Blechmann M (ed) Revolutionary romanticism. City Lights Books, San Francisco

Ache P (2015) The future of the city—the future of the metropolis? In: Hospers G-J, Van Melik R, Ernste H (eds) Visies op de stad. NN, NN

Ache P (2017) Metropolitan visions: instant, concrete, and conflict free futures? Territorio:7–14

Alexander E (1992) Approaches to planning: introducing current planning theories, concepts and issues. Gordon and Breach Science Publ. SA, Luxembourg

Bloch E (1985/1954) Das Prinzip Hoffnung Berlin, Suhrkamp

Bo D (2012) Our posthuman futures and education: Homo Zappiens, Cyborgs, and the New Adam. Futures 44:55–63

Byrne D (2003) Complexity theory and planning theory: a necessary encounter. Plan Theory 2:71–178

Connell DJ (2009) Planning and its orientation to the future. Int Plan Stud 14:85–98

Davidoff P (2003) Advocacy and pluralism in planning. In: Campbell S, Fainstein SS (eds) Readings in planning theory. Blackwell, Oxford, UK
Dick P (1968) Do androids dream of electric sheep? Doubleday
Dick PK (1969/2010) Ubik. Orion House, London. eBook Edition Gollancz (2010)
Faludi A (1973) Planning theory. Pergamon Press, Oxford a.o.
Forester J (1989) Planning in the face of power. University of California Press, London
Friedmann J (2002) The prospect of cities. University of Minnesota Press, Minneapolis
Ganis M (2015) Planning urban places: self-organising places with people in mind. Routledge, London and New York
Glaeser E (2011) Triumph of the city: how our greatest invention makes us richer, smarter, greener, healthier, and happier. Penguin, New York
Gottmann J (1961) Megalopolis: the urbanized Northeastern seaboard of the United States. M.I.T. Press, Cambrigde (MA)
Harvey D (2012) Rebel cities: from the right to the city to the urban revolution. Verso, London (Kindle)
Healey P (1996) The communicative turn in planning theory and its implications for spatial strategy formation. Environ Plann B-Plann Des 23:217–234
Herrschel T (2013) Competitiveness and sustainability: can 'smart city regionalism' square the circle? Urban Studies 50:2332–2348
Hillier J, Healey P (eds) (2008) Foundations of the planning enterprise: critical essays in planning theory, vol 1. Ashgate, Aldershot
Hodge G (2003) Planning Canadian communities: an introduction of the principles, practice and participants. ITP Nelson, Scarborough, ON
Huxley M (2002) Governmentality, gender, planning: a foucauldian perspective. In: Allmendinger P, Tewdwr-Jones M (eds) Planning futures: new directions for planning theory. Routledge, London
Lefebvre H (1968) Le Droit à la ville, Paris, Anthropos
Levitas R (2013) Utopia as method: the imaginary reconstitution of society. Plagrave Macmillan, Basingshoke and New York
Miller R, Rossel P, Jorgensen U (2012) Future studies and weak signals: a critical survey. Futures 44:195–197
Morus T (1516) De optimo rei publicae statu deque nova insula Utopia (Projekt Gutenberg), Leuven, Martin van Aelst
Myers D, Kitsuse A (2000) Constructing the future in planning: a survey of theories and tools. Plann Educ Res 19:221–231
Novy A, Swiatek DC, Moulaert F (2012) Social cohesion: a conceptual and political elucidation. Urban Studies 49:1873–1889
Paris C (1982) A critique of pure planning: Introduction. In: Paris C (ed) Critical readings in planning theory. Pergamon Press, Oxford, UK
Pinder D (2013) Reconstituting the possible: Lefebvre, utopia and the urban question. Int J Urban Reg Res 39:28–45
Pløger J (2004) Strife: urban planning and agonism. Plan Theory 3:71–92
Sleight G (2009) Introduction. In P. K. Dick, Do Androids Dream of Electric Sheep? London: Orion Publishing
Smith MM (2006) Introduction: Ubik. Orion, London
Shanahan T (2014) Philosophy and Blade Runner New York: Palgrave Macmillan
Swyngedouw E (2009) The antimonies of the post political city: in search of a democratic politics of environmental production. Int J Urban Reg Res 33:601–620
Throgmorton J (1992) Planning as persuasive storytelling about the future: negotiating an electric power rate settlement in Illinois. J Plann Educ Res 12:17–31
Townsend AM (2013) Smart cities: big data, civic hackers, and the quest for a new utopia. WW Norton, New York
United Nations (2014) World urbanization prospects: highlights. United Nations, New York
Vanolo A (2014) Smart mentality: the smart city as disciplinary strategy. Urban Studies 51:883–898

Peter Ache, Prof. Dr. works at the Department of Geography, Planning, and Environment at the Faculty of Management Science, Radboud University, Nijmegen, the Netherlands. He graduated and received his PhD from TU Dortmund, Germany, as a spatial planner. His research interest focuses on urban futures with a particular concern for vision making processes and strategic spatial planning. These topics also inform his teaching, f.i. on the 'beauty of transition' (MA). He is a member of the German Academy for Territorial Development (ARL) and was, from 2006 to 2008, President of AESOP, the Association of European Schools of Planning.

Chapter 16
Max Weber, *Die Stadt* (1922), English Edition, Max Weber, *The City*, Edited and Translated by Don Martindale and Gertrude Neuwirth, the Free Press, 1958

Paolo Perulli

Abstract Max Weber's "The City" has been the first and foremost masterpiece of sociology of the city in Western culture, yet it is neither a text of urban sociology nor a text of urban history. It is rather a text on *power*, as part of the chapter on "Types of power" in *Economy and Society*, the unfinished Weberian summa opus. It is a chaper devoted to "not legitimate power". The city is the prototype of a power created through usurpation of Empire's power by groups of free men, the future bourgeois class, in Middle Ages's Europe. Types of ancient cities, from elitist democracy to the plebeian city, are sketched. Their importance and significance are, since then, universal and still alive. Not in the East, however, from China to india, where the concept of "citizenship" never took place and State's domination is still the negative feature of Eastern societies. The future of cities in Western nations is therefore linked to the long-lasting struggle of competitive world conceptions and views.

Keywords City · Power · Empire · Freedom · Western culture

16.1 Weber's Model of City

In the era of global capitalism and planetary urbanization, is it still useful to refer to a text like Max Weber's *The City*? It is a text well known in the twentieth-century history of urban thought but hardly quoted today, and even forgotten by most urban scholars. I believe that Weber's text is necessary, and that it is a wrong approach to urban studies to ignore the Weberian theory of the city, and the intertwined theories of capitalism and democracy as well as well explain in this article. For Weber in fact the 'city' is essential part of a web of relations linking autonomous cities, free labour, rational law and state command to capital accumulation and circulation in a social system that we continue to define 'capitalism'. The changing scale of such relations

P. Perulli (✉)
Economic Sociology, University of Piedmont Orientale, Vercelli, Italy

Urban Sociology, Architecture Academy of Mendrisio, Mendrisio, Switzerland

C. Perrone (ed.), *Critical Planning and Design*, The Urban Book Series,
https://doi.org/10.1007/978-3-030-93107-0_16

and the passage from national to global assemblage don't cancel the strength of the Weberian model which combines structures and agencies in an unrivalled manner. Any critical approach to cityness, civicness and the city itself should reconsider this web of relations. This is why revision and expansion of Weberian analysis of types of city in the era of world urbanization is necessary. The idea which has to be made explicit is that planetary urbanization makes the different types of city closer, instead of doing these types unrelated. This takes place in distinct evolutionary paths. We can then elaborate on the connection between distinct spatial–territorial forms, a term which includes economy and society in their variable matching. The idea of assemblage is let aside here, although relevant in recent urban theory like in Saskia Sassen's *Territory, Authority and Rights* (2007). The Weberian category of urban political economy can be still proposed, based on comparison of different types of city and the range of actors and economic-territorial systems taken into account.

Weber's text has been the first and foremost masterpiece of sociology of the city in Western culture, yet it is neither a text of urban sociology nor a text of urban history. It is rather a text on power, and actually it was a part of the chapter on 'types of power' written by Weber in *Economy and Society*, the unfinished Weberian summa opus. It is more precisely a chapter devoted to 'not legitimate power'. The city is for Weber the prototype of a power created through usurpation of legitimate power by groups of free men opposed to the Imperial power, in Europe during the Middle Ages. But its importance and significance are, since then, universal.

For Max Weber, the city is a large, inhabited centre that lacks mutual and personal knowledge of the inhabitants. Therefore, indirect and complex forms of *political* governance are necessary. As well complex and—in the same sense—'modern' are the manufacturing activities carried out in the city: it is the other side of the discourse, the *economic* one, about the city.

There are two ways of forming the city:

1. seat of landlordship or principality, *oikos* of the landlord and production centre for the princely court: the 'city of the prince';
2. place of settlement of an exchange of products, where an essential element is the profit and the coverage of the needs of the inhabitants with the existence of a market: the 'market cities'.

The economic centre of both types, oikos and the market, stand side by side:

(i) the prince or landlord who extracts taxes, duties, land rents, procedural rights (this is the case of the cities of Eastern and Central-Northern Europe);
(ii) concessions to entrepreneurs who founded cities—the union of foreign invaders, pirates, commercial settlers or local actors interested in transit trade (as historically in the Mediterranean Europe).

On one side is the city of the prince who develops revenues: the city of officials (Beijing), the city of landowners (Moscow) are 'cities of consumption'. On the other hand, the city's land rents flow into the hands of an urban aristocracy both in antiquity (Rome, Byzantium), and in the Middle Ages: it is the 'commercial city' or 'city of production'.[1] But in long-distance trade profit was realized outside the city: today the same phenomenon occurs in finance as in London, New York, and other 'global cities' (Sassen 1991). The town's economic policy develops the regulation of the economy. Distinct from it, the political-administrative sphere is linked to the ancient fortress-city, in the Middle Ages to the city of the burgesses, citizens who have the obligation to defend and maintain the walls. Weber underlines the plastic dualism between urban economy and urban politics. In Rome next to the comitium and the campus martins (assembly and military garrison) are the fora or economic markets. Here is born a citizen status group or Stand (the bearers of privileges of estate): it is the characteristic of the city in a political sense.

Who governs? a patriciate citizen (like the gens Claudia in Rome): the patrician families, then in the Middle Ages the families of the knights who hold military power. The Roman *civitas* is the prototype of the 'modern' city founded by the common settlement of natives and individuals of foreign origins (Tito Livio), fraternized with ritual ceremonies in religious communities (civil religion). And then divided into families, curias, tribes that make their aggregation 'legitimate'. Instead in the medieval city for the first time, one becomes a citizen as a single individual. "The air of the city makes free" those who were serfs: the aggregation as a brotherhood gives rise to the municipality which is the new institution. For the first time the political equation of the inhabitants, the election of city officials, is realized, while the differentiation moves to the class (a class of 'notables' emerges). This prepares

[1] The following citation of Weber's text gives us back the richness of his contrasting models, whose actuality is very important: the Hanse has been recently rediscovered as a model of contemporary networked cities, and the commenda has been applied to explain how the start-up enterprises of Silicon Valley act today to collect risk capital.

"The large consumers can be rentiers spending their business incomes (today mainly interest on bonds,

dividends or shares) in the city. Whereupon purchasing power rests on capitalistically conditioned monetary rentier sources as in the city of Arnheim. Or purchasing power can depend upon state pensions or other state rents as appears in a "pensionopolis" like Weibaden. In all similar cases, one may describe the urban form as a consumer city, for the presence in residence of large consumers of special economic character is of decisive economic importance for the local tradesmen and merchants.

A contrasting form is presented by the producer city. The increase in population and purchasing power in the city may be due, as for example in Essen or Bochum, to the location there of factories, manufactures, or home-work industries supplying outside territories thus representing the modern type. Or, again, the crafts and trades of the locality may ship their goods away as in cities of Asiatic, Ancient, and Medieval types. In either case the consumers for the local market are made up of large consumers if they are residents and for entrepreneurs, workers and craftsmen who form the great mass, and merchants and benefactors of land-rent supported indirectly by the workers and craftsmen. The trade city and merchant city are confronted by the consumer city in which the purchasing power of its larger consumers rests on the retail for profit of foreign products on the local market (for example, the woolen drapers in the Middle Ages), the foreign sale for profit of local products or goods obtained by native producers (for example, the herring of the Hansa), or the purchase of foreign products and their sale with or without storage at the place to the outside (intermediate commercial cities). Very frequently a combination or all these economic activities occurred: the commenda and societas maris implied that a tractator (traveling merchant) journeyed to Levantine markets with products purchased with capital entrusted to him by resident capitalists.".

modernity: Pierre Bourdieu explains the social 'distinction' in contemporary societies as a condition of existence that creates a habitus and a lifestyle and is linked to a habitat.

In history, Italy was the homeland of conjuratio, a common oath between different groups and classes. It is the Weberian 'city of the plebeians': the genesis of the people, entrepreneurs and artisans against knightly families (a process that took place in Milan in 1198, then in Pavia, Lodi, Lucca, Bologna, Verona, Siena). Who governs? the people elect one of its captains and organize themselves into corporations and guilds.

Another feature of the city in its Western form is to be a *unitary community*, as Weber underlines in 'Die Stadt' (and also in his last lessons on world economic history in Munich). Decisive is the autonomy of the city, its nature of self-governing and self-regulating entity. To constitute a full urban community—Weber observes—a settlement must display a relative predominance of trade and commercial relations with the settlement as a whole displaying the following features: (1) a fortification; (2) a market; (3) a court of its own and at least partially autonomous law; (4) a related form of association; and (5) at least partial autonomy and autocephaly, thus also an administration by authorities in the election of whom the burghers participated.

This complex of institutions of self-governance (as today we call it) is of the utmost importance as they prepare the 'state form', and are different from other forms of political domination, as they are self-determined and autonomous (*autòs nomos* = holding its own law). Therefore, the 'air of the city' is the atmosphere of community relations no longer based on family, lineage or traditional powers but on parity of rights and moral norms administered by the city itself.

In Barrington Moore's research on the 'moral aspects of economic development', a clear Weberian attitude is traced and greatly enriched. In medieval cities, for the first time in history courts are created to protect and regulate merchants' activities. The judiciary power of courts creates and applies norms of reputation, good faith, business ethics, prudence and obedience that are still the basis of modern capitalism, although under attack of neoliberalism and deregulation today.

These moral norms are created in medieval urban courts to guarantee markets' supply and standards' respect, to prevent the formation of monopolies, to control rates of intermediation. These norms were born from interaction between merchants' elites and city governments, which overlap and become intertwined.

Medieval citizen was such as he participated in tribunals and in the choice of authorities. This is not at all new in Western culture, as the Greek *polis* was already a form of brotherhood, a synekism (*sun oikos* means the union of dwellings in a single unity: "Theseus united them in today's city", as Tucydides wrote).

The polis is already, Weber underlines, the product of a legal association and its symbolic, identitarian meanings: a common ritual banquet is prepared, a religious association is created destined only to people owning their tombs and dwellings within city limits. In Medieval cities, the act of brotherhood is *coniuratio*, an armed confraternity offensive and defensive which usurps the Imperial political power. Venice in 726 A.D. was the first to make such detachment from Byzantine Empire a matter of fact.

Therefore, religious brotherhood and military equipment make the city possible. But then it is the force of the economy, of the market to make it grow. What determined knights' families to become city residents was the possibility to take part in commercial activities. But essential to creating the Medieval city was the birth of democracy. *Demos, plebs, people, Burghershaft*: the irruption of a mass of citizens which don't live as knights.

Weber's lesson is therefore also an extraordinary theory of the origins of democracy based on military force, to be read and reread at the light of contemporary times. The Medieval noble, the knight—Weber observes—are placed under surveillance, deprived of their right to vote and of their rights, like the Russian bourgeois class was deprived under Lenin. The reason of democracy is always military in nature: as the masses of not-knights were to be enrolled, the arms were placed in their hands and therefore the *political power. But again, the economic side has to be underlined: they were the merchants, the artisans.*

This is also the fascinating origin of the city of the bourgeois class in Pirenne's history of Medieval cities: cities are agglomerations of merchants. They are social outsiders: masses of adventurers without roots are the first adepts of the future bourgeois class. They locate their commodities' deposits around the walls of the roman cities or out of the walls (*forisburgus, foubourg*) or in *portus* (in England and Flanders). They become the burgenses (inhabitants of the new borough).

In this context, a political economy of cities is developed, giving rise to a 'city-state' (in Italy and Germany, not in England). In this framework, also a public sphere is created again, after the long interruption of their feudal world. This public sphere is no longer—as it was in the ancient *polis*—based on separation between domestic and political sphere, but it assumes the modern form of 'society', the strange 'hybrid'—according to Hannah Arendt—where private interests reach a public significance. In the past ancient times, the concept of private life was the equivalent of a loss or deprivation, as the public sphere was the only place granted for value (the political action, arts and public honours). The public sphere is the common world which puts all in connection with all, the *koinon* (common). The communal city is then the prototype of this public sphere. Although the organization in guilds and corporations is limited to a stage of "enlarged domestic sphere", a civil society will be born only later, in the seventeenth century.

The public spaces of square, of communal assembly, of university (Universitas), of library (the first public library was born in Florence in the fifteenth century) are activities of the *vita activa* (the social world of free 'agency' according to Arendt) which take place in the public sphere. The very nature of the city 'to make free' its inhabitants, hence a parity equilibrium of strata and the freedom for all, is the decisive feature of the Medieval city according to Weber. This leads to revision of the idea regarding the condition necessary to be admitted in the public sphere. Originally the artisan was neatly distinct from the free man in city, as only the latter was wealthy enough to devote himself to the public sphere (this was also the case in the antiquity: the artisan lived in a condition of limited serfdom according to Aristotle). The *Book of Customs* in England (in the Early Middle Ages) still made a distinction between artisan and free man (*franke homme*) in cities. Only later the artisans became so wealthy that the citizenship started to coincide with the membership of a corporation or artisans' company.

16.2 Legacies and Interpretations of Weber's Model and Its Actuality

The modern State, the nation-state is the product of processes based on the progressively intensified elimination of the autonomy of cities. The history of this long-lasting conflict—actually an entire millennium- has been written by authors like Tilly (1990) giving to the city the role of 'capital' and to the state the role of 'coercion'. In other terms the city is the node of attraction of the various and different functions of capital accumulation and circulation, whereas the state is the crystallization of the coercive power. Why regional Empires, federations and other weaker forms of authority were not successful in the European case, and the nation-state form finally prevailed, is the main contribution of Tilly's historical sociology. His work is a creative development of previous questions made by the works of Barrington Moore, Stein Rokkan and Lewis Mumford. Tilly is able to show how the capitalist forms of production were successfully linked to cities, and later to nation-states, whereas forms of agrarian production regimes were destined to survive in different regions (Asia, Africa, and the Americas). However, in Tilly's work the contribution of Weber about the state formation is given for granted (as well as the theories of Marx, Schumpeter, Rokkan, Moore and others). My doubt is about the separation between capital and coercion: on one side the city, on the other the state. The Weberian approach is clearly different: economy and polity cannot be divided, and they coalesce in the city, and later in the state form. Also, we can observe that also capital is coercion (forced submission of people to industrial discipline, expropriation of lands and estates, enclosures etc.) and that the state is not only coercion (but also the owner and dispenser of cultural capital, symbolic capital, social capital and so on). Therefore, the historical link between capital and coercion is stronger than the historical opposition between city and state forms.

To be true, Tilly's model is based on two spatial logics: the bottom-up spatial logic of capital and the top-down spatial logic of the state. These logics are intertwined: in Europe, and elsewhere (China) the power of Empire and the state power are descendent, the power of capital (commerce) is ascendent. The network of capital is essentially commercial (between and among cities of different size, small towns and villages), whereas the state network is coercive (through army and taxes administration). The two networks of commerce and coercion can be differently articulated: strong central power and strong capital can be linked, as in the case of capital cities or strong commercial cities like Amsterdam, Venice, Genoa; strong central power can be also confronted with dispersed capital, as in seventeenth to nineteenth-century Eastern Europe. But strong central power can also be incompatible with dispersed capital and fragmented spaces, as in the case of the network of cities of the Hanseatic League in Northern and central Europe between thirteenth and seventeenth centuries.

Weber has been neglected by most students of urban studies in the link between territory and power that is his main contribution. The famous definition of the state contained in Weber's *Politik als Beruf* conference of 1919: *"The state is that human community, which within a certain area or territory has a successful monopoly of legitimate physical violence"* has been largely used in the final part of the sentence regarding the state power, but largely forgotten in the first part of the sentence regarding the territorial dimension. The point has been rightly made by Stuart Elden in his excellent text on The Birth of Territory in Western political culture (2013). Yet Elden is not coherent with the assumed neglect of Weber as a critical theoretical point, as he himself forget Weber in the rest of his book and never quote Weber's Die Stadt.

It is not surprising that Weber is not even quoted in the influential book edited by Neil Brenner, *Implosions/Explosions. Towards a Theory of Planetary Urbanization* (2014). Brenner's aim is to "advance a hihertho largely subterranean stream of urban research that has, since the mid-twentieth century, cast doubt upon established understandings of the urban as a bounded, nodal and relatively self-enclosed sociospatial condition in favour of more territorially differentiated, morphologically variable, multiscalar and processual conceptualizations". No doubt that Weber is included (although implicitly) into the established literature about the city as a closed system. But this is a mistake, as Weber's city is as open and variegated, multilayered and dynamic as the types of cities that are included in the text (from polis to civitas, from Western Medieval city to Oriental city, to early modern city). It is the method of understanding the variable linkages connecting various types of city and various socio-economic, political and cultural dimensions that is at stake here.

When Weber is quoted in urban studies literature is sometimes to criticize his theory. This is the case of Peter Taylor's criticism in his important book on *Extraordinary Cities*, a study of moral syndromes, world system and city/state relations (2013). As I have already mentioned, the plastic dualism of economy and polity is illustrated in Weber's theory of the city but is not understood by Taylor. As I explain in my text *The Urban Contract* (2017), in the Western medieval type of city the urban contract was written by the new emerging force of merchants and of free citizens. This is the seed of a "politically-oriented capitalism", or better of a "regulated capitalism" developing only in the Western cities, and later in Western states (see the excellent text of Patrick Le Galès, *European Cities*, 2002, for a strong connection to the Weberian legacy). Western rationality is based on this pact between the Economic (market) and the Political (city, then state: both expressions of polis) considered by Weber as distinct, separate spheres. The pact gave rise to rational capitalist enterprise, the monetary economy and the circulation of negotiable financial instruments. This link between economy and society is presented in a different way by Taylor, where a cityness model based on the agglomeration of the city's economies, and its circulation in a wider city network is proposed. Taylor criticizes the "Occidental Cities" thesis of Weber but fails to grasp the core of Weber's argument: the *plastic dualism* between economy and society and the role of the Political. "Weber's theory relates to guardian process, but what is important is how such institutional freedom is used commercially. It is commercial autonomy that is the really important process" Taylor writes (2013, p. 236).

However, the Political is a key partner of the urban contract, in direct relation with the Economic: they are not separate entities. As I have elaborated in more detail in a recent book (Perulli, 2017), The Merchant is the other fundamental partner in the urban contract. From the early stages of commercial guilds in the medieval city to contemporary forms of geo-finance in the global cities there is a strong essential continuity, but also a significant discontinuity: the guilds were incapsulated in the city and controlled through its political regulation, whereas geo-finance today is deregulated, open to the globe and beyond the political control of the nation-states. Even unions of nation-states, like the European Union, are unable to cope with the major problem of regaining control of the geo-financial "anomic" will.

In this sense we can still use Weber in a normative perspective. For Weber capitalism is a system in which the rational ethics has driven and controlled the individual capitalists' impulse toward capital accumulation. Cities have been the loci where capitalist accumulation and economic ethics have coalesced, from the Lutheran Reform to the growth of capitalism in the Unites States. Today' capitalism has forgotten or refused such linkage and has expanded its freedom at the global level. The state is the main victim of such an epochal shift.

The state has been created as autonomous from any external and superior power, and free from the acknowledgment of any rival jurisdiction over its territory. The invention of the concept of State-which has Medieval origin-consists in the transfer to the State and its territorial authority of the sacred nature which was the essence of religion. This has been largely explained by Schmitt (1922) in his *Political Theology*, which is Weberian in its essence (Schmitt himself being a student of Weber in the Munich inner circle). The State presents itself as immutable and eternal, its Body-the political Body which is sacred like the corpus mysticism of religion (Kantorowicz 1957)-*allows* to give identity to its subjects, both in positive and negative senses (e.g. the state has the monopoly in defining its enemies, and perhaps the modern state was born when this monopoly power to define enemies has been obtained; the state has the power to orient and control its civil society; etc.).

Two main direction, two different models of relationship between the state and civil society can be elaborated after Weber.

The first is the model of the "*Absolute Politics*", a term coined by Pizzorno (1993) and fully schmittian and Weberian in nature, as Pizzorno himself explicitly acknowledges in his text (1993, p.14). Absolute politics is the 'high' politics, whereas interest politics is the 'low' politics'. The former term comes from religion (political theology) while the latter term comes from economics (the interest is an economic concept). National territorial governments needed a normative concept to differentiate their own power from the Church's power and found it in the concept of State. In this perspective the civil society is the 'subject' where the interests' intermediation is conducted, in its 'low politics' dimension. The State holds a superior power and can give some of its power to decentralized subjects in the civil society sphere.

The second direction is a different one and comes from a non-European theory of the state, the North American one. Tocqueville had explained in "Democracy in America" that in the United States' political culture the state doesn't detach itself from the civil society; it is composed by different societies united and embedded within each other and formed by governments which are separated and almost independent from each other, the federal, state and local ones.

The city here is the township, whose origin is one of a small independent nation; it has not received its power from the state, on the contrary it has given a small part of its independence to the state. One should observe how skilfully in the American township the power has been dispersed in such a way to involve the more people as possible in the public sphere. The county is a wider administrative unit; however, its administrative power has no centralized nor hierarchical nature; that's why nobody can see it, it is dispersed in many hands. The legislation itself is referred to the particular interest: the individuals interested in the state for a kind of selfishness. As far as he defends the Union, he also defends his own prosperity. The Federalist Papers assert that the powers delegated to the federal government are well defined and few, whereas the powers which remain in the hands of the single State are numerous and independent.

This striking difference makes the American model of township less able to be defined by European categories. The Pizzorno model (1993), largely elaborating on a Weberian legacy, distinguish between four types of social linkages: based on kinship, religious, territorial, and associative. Kinship and religious are predetermined linkages: one can choose religion but cannot choose its kinship. Territorial and associative are voluntary linkages: one cannot choose its territorial identity but can choose the associations to be associated with. However, the model fits less in the American model, as Tocqueville described it. In fact, the territorial and associative linkages are both so interlinked and intertwined that it is quite impossible to trace a clear distinction between them.

The territorial linkage, it's true, is more individualistic and universalistic in nature (I belong to a territory where standard universal norms can be applied to all), whereas the associative linkage is strictly voluntary and collective (I belong to an association sharing some specific values and 'commons' with others). However, the limit is not so easy to be traced in the American democracy of Tocqueville, where the territorial and associative lives are almost made of the same fabric.

Weber has asserted, in a rather cryptical phrase of his last lessons devoted to the world economic history, that as long as the closed nation-state will exist also capitalism will survive (1993, p. 236). The reason is that capitalism and the State support each other, and the nation-state is in competition with the other nation-states to attract and maintain capital within its territory. *But not so in presence of a world Empire.*

What Weber exactly wants to say? Maybe he is saying that capitalism would not survive in a world State, given that the competitive game of capitalism and the nation-states will no longer exist. The animal spirits of capitalism would finally be appeased, and we would enter the phase of postcapitalism?

And what about the city? Its life is much longer than capitalism: more or less 3800 years ago Hammurabi, the lord of a small city-state in Mesopotamia, conquered other city-states of the region and created the first normative state, with laws and codes. Capitalism is a recent, much younger regime. Will cities survive capitalism, and will they be again the loci of creative destruction, of novel forms of social integration, of new forms of social power? Any answer to such questions will not avoid the Weberian lesson. In the last pages of Die Stadt, Weber notes that, whereas the political condition of the Middle Ages addressed the citizen to be a *homo oeconomicus*, the ancient polis maintained its characteristic of perfect military association, and the citizen was a *homo politicus*. So, what about the next phase of globalization? Will a new kind of *homo*, a new kind of humanity emerge? Will it be a cosmopolitan humanity, as some philosophers, from Jurgen Habermas to Kwame Anthony Appiah, try to redefine? In any case, as Enzo Paci, the Italian philosopher, has written at the end of his 1950 Preface to the Italian edition of Weber's *The City*—studying the sociology of the city was for Weber, and remains true for us, looking in history for the signs of our destiny".

References

Brenner N (ed) (2014) Implosions/explosions: towards a study of planetary urbanization. Jovis, Berlin

Elden S (2013) The birth of territory. University of Chicago press, Chicago

Kantorowicz EH (1957) The king's two bodies. a study in mediaeval political theology. Princeton University

Le Galès P (2002) European cities: social conflicts and governance. Oxford University Press, Oxford

Paci E (1950) La Città. Introduzione di Enzo Paci, Max Weber, Bompiani, Milano

Perulli P (2017) The Urban Contract. Community, governance and capitalism. Routledge, London and New York

Pizzorno A (1993) La politica assoluta e altri saggi. Feltrinelli, Milano

Sassen S (1991) The global city: New York. Tokyo, Princeton University Press, Princeton, London

Sassen S (2007) Territory, authority and rights: from medieval to global assemblage. Princeton University Press, Princeton

Schmitt C (1922) Political theology: four chapters on the concept of sovereignty. MIT Press, Boston

Taylor PJ (2013) Extraordinary cities: millennia of moral syndromes, world-systems and city/state relations. Edward Elgar, UK

Tilly C (1990) Coercion, capital, and European states, AD 990–1990. Basil Blackwell, Oxford, UK

Paolo Perulli has been a full Professor of Economic Sociology at the University of Piemonte Orientale (Italy), visiting scholar at MIT, visiting professor at Université de Paris Sud, and a Professor at the Academy of Architecture-Mendrisio (Switzerland), University Institute of Architecture-Venice and University of Molise. He is the author of The Urban Contract. Community, Governance and Capitalism (Routledge 2017), Il debito sovrano. La fase estrema del capitalismo (La nave di Teseo 2020), Nel 2050. Passaggio al nuovo mondo (Mulino 2021). He is an Editorial Board member of City, Culture and Society.

Chapter 17
Antonio Gramsci and the *Prison Notebooks*

Stefan Kipfer

Abstract Planning was not a central and explicit concern for Antonio Gramsci. In a few of his *Prison Notebooks*, Gramsci did however suggest that planning broadly understood (as well as architecture) be understood within a larger process of social rationalization that he called Americanism and Fordism and investigated in relationship to the *longue durée* of Italian history. In order to situate these rare comments—and erect a few signposts for future Gramsci-inspired research on planning practice—this paper discusses the complex historical geographies permeating Antonio Gramsci's work. It does so by concentrating on Gramsci's multi-temporal and multi-scalar understanding of city, country and urbanization. This understanding helps us grasp spatial questions (from processes of urbanization and unevenly developed social struggles to intellectual claims to urbanity and rurality) as active elements within longer term historical dynamics, the relations of force that shape particular historical conjunctures, and, of course, hegemonic projects oriented towards building a socialist future.

Keywords Antonio Gramsci · City · Country · Urbanization · History · Geography · Politics

Abbreviations of Works by Antonio Gramsci

PNI	*The Prison Notebooks*, vol. 1, ed. and intro. J. A. Buttigieg, trans J.A. Buttigieg and A. Callari, Columbia University Press, New York, 1992.
PNII	*The Prison Notebooks*, vol. 2, ed. and trans. J. A. Buttigieg, Columbia University Press, New York, 1996.
PNIII	*The Prison Notebooks*, vol. 3, ed. and trans. J. A. Buttigieg, Columbia University Press, New York, 2007.
SCW	*Selections from Cultural Writings*, ed. D. Forgacs and G. Nowell-Smith, trans. W. Boelhower, Lawrence and Wishart, London, 1985.

S. Kipfer (✉)
Faculty of Environmental and Urban Change, York University, Toronto, Canada
e-mail: kipfer@yorku.ca

C. Perrone (ed.), *Critical Planning and Design*, The Urban Book Series,
https://doi.org/10.1007/978-3-030-93107-0_17

SPN *Selections from the Prison Notebooks*, ed. and trans. Q. Hoare and G. Nowell-Smith, Lawrence and Wishart, London 1971.
SPWI *Selections from Political Writings, 1910–1920*, ed. Q. Hoare, trans. J. Matthews, Lawrence and Wishart, London, 1977.
SPWII *Selections from Political Writings, 1921–1926*, ed. and trans. Q. Hoare, Lawrence and Wishart, London, 1978.

Planning was not a central concern for Antonio Gramsci. During his writing years, from 1910s to the mid-1930s, the apparatuses of planning that generate the need for planning education in bourgeois society had not yet been consolidated. Moving within the early networks of the Communist International, Gramsci was of course exposed to early debates about planning and socialism. But he could not have been preoccupied with what we now call planning theory: a normative concern centred on the practice of specialists whose functional understandings of planning are intimately shaped by the hierarchies and divisions of labour of the extended state. Instead, Gramsci's stated concern was analytical and hermeneutic; he wanted to understand planning-like practices as elements of larger processes. In the few sections of the *Prison Notebooks* devoted to the topic, Gramsci relates planning to rationalism and rationalization in art, architecture and society. In general, he warned against treating plan-making efforts (generally understood) in formalistic, bookish ways instead of emphasizing their uncertain effects on the broader historical process (Gramsci, Q3, §33, PNII, p. 32). He commented on the aesthetic of functionalism in literature and discussed the artistic dimension of architecture (the drawing) and its alienation (in the building). Further, he placed architecture in a wider "urban panorama", a whole "architectural complex" of forms and activities that he saw being restructured in his short lifetime (Q3, §155, PNII, p. 125). Gramsci's rare notes thus situate planning and architecture within a process of social rationalization that he called Americanism and Fordism and investigated in relationship to the *longue durée* of Italian history (Q14, §65, SCW, pp. 129–31).

Gramsci urges us to embed planning praxis within larger historical dynamics, which for Gramsci were also profoundly spatial, as the following pages show.[1] For example, Gramsci understood city, country and urbanization in all their historical depths and social specificities. He related all three to regional questions (notably the North–South divide in Italy), national unification and international relations (colonialism and imperialism included). He understood well that communist hegemonic projects would have to transform both city and country. He therefore insisted that claims to urbanity/rurality represent moments of struggle (not obvious expressions of place-bound environments). In this broader historical–geographical context, investigating planning practice may take two, inter-related forms. First, it can mean shedding

[1] This paper is a significantly shortened, simplified and reframed version of chapter in a volume co-edited with Mike Ekers, Gillian Hart and Alex Loftus that explores the geographical aspects of Antonio Gramsci's work (Kipfer 2013).

light on the 'place' of planning within a particular historic bloc, that is the relations of force that define a particular historical conjuncture and the deeper historical–geographical patterns converging in this conjuncture. Second, it can mean studying planning as a practice of organizing the foundations of a non-capitalist social order. This we can extrapolate from Gramsci's involvement in Italy's red years (the project of building Soviets out of councils of workers, peasants and agricultural workers in 1919 and 1920) and his experience building the Italian Communist Party (which informed his understanding of organic intellectuals as those leading permanent organizing efforts). In short, Gramscian scholars of planning may ask: what is the role of planning practice in (dis-)organizing political rule and social relations?

17.1 Cities and Urbanization

Antonio Gramsci constructed key concepts through analyses of specific geographical situations and particular historical moments. Urban questions were important for Gramsci in this larger context, In Gramsci's pre-prison writings, the city-countryside problematic emerged out of a double movement. His intellectual biography took him from "country boy" to big-city dweller, from a "Sardist" critique of the Italian continent and follower of Gaetano Salvemini's "Southernist" socialism to a communist revolutionary who asserted the leadership of the urban–industrial proletariat (Davidson 1977; Santucci 2010). In a second movement, the October revolution and the isolation of the Turin workers during the red years of 1919 and 1920 (which led to their defeat and made them vulnerable to fascist encirclement) converged in Gramsci's political preoccupation with the imperative of organizing an alliance between the Northern proletariat (industrial and agricultural) and the poor peasants and landless labourers of the South. This preoccupation with revolutionary unity (which gave the name *Unità* to the paper of the *PCI*), permeated his writings in *Ordine Nuovo* (from 1919) and his contributions to party congresses from Livorno (1921) to Lyon (1926). In this context, Gramsci treated city and country in relation to historically specific social forces and geographically uneven, often shifting political dynamics. In 1926, Gramsci systematized his insights by placing city and countryside at the heart of the modern Italian situation, which he characterized, in the *Lyon Theses* (with Togliatti) and in *Some Aspects of the Southern Question*, with late industrialization, a fragmented ruling class, a minoritarian industrial proletariat and a semi-colonial relationship between North and South that was permeable to international (French, Austrian and Anglo-American) influence and tied to Italy's late but frenzied imperial adventures (SPWII, pp. 340–75; pp. 441–62).

In the *Prison Notebooks*, Gramsci added depth to this discussion. For him, the Risorgimento refracted Italian urban history and its relationship to culture, social structure and intellectual mentalities. Originally, the communes or city-states of the eleventh, twelfth and thirteenth centuries signalled a progressive epoch. They weakened feudal social relations. However, the communes ultimately failed to undo feudalism as the dominant urban classes became reintegrated into a system of feudal privileges. They did not break with the cosmopolitanism of the Roman Empire, the Catholic Church and the Latin language; they thus failed make the Renaissance and Humanism into a national-absolutist force analogous to the French monarchy (Q17, §3, §8, Q5, §104, §123; SCW, pp. 216–36; Q5, §127, SPN, p. 249; Q5, §55; PNII, pp. 313–6). Italian urban culture stalled in a frozen state made up of rural towns and villages in Puglie and Sicily (the "100 Cities"), the large cities of the Mezzogiorno (Rome and Naples) and the small cities in Central and Northern Italy which had shrunk to mere shadows of their former glory as city states (the "Cities of Silence") (Q19, §26; SPN, pp. 90–3). As a result, Italy's urban intellectuals failed to have a "national-popular" impact. Italian stagnated as a vernacular language (Q3, §76; SCW, pp. 167–70; Q29, §3, SCW, pp. 183–4; Q21, §5, SCW, pp. 206–12). Gramsci's observations resonate with those of Ibn Khaldoun on the fourteenth-century Maghreb. While the latter traced the impasse of Islam to the failure of city-based Arab–Muslim dynasties to incorporate rural—nomadic and pastoralist—social forces into consent-like forms of rule (*Asabiyah*) (2005, pp. 263–94), Gramsci described the failure of Italian cities to provide the basing points for integrative absolutism.

Gramsci suggested that the Risorgimento failed to overcome the *longue durée* of Italian urban history. Since the early 1800s many political crises originated in the Southern countryside (in the form of land occupations and revolts). But these political energies were mitigated by "transformism": the "decapitation" and molecular absorption of Southern intellectuals into an agrarian bloc shaped by Northern political forces (Q19, §26; SPN, pp., 90–2, p. 94, p. 96; Q19, §27; SPN, pp. 102–4; Q15, §59; SPN, pp. 105–6; Q19, §28; SPN, pp. 58–59). Only with rural land reform and a break between parasitic urban forces in the South and the peasantry (Q19, §24, SPN, pp. 63–4, p. 74, pp. 80–1) could the Northern urban bourgeoisie have followed the French example, created a dynamic internal market, unblocked industrial urbanization and led Italy's diverse socio-spatial forces (Southern urban strata and rural classes of the South and the Centre) into an organic national bloc (Q19, §26; SPN, pp. 98–100). Parallel to these failures, Italian intellectuals contributed to the lack of popular support of the Risorgimento. More in tune with European than Italian cultural trends, they could not build a truly national hegemonic romantic literary movement because they failed to overcome Italian urban history and the traditional cosmopolitan role Italian intellectuals played therein (Q21, §5; SCW, pp. 206–12).

The political impasse of the Risorgimento were accentuated by Mussolini's fascism. One example Gramsci provided in this regard was the literary debate between exponents of "supercountrymanism" (*strapaese*) and "supercitymanism" (*stracittà*) in the 1920s (Q22, §6; SPN, pp. 287–8; Q15, §20; SCW, pp. 117–9; Q23, §8, §30; SCW, pp. 213–5, p. 329; Q1, §17, PNIII, p. 223, p. 379, pp. 514–5; Q6, §27, PNIII, pp. 22–3, pp. 404–5). Coined by Curzio Malaparte, the terms of this debate pitted two claims to the Italian nation against each other. While one (*strapaese*) equated *Italianità* with rural village life and its imputed non-modern values of balance and simplicity, the other saw Italy in futurist fashion as part of the techno-centric dynamism of modern Europe (*stracittà*). Although both currents were pro-fascist in orientation, they mobilized competing imaginaries (of urbanity and rurality) to buttress Italian nationalism. In this, they both failed because they remained inorganic claims to hegemony. They reproduced the already widespread intellectual disdain for the lived experience of common Italians by reducing rural popular culture to picturesque folklore or elevating urban life to superficial cosmopolitanism (Q15, §20; SCW, p. 118; Q21, §1; SCW, pp. 201–2; Q23, §8; SCW, pp. 213–5). Gramsci's biting satire of these reactionary imaginaries of city and countryside was understandable. In his Pre-Prison Writings, he had already sketched a differentiated *tableau* of cities, which undermined any simple analogy between progress/reaction and city/countryside. Italian cities played a progressive role only in certain eras and were often prevented from doing so by the very historical structure of Italian urbanism. For example, Gramsci's optimism about industrial urbanization and the progressive role of the bourgeoisie in developing the productive forces was Engelsian in tone (SPW1, pp. 150–54, pp. 262–3; SPWII, p. 139, p. 376) but tempered by the observation that industrial urbanization in Piedmont and Lombardy had failed to transform Italy's urban structure as a whole.

Gramsci's hedged optimism about modern, industrial urbanization alerts us to the way in which he treated urban questions in multi-scalar, transnational way. If Italian urban history from the Renaissance to the nineteenth century exposed Italy to the hegemonic influence of international intellectual forces, and if fascism paradoxically perpetuated Italy's permeability to foreign forces, Gramsci also hoped that American Fordism would help Italy move to a new urban era. Observing the beginnings of comprehensive urban planning in Milan, Gramsci linked Fordism to the "exaltation of the big cities", "grandiose [urban] projects" (Q22, §2; SPN, p. 286) and superhighways (Q1, §95; PNI, p. 335). He left no doubt that, after the defeat of revolutionary politics, Fordist urbanization and planning would provide the basis upon which to reshape Italy's social and spatial structure. It posed a welcome demographic threat to parasitic city-countryside relationships in Italy and what he saw, conservatively, as backward forms of sexual morality in the countryside (Q22, §3; SPN, pp. 294–7). Functionalist architecture and the advances in city-building techniques in America were signs of a collectivization of capitalist accumulation (Q14, §1, §2, §65; SCW, pp. 128–32). Together with the transformation of the state, the labour process, household structures and workers' subjectivities, urbanization was key to what Gramsci saw as a positive, if contradictory rationalization of social life.

These passages about fascism and Fordism highlight the degree to which Gramsci linked hegemony to urban questions. On the one hand, Gramsci saw modern(-ist) urbanization and planning as a way to demographically reorder the "terrain" of hegemony (Q22, §3; SPN, p. 296) and interpreted urban space (architecture, the layout and names of streets) as key "ideological material" for bourgeois rule (Q3, §49; PNII, pp. 52–3). On the other hand, Gramsci was unambiguous about the positive role urban transformations could play in the multi-scalar, spatially and temporally differentiated project of communist wars of position. While he saw hegemony as a modern form of cosmopolitanism that would build on the national bases of proletarian movements (Q4, §68; SPN, pp. 240–1), Gramsci's view of communist hegemony was both multi-scalar and deeply shaped by urban considerations.

The most developed revolutionary forces Gramsci detected in the industrial city: Turin. Gramsci contrasted Turin to Milan, Italy's corporate control centre with its petty and grand bourgeois leanings and reformist trade unionism, to Rome, the parasitic city of bureaucrats, and to Bologna, Florence, Naples, Bari, the cities of landowners or the agricultural bourgeoisie (SPWI, p. 240; SPWII, pp. 49–50, p. 130). Like Engels in his writing about Manchester, he thought that Turin, where capital and labour were spatially separated from each other with little interference by intermediary strata, provided the most fertile conditions for the revolution (SPWI, pp. 110–11, pp. 183–4, pp. 312–3). In the integrated "productive unit" of Turin, the proletariat could grasp the division of labour as a whole (SPWI, p. 120) and begin to see itself as a subject, not an object of history (ibid. 345). Key in this regard were organizations that could revolutionize workers' subjectivities (SCW, p. 33, p. 37; SPWI, pp. 65–8). On the basis of these revolutionary urbanities in factories and working-class neighbourhoods, Turin workers demonstrated their capacity (during the red years of 1919 and 1920) to appropriate productive capital through economic self-management (factory councils) and build embryonic forms of a communist state (in soviet-style municipal councils) (SPWI, p. 220, p. 224, p. 226, p. 250; pp. 326–7). Gramsci hoped that industrial action and political self-organization in Turin would join up with the land occupations that swept through Northern and Southern agricultural zones (SPWI, p. 180, pp. 188–91, p. 335), prepare the ground for a post-capitalist, planned social order while building the political bases for a revolutionary war of movement—the takeover of the heights of bourgeois economic and political power in Milan and Rome.

Much more than Engels in his discussion of urban England, Gramsci recognized however that revolutionary politics could not be built by extrapolating from urban industrial struggles. Having failed to build a Piedmont-wide party organization and link up with Northern agricultural workers and Southern peasants and landless labourers (SPWII, pp. 68–9, pp. 207–8), Turin workers remained as isolated as the Paris communards in 1871 (SPWII, p. 253). They were defeated and had to confront fascist thugs imported from the central Italian countryside (SPWII, pp. 5–9, pp. 59–67). Armed with his experience of the Italian situation (and his knowledge of the Russian revolution), Gramsci argued that a national hegemonic project must itself be grounded in a spatially differentiated constellation of subnational (but transnationally articulated) political forces. In this view, Turin represented the kernel for a nation-wide web of class alliances (among workers, agricultural workers and poor peasants) and an intellectual and organizational struggle to bind city and countryside into a hegemonic bloc (SPWI, p. 360, p. 376; SPWII, pp. 354–5). To achieve this ambitious goal was impossible without a transformation of the racist leanings of urban workers against Southerners (SPWII, pp. 444–9; Q1, §50; PNI, pp. 161–2). If socialists needed to promote "genuine independence" of women from men (Q22, §3; SPN, p. 296), they also had to help Northerners emancipate themselves from their prejudices.

17.2 The Country and Rural Questions

The merits of Gramsci's geographical–historical approach are best demonstrated by his differentiated treatment of urbanization, which ran counter to "urbanist" interpretations (in Max Weber, Adam Smith or Henri Pirenne) that see Europe's late medieval city states as embryonic forms of modern capitalism (Holton 1986). For Gramsci, studying urbanization required that one specify the relationships between cities and their internal social structure, their relationship to the countryside and their links to the state. The lack of success of Italian communes he explained with reference to their failure to become the *socio-spatial core* of an Italian absolutist state. He was critical of the urbanist celebrations of cities as loci of progress, liberal-cosmopolitan modernity and the capitalist spirit. Conversely, he also rejected urbanism's mirror image: the anti-urban, romantic rejection of cities as symbols of the evils of industrialism, civilization or human arrogance. While he saw the progressive role of cities as contingent, not prescribed, Gramsci did think that industrial urbanization provided a fertile ground for revolutionaries by re-ordering the terrain for hegemony under Fordism. Capitalist urbanization thus represents one spatial medium of the revolution. Hegemony is tied to a claim to a revolutionary proletarian urbanity defined in relationship with, not in opposition to the countryside.

If Gramsci's treatment of urbanization—and its complex relationship to emancipatory politics—was both nuanced and visionary, so was his treatment of the countryside and rural social groups. In his discussion of the Renaissance, the Risorgimento, fascism and Fordism/Americanism, Gramsci did not shy away from using jarring language ("backward", "immature"). He deployed this language to both urban and rural phenomena, however. As a crucial microcosm of his analysis of the relationship between North and South in Italy, his treatment of city and countryside escaped the diffusionism that lies at the heart of Euro-centric conceptions of the development of capitalism (Morton 2007, p. 50, p. 72). If, for Gramsci, the city was not automatically synonymous with progress, modernization and the revolutionary working class, the 'countryside' per se did not connote backwardness, tradition, and the political conservatism of peasants. Gramsci stressed that industrial-capitalist expansion in the age of empire required exploiting agricultural and resource hinterlands in colonies and peripheral regions in imperial countries (SPWI, pp. 59–60, pp. 301–4). Indeed, for Gramsci, the relationship between city and countryside in modern Italy had "colonial" aspects because it was tied to Italy's African expeditions and the role of the Italian South as an "internal colony" (Q8, §193; PNIII, p. 345; SPWI, pp. 3–5, p. 375; SPWII, pp. 343–8, pp. 396–7; Q19, §24; SPN, p. 68).

While overdetermined by the schism between North and South (within which "the whole zone of the South and islands functions as an immense countryside in relationship to northern Italy" (SPWII, pp. 396–7)), the Italian countryside was not reducible to this division. As his analyses in both Pre-Prison and Prison Writings made clear, Gramsci's Italian countryside was highly differentiated (SPWII, p. 343; see also Arnold 2000). It was unevenly developed, with qualitative variations between the agricultural capitalism of the Po Valley in the North, the cooperative farms in

Central Italy (Reggio Emilia, Tuscany), and the semi-feudal agricultural zones in the continental South, Sicily and Sardinia. The social composition of Italy's countryside was equally complex, with regionally distinct relationships between share croppers, small holders, landless agricultural workers and tenant farmers, one the one hand, landlords, agricultural capitalists and intellectuals on the other (SPWI, p. 164, p. 170, p. 335, p. 345; SPWII, pp. 63–4, p. 130, p. 393, pp. 449–60; Q19, §24; SPN, pp. 75–6; Q13, §23; SPN, pp. 213–4; Q3, §46; PNII, pp. 44–6). As with industrial workers, the consciousness of rural subalterns was a contradictory, unstable combination of apolitical passivity, mental tenacity and spontaneous revolt (Q19, §24, §26; SPN, pp. 74–6, p. 97, p. 102; Q3, §48; PNII, pp. 51–2; Q13, §29; SPN, p. 203; Q3, §46; PNII, pp. 44–7; Q11, §12; SPN, pp. 325–6, p. 337, p. 340). On this basis, Gramsci's remarks about Italian politics in the 1910s and 1920s detected not only regional distinctions in political culture (reformist socialism in Central Italy, democratic republicanism in the Po Valley, specific socialist traditions in rural Sicily) but also rapid shifts in how these regions were affected by the factory and land occupations in 1919–20 and the subsequent wave of fascist counter-mobilizations (SPWI, p. 180, pp. 188–91, p. 335; SPWII, pp. 59–61, p. 63–7, p. 152, p. 259, p. 360, p. 376, p. 393, p. 449, pp. 453–4). In this complex context, a hegemonic alliance of urban and rural forces needed to transform the spatial relations between industrial and agricultural workers, peasants and intellectuals.

The way in which Gramsci linked city-countryside questions to imperialism and internal colonialism, the care with which he linked rural questions to a nuanced view of the consciousness and political potential of subaltern groups (including the peasantry), and the quality of his analyses of incomplete ("passive") revolutions from above (in the Risorgimento and fascism) explains the appeal of his work for analyses of imperialism, colonization and racialization (Hall 1980, p. 333, 1996, p. 416; Bannerji 2001a; Kiernan 1995; Ahmad 1996). This has been the case forcefully in India, where Gramsci played an important role (next to Maoism, British marxist historiography, and debates about Gandhi's legacy) in the formation of subaltern studies. In this context, Gramsci provided analytical tools for a highly innovative "historiography from below" that would pay appropriate attention to rural life and peasant agency (Guha 1982) and the failure of India's independence movements to develop national liberation to a moment beyond "passive revolution" (Chatterjee 1986; Kaviraj 1988). Key to these Gramsci-inflected analyses was the insight that in colonies, the reach of hegemony remained weak to such an extent as to limit the capacity of nationalist independence movements to make inroads into the "autonomous" worlds of subaltern forces (Guha 1982, 1983; Arnold 2000).

The emphasis subaltern scholars in India placed on the autonomy of (mostly rural) subaltern social forms from other aspects of the Indian social formation took them a considerable distance away from Gramsci's conception of hegemony as a boundary-crossing, dialectical transformation of rural and urban forces. Far from a simple reflection on the comparative differences between Italy and India (as qualitative as these are), the argument about peasant autonomy took on a life of its own in Guha's detection of a "structural split" between elite and subaltern political domains (1997), Chatterjee's distinction between "inner/spiritual" and "outer/material" sides of nationalism (1993), and the ultimate move by Chakrabarty (2000), who treats the peasant-subaltern as a counterpoint to the elite realm of colonial, Western, rationalist modernity. In Chakrabarty, Gramsci's historico-geographical sensibilities are displaced by a round-about anti-historicism, within which "the 'peasant' acts ... as a shorthand for all the seemingly nonmodern, rural, nonsecular relationships and life practices...The peasant stands for all that is not bourgeois (in a European sense) in Indian capitalism and modernity" (2000, 11). Here, the peasant gets lost in a grand philosophical confrontation between Marx and Heidegger, reason and experience, analysis and understanding, West and East, History I and History II, science and spirituality/religion. Such an approach is difficult to reconcile with Gramsci's relational view of rural and peasant life (Sarkar 2000; Bannerji 2001b, 43).

17.3 Urbanity, Rurality, Hegemony

The power of Gramsci's own formulation about subalternity lies precisely in his refusal to treat the city and the countryside divide as equivalent to social distinctions (peasantry versus worker or intellectual, subaltern versus elite) or transhistorical conceptions of modernity/tradition, progress/stagnation and civilization/barbarism. His socially nuanced treatment of geographical differentiation (as well as his spatially fine-grained treatment of social difference) helps Gramsci escape the pitfalls of urbanism (in classical sociology, bourgeois economics and modernization theory) and its perennial anti-urban opposites (rural romanticism, neo-traditionalism, deep ecology). Gramsci recognized that hegemonic politics required much more than a recombination of "autonomous" social forces—some rural, some urban—in an additive and instrumental project of coalition building. Linking urban with rural forces required a transformation of both, and thus also a break with "the aversion of the country for the city" and the "hatred and scorn for the 'peasant'" he observed in his own time (Q19, §26; SPN, p. 91). Against such inorganic urbanisms/ruralisms, Gramsci's hegemonic project included a claim to an expansive urbanity that could incorporate instead of shunting aside the countryside.

In his quest to transform urbanism and anti-urbanism, Gramsci had much in common with Frantz Fanon and Henri Lefebvre (see also Hart and Sitas 2004). Endulging in rural populism only for a moment in a larger emancipatory quest, Fanon saw national liberation struggle as an internationally articulated national alliance of social forces in both city and countryside (albeit with the weight clearly on the latter rather than the former). In a manner reminiscent of Gramsci's "precocious Fanonism" (Sekyi-Otu 1996, p. 118), Fanon realized that both urban bias and rural spontaneism must be countered if national liberation was to escape the traps set by colonial empires and if liberation struggles were to appropriate and transform, not magnify, pre-colonial and colonial elements of colonized societies (Gopal 2004, pp. 160–1). Lefebvre's rural sociological remarks (1970, 1991) about land rent and rural class relations in Tuscany and festivals in Southern France provided complex insight into analyses of rural social relations-in-transformation not dissimilar from

Gramsci's own. Lefebvre, too, argued against transhistorical conceptions of rurality to emphasize the historically varied and geographical uneven ways in which rural social relations, cultural forms and political struggle undergo structural change in long transitions to capitalism. Combining rupture with continuity, this change is not a linear shift from rural tradition to urban modernity. What he called the "urban revolution" (2003) was an incomplete, uneven transformation of the meanings of city and countryside.

Following Lefebvre's hypothesis of the "complete urbanization of society" superficially, one is tempted to dismiss Gramsci's treatment of city and countryside as anachronistic. It is true that if one uses city and countryside as objective markers of settlement forms that embody qualitatively distinct socio-economic and cultural social relations and adequately describe the sociopolitical forces that emanate from these relations, one cannot hold onto them today without elaborate qualification, if at all. But if one follows Gramsci and treats city and countryside *also* as intellectual and linguistic-metaphorical claims to hegemony, he helps us illuminate Lefebvre's own insights. Gramsci's emphasis on social reality, including language, metaphor and ideology (Ives 2006), as a confluence of multiple, uneven temporalities allows us to understand why competing claims to the city and the countryside may persist in an urban context. Far from arbitrary (Thomas 2009) imaginaries, metaphors and ideologies can outlive their original contexts and take on new meaning once they become articulated with new struggles. This allows us to understand how Lefebvre managed to see the *Commune* and 1968 as revolutionary movements that laid claim to "the city" (urban centrality, democracy, the social surplus) even as they radically transposed traces of the rural festival, which Lefebvre had analysed earlier (1965, 1968).

In a rapidly urbanizing world, city and countryside may be both outdated (as markers of qualitatively distinct social spaces) and rejuvenated (as intellectual, linguistic and political claims to hegemony), as Raymond Williams has demonstrated in British literature (1973). Gramsci's fleeting insight (that claims to urbanity and rurality can mediate hegemony) holds even in an urbanizing context, particularly if meanings of city and countryside become reinscribed in material processes (urbanization, state formation) that appear to make these terms redundant in the first place. In Anglo-America, for example, non- or anti-urban ideologies of nationhood (the English countryside, the American frontier, the Canadian Great White North) have combined powerfully with an anti-urban ethos of suburban and small town life. Fused with vulgar modernist urban planning and neo-romantic landscape design modelled on pastoral or wilderness imaginaries, these twin ideologies became powerful lived forms of Fordism (Wilson 1991). Particularly in the privatized realm of North American suburbia, these anti-metropolitan lived worlds help explain why the standardized urbanisms of the Fordist era managed to rationalize the bourgeois world in a more systematically counter-revolutionary fashion than Gramsci anticipated. There, agoraphobic urbanisms continue to wield tremendous influence over claims to "urbanity" in today's sprawling metropolitan regions, despite the rise to prominence of liberal-cosmopolitan urbanisms and their proponents—gentrifiers and fractions of transnational capital.

In Algeria and India, claims to rurality and urbanity were intimately tied up with the passive revolutions of independence. After being forced into Tunisian exile after 1957, the Algerian FLN presented itself as a movement centred in the countryside, irrespective of the urban origins of many of its cadres (Harbi 1980) and notwithstanding the fact that French colonialism had de-structured city-countryside relations to the point of where the meaning of urban and rural was less than obvious (Bourdieu and Sayad 1964). The FLN's rural shift was part of an attempt to assert its dominance over resistance forces in Algeria after the preliminary defeat of the urban movements in 1957. This shift (in which Fanon participated even though he recognized its pitfalls) had important implications for how the FLN consolidated its rule (Harbi 1980) and redirected colonial spatial development strategies after 1962 (Popelar and Vannier 2010). In India, finally, the "village" functioned as a powerful socio-spatial reference point for nation-building before and after independence, first positively (for Gandhians), then more negatively (under Nehru), and then again affirmatively with the Maoist-inflected challenges to Nehruvian and Communist modernization strategies in the 1960s (Prakash 2002). Subalternists were shaped by this third attempt to lay claim to the Indian countryside. But the argument about peasant autonomy is now more difficult to defend not only because of the recent revisionist "urban turn" proclaimed by former subaltern scholars (ibid.), but also because of the way in which structural adjustment, partial depeasantization and new bourgeois urbanisms have reconfigured, once more, the meaning of city and countryside in Indian politics (Chatterjee 2004, 2008).

17.4 Conclusion

The political revolutions in North Africa in 2011 invite us to put Gramsci to work (and hint at the possible place of planning practice in wider political dynamics). The events in Tunisia and Egypt, for example, articulated multiple temporal rhythms and geographies. In long-term perspective, popular demands for democracy, freedom, equality and dignity showed how decolonization had remained incomplete. More immediately, popular anger led to an organic crisis of the authoritarian, neo-colonial regimes, which since the 1980s had transformed post-independence historic blocs with structural adjustment. Sparked by events such as the self-immolization of Mohamed Bouazizi in Tunisia, popular revolt intensified conflicts within the ruling class after increasingly naked, personalized forms of corruption had undermined the last vestiges of the regimes' legitimacy, and after years of intermittent protest, riots and strikes had encouraged people to claim their dignity irrespective of relentless state repression. The uprisings were also products of uneven development. While the regional influence of the revolution in Tunisia was widely commented, the uneven processes of mobilization that converged on and produced globally visible 'faces' of the revolutions—L'Avenue Habib Bourgiba in Tunis, Tahrir Square in Cairo—received originally little attention. These were urban revolutions not because they grew out of central city spaces in the capitals but because they were produced by spatial dynamics of mobilization that articulated various points in the countries' differentially urbanized landscapes. Most strikingly in Tunisia, the revolution reached across deep socio-spatial divides, as the Tunisian Planning Association pointed out early (Association Tunisienne des Urbanistes 2011). It began in peripheral mining and agricultural towns in the country's arid centre before reaching working and middle-class quarters in coastal cities like Sfax and Tunis.

The North African uprisings cannot be explained in civilizational terms (and their urbanist or ruralist assumptions). They in fact signal the intellectual bankruptcy of recent Orientalist formulations. They fly in the face of racist images of the "Arab street" as a volatile mix of fatalist passivity, male chauvinism and religious fanaticism. They also contradict attempts to buttress American empire by identifying democracy, freedom and secularism with Western urban civilization (Kipfer and Goonewardena 2007). Gramsci's analyses of conjunctures as moments of struggle that articulate and produce a multiplicity of temporal rhythms, spaces and scales provide us with a powerful methodological counterpoint to civilizational reification. They invite us to investigate urbanity and rurality alike not as civilizational markers but as material grounds of historic blocs, products of the interaction of socio-political forces, and cultural–ideological component parts of hegemonic claims. In turn, the search for spatial forms of a new social order (and the possible role of spatial planning in that search) is a 'terrain' one should not leave to bourgeois voices. Certainly, considerations of spatial organization (planning-related or otherwise) are vital for the prospect of both communist hegemony and the future of the planet. In this regard, Gramsci's emphasis on hegemony as a project to *transform* spatial divides (rural, urban or otherwise) is more crucial than ever.

References

Ahmad A (1996) Lineages of the present. Tulika, New Delhi

Arnold D (2000) Gramsci and peasant subalternity in India. In: Chaturvedi V (ed) Mapping subaltern studies and the postcolonial. Verso, London, pp 24–49

Association Tunisienne des Urbanistes (2011) "Manifeste Ville de demain" http://www.tunizien.com/145853-tunisie-style-de-ville-manifeste-pour-la-ville-tunisienne-de-demain.htm

Bannerji H (2001a) Inventing subjects. Tulika, New Delhi

Bannerji H (2001b) Pygmalion nation. In: Bannerji H et al (eds) Of property and propriety. University of Toronto Press, Toronto, pp 34–84

Bourdieu P, Sayad A (1964) Le déracinement: la crise de l'agriculture traditionelle en Algérie. Editions de Minuit, Paris

Chakrabarty D (2000) Provincializing Europe. Princeton University Press, Princeton

Chatterjee P (2008) Democracy and economic transformation in India. Econ Polit Wkly (April 19), 53–62

Chatterjee P (2004) Are Indian cities becoming bourgeois at last?" In: Chatterjee P (ed) The politics of the governed. Columbia University Press, New York, pp 131–147

Chatterjee P (1993) The nation and its fragments. Princeton University Press, Princeton

Chatterjee P (1986) Nationalist thought and the colonial world. University of Minnesota Press, Minnesota

Davidson A (1977) Antonio Gramsci: towards an intellectual biography. Merlin, London

Gopal P (2004) Reading Subaltern History. In: Lazarus N (ed) The Cambridge Companion to Postcolonial Literary Studies. Cambridge University Press, Cambridge, pp 139–161
Guha R (1997) Dominance without hegemony: history and power in colonial India. Harvard University Press, Cambridge, Mass
Guha R (1983) Elementary aspects of peasant insurgency in colonial India. Oxford University Press, London
Guha R (1982) On some aspects of the historiography of Colonial India. In: Guha R (ed) Subaltern studies I: writings on South Asian history and society. Oxford University Press, Delhi, pp 1–8
Hall S (1996) Gramsci's relevance for the study of race and ethnicity. In: Morley D, Kuan-Hsing C (eds) Stuart Hall: critical dialogues in cultural studies. Routledge, London, pp 411–440
Hall S (1980) Race, articulation, and societies structured in dominance. In: UNESCO (ed) Sociological theories: race and colonialism. UNESCO, Paris
Harbi M (1980) Le F.L.N.: Mirage et réalité. Editions j.a., Paris
Hart G, Sitas A (2004) Beyond the urban-rural divide. Transformation 56:31–38
Holton RJ (1986) Cities, capitalism, civilization. Allen and Unwin, London
Ives P (2006) Language, agency and hegemony. In: Bieler A, Morton AD (eds) Images of Gramsci. Routledge, London, pp 61–74
Kaviraj S (1988) A critique of the passive revolution. Econ Polit Wkly (November), 2429–2444
Kiernan VG (1995) Gramsci and the other continents. In: Kiernan VG (ed) Imperialism and its contradictions. Routledge, London, pp 171–86
Kipfer S (2013) City, country, hegemony: Antonio Gramsci's spatial historicism. In: Gramsci: nature, space, politics (Co-edited with Mike Ekers, Gillian Hart, and Alex Loftus). Blackwell, London, pp 83–103
Khaldoun I (2005) The Muqaddimah: an introduction to history (trans: Rosenthal F). Princeton University Press, Princeton
Kipfer S, Goonewardena K (2007) Colonization and the new imperialism: on the meaning of Urbicide today. Theory Event 10(2):1–39
Lefebvre H (1965) La Proclamation de la Commune. Gallimard, Paris
Lefebvre H (1968) L'irruption. Anthropos, Paris
Lefebvre H (1970) Du Rural à L'Urbain. Anthropos, Paris
Lefebvre H (1991) The critique of everyday life: volume I. (trans: Moore J). Verso, London
Lefebvre H (2003) The urban revolution (trans: Bononno R). University of Minnesota Press, Minneapolis
Morton A (2007) Unraveling Gramsci. Pluto, London
Popelar A, Vannier P (2010) Urbanisme, une autre bataille d'Alger. Le Monde Diplomatique, Dec 10
Prakash G (2002) The urban turn. In: Sarai Reader: the cities of everyday life, 2–7
Santucci A (2010) Antonio Gramsci. Monthly Review Press, New York
Sarkar S (2000) The decline of the subaltern in *subaltern studies*. In: Chaturvedi V (ed) Mapping subaltern studies and the postcolonial. Verso, London, pp 239–255
Sekyi-Otu A (1996) Fanon's dialectic of experience. Harvard University Press, Cambridge
Thomas P (2009) The Gramscian moment. Brill, Leiden
Williams R (1973) The country and the city. Oxford University Press, New York
Wilson A (1991) The culture of nature. Between the Lines, Toronto

Stefan Kipfer teaches and researches urbanization, comparative urban politics and planning, and critical social theory in the Faculty of Environmental and Urban Change, York University, Toronto, Canada.

Chapter 18
Le droit à la ville, 1968: Reading Lefebvre's *The Right to the City* in Planning Perspective

Barbara Pizzo

Abstract The present reading of 'The Right to the City'—one of the most discussed and quoted writings in the field of urban studies—has a limited scope, focusing on two main concepts and issues: 'city' and 'urban planning', the fundamental objects and fields of research and action of planners. With the aim of extrapolating a 'planning lesson' from a classical text, this paper offers a critical discussion of the meaning of the city as a *mediation*, and therefore an analysis of L.'s idea of an 'intermediate domain' to which the city belongs, in between what in Gramscian terms could be called 'structure' and 'agency'. The paper also tries to highlight the implications that this peculiar position of the city may have for urban planning.

Keywords The right to the city · Urban planning · Urban studies · City · Urban renewal · Lefevbre · Gramsci

It is inevitable, to approach one of the most discussed and quoted writings in the field of urban studies with a certain timidity.[1]

[1] In fact, thousands of works have been written, which are based on Lefebvre's 'The right to the City', or are somehow referred to it, so much so that even to provide a 'simple' literary review on the topic would require an enormous work—and probably more, if the aim is to piece the contributions together to get a history (or also a geography) of its influence. Just checking on Scholar, one can finds about 3,800,000 writings which are directly or indirectly related to Lefebvre's Right to the City, and more than 70,000 those which explicitly quote it—10,000 only in 2019 (while over a million those referring to David Harvey on the same topic). We could ask ourselves if really there is still something that has not been said, if our contribution is really necessary. I myself already wrote a short paper on 'The right to the city' (see Pizzo 2013), although with a very limited scope also in that case (Harvey's essay with the same title, 40 years after Lefebvre's), as well as on the concept of 'self-organization' (2019).

B. Pizzo (✉)
Urban Planning Studies Division, Department of History, Design and Restoration of Architecture - DSDRA, Sapienza University of Rome, Rome, Italy
e-mail: barbara.pizzo@uniroma1.it

C. Perrone (ed.), *Critical Planning and Design*, The Urban Book Series,
https://doi.org/10.1007/978-3-030-93107-0_18

The specific task we got, that of extrapolating 'planning lessons'[2] from a classical text, and to re-read it in a disciplinary, even *technical*, perspective, providentially helps in limiting the scope of the commitment. Therefore, this contribution is articulated following that basic indication. After a brief introduction, I will focus on two main concepts and issues: 'city' and 'urban planning': the fundamental object and field of research and action of planners.

The book appeared for the first time in 1967, as a peculiar 'celebration', that of the centenary after the publication of Karl Marx Capital. In France, as well as in Italy and elsewhere, it had various editions. It has been the first important work dedicated to urban topics by Lefebvre: a couple of years later, he published *La Production de l'Espace*,[3] thus becoming a reference point for urban scholars, although he was not among them by education.[4]

18.1 A Striking Outdatedness

Among the manifold reasons to argue that this book is still worth to be read and discussed although it is fifty years old, the *impact* that it had[5] and still has on contemporary research[6] and the *content* should be considered.

But, if usually the interest for 'classical' books is justified through their 'extraordinary topicality', I will start here with what seems a striking outdatedness.

In fact, there are some passages and expressions that sound anachronistic, even naïve, while in my view they are extremely helpful in unveiling fundamental limits and contradictions of our time. Here are two examples.

[2] Unfortunately, it is not possible to get in English the nice pun Giancarlo Paba and Camilla Perrone used. They proposed to us to deliver 'Lezioni di *Piano*' [*Planning* lessons], using the word 'Piano' which in Italian is both the music instrument, and the major planning instrument.

[3] See C. Schmid essay in this same collection.

[4] It is rather difficult to 'label' Lefebvre or to encapsulate his work within one disciplinary field. By education, he is considered a philosopher and a sociologist, one among the most important French intellectuals, of the '60s and '70s in particular. He vividly represents (the need of) contamination and scouting through disciplinary boundaries, when tackling urban phenomena in particular. On this regard, see also Pizzo (2019), with Gribat et al. (2019).

[5] Although it risks to sound obvious, and applicable to many other books, indeed it lays at the root of an entire literature: it can be the starting point of a journey through an immense library, where each book, essay or author brings to another.

[6] Which does not mean only that to have an in-depth understanding of what Harvey or Brenner, or Schmid (just to name the more prominent ones) wrote we could need to go back to one of their main sources of inspiration: in similar cases, going back to the original text is imperative, since the literature referred to this work is so vast and variegated (Pizzo and Rolnik 2019). Let us just think about the relationship between Lefebvre and Harvey; the acknowledged debt of Brenner (e.g. 2000, 2014), Schmid (e.g. 2018) and also others (see in particular: Buckley and Strauss 2016) for the concept of 'planetary urbanization', which occupied a major part of planning and geographic debates in the last decade or so.

As we all probably remember, Lefebvre's proposal is oriented towards "urban renewal" (which is not, or just indirectly, what planners usually mean, a physical–spatial project, but an "inevitably revolutionary" strategy instead—p. 154), and this proposal is entirely based on the 'capacity of playing' ("from play, *ludo*, a term which must be understood here in its broadest and deepest meaning"—p. 171) of the 'working class', since "only the working class still knows how to really play, feels like playing over and above the claim and programmes of economicism ..." (p. 172).

If one shares Lefebvre's transformative perspective (or *prospective*, p. 160ff.), thinking and trying to translate that sentence into today socio-economic reality will turn a certain initial smile into grimace.[7]

The idea that a project, which is a profound renewal process, is based and *depends* on the ability and propensity to play of the working class, dramatically evidences a reality that almost does not exist anymore, and the fundamental categories Lefebvre's used to understand it have become contentious: we do no longer even have a 'working class', the concept of class is contested—putting into question *who* could be the subject capable of playing and thus of leading to that renewal. If it is no longer the working class, who could play that role now? In L.'s view, it is not the technicians, neither sociologists nor philosophers, nor urban planners, who can find the 'synthesis' that leads to urban renewal. Thus, the initial smile vanishes while the question remains open: who could play that role *now*? The 'synthesis' that Lefebvre appeals to cannot be provided technically or through expertise: as urban scholars, such question provokes us in many different ways.

The second example is taken from the very beginning of the book (from the first page of the introduction).

Lefebvre claims that "*Urbanism*, almost as much as the system, is fashionable" (p. 63). This claim makes us smile too, because certainly urbanism is *not en vogue* today: on the contrary, it never has been so much out of fashion as nowadays, and this makes us conscious that we are living in a different world indeed.

What Lefevbre took 'for granted' (a certain societal structure and organization, in a certain co-productive relation with the urban space; a certain meaning and role of planning, as a discipline devoted to understand and shape that relation), for us is no longer so; despite this, his interpretation and proposal still have meaning for us. Can it be reduced to a problem of *translation*?

I propose here to consider the content of the book as an array of 'known' and 'variable' elements that can be combined and thus lead to solutions that are probably different from Lefebvre's ones but consistent with his 'transformative proposal'. For instance, going back to the previous example: if it is not the 'working class' (as we knew it from the '60s and '70s) anymore, who plays or could play *that* role today? Subordinate classes still exist—the forms and ways of oppression, marginalization and segregation increased—however they do not show the same 'playing capacity' that L. acknowledged to the 'traditional' working class. I fear that they are

[7] It is important to underline also the radical difference from this 'capacity of playing' and the idea of free-time (*leisure*), which is the object of a strong critique (see e.g. pp. 168–171).

increasingly turning into what L. himself called the 'masses': "The mass, under pressure from many constraints, spontaneously houses itself in satellite cities, planned suburbs and other more or less residential ghettos. There is for it only carefully measured space. Time eludes it. It leads its daily life bound (perhaps unwittingly), to the requirement of the concentration of powers. But this is not a concentrationary universe. All this can quite do without the ideology of freedom under the pretence of rationality, organization and programming. These masses who do not deserve the name of people, or popular classes, or working class live relatively well. Apart from the fact that their daily life is remote-controlled and the permanent threat of unemployment weighs heavily on them, contributing to a latent and generalized terror" (p. 162).

It is this "latent and generalized terror", derived from "the permanent threat of unemployment" with all the related threats (e.g. eviction or foreclosure), which keep them subjugated, limit their opportunities and availability to engage in other activities,[8] and of course prevent them to *play*. In the void that is left, market actors became even stronger. Discussing together the last edition of a key work on the planning history of Rome,[9] Italo Insolera attributed that fundamental social *energy* to migrants and to gypsies, whose community relationships and life styles tend to and often actually escape from commodification.

Concerning the second example, we could ask ourselves: If urban planning is no longer in fashion, why is it so, what does this mean and what does it imply? What replaced it (if anything), why and how? Who has become the bearer of that technical knowledge which anyway is not sufficient to produce the awaited 'urban renewal'? Or, better: If, when it was fashionable, urban planning was said of being not capable of producing 'urban renewal', now that it is out of fashion, has it left room for such 'renewal'? Unfortunately, I do not think so. Thus, does it demonstrate planning actual *uselessness*? Formally, planning still exist, and regulatory planning is still considered as a main market and private initiative constriction, while it is criticized for its constant adaptation to the existing order, which it contributes at reproducing notwithstanding a number of alleged 'innovations' (Savini 2019). Thus, which conceptualization of planning had Lefebvre in mind (the rational-comprehensive planning of the '60s?) and how has it changed, and how could it change in order to be not necessarily perceived as either negative or useless?

Summarizing: we have a theoretical, philosophical and political proposal with its own still convincing inner consistency, which however must be translated into today's reality; the difficulties of this operation seem to increase considering a series of expressions, concepts and seemingly 'bizarre' questions, that risk to be labelled and quickly dismissed as outdated or even naïve, while they have the power to shed light to the trajectory that cities and urban societies followed, and their further possible directions, shaping a vast space for critical reflection.

As I said, to respond to the request to extrapolate 'piano [planning] lessons' from the text, I think it is essential to start with a few key words, namely that of *city* and

[8] See also Madden and Marcuse (2016), pp. 131–133.

[9] Insolera 2018 (2011 [1962]).

that of *urban planning*. This limited task is not easy anyway since arguments are not systematically treated.

18.1.1 About the City and its *Domain*

There is not one unique definition of what a city is, but three main ones, and its conceptualization is developed discursively throughout different chapters.

What interests me mostly is the fact that L. places the concept of city in a precise 'domain', which is an 'intermediate' domain between other two, which recall Gramsci's 'structure' and 'agency' interplay: thus, aware of the risk of conceptual stretching, what I adopted is a 'Gramscian reading' of L's work. If Marx's legacy is acknowledged by Lefebvre[10] but has been object of contestation within more 'orthodox Marxism'; that of Gramsci is, at least, more uncertain.[11]

However, this is my starting point, and my view angle.

Alongside the structural issues Marx deals with, which explain the relationship between economy and society through the socio-economic structures of the modes of production, Gramsci introduces a series of issues that have to do with behavior, with habits, with practices, with daily life, which are obviously connected to *structure*, often shaped by it, but which also challenge it constantly: it is the *agency*, its apparent weakness compared to structures, its potential transforming power.

For instance, L. (p. 100f.) claims that: "The city always had relations with society as a whole, with its constituting elements (countryside and agriculture, offensive and defensive force, political power, States, etc.), and with its history (...). Yet, the city's transformations are not the passive outcomes of changes in the social whole. The city also depends as essentially on relations of immediacy, of direct relations between persons and groups (...). It is situated at an interface, half-way between what is called the *near order* (relations of individuals in groups of variable size, more or less organized and structured and the relations of these groups among themselves), and the *far order* of society, regulated by large and powerful institutions (Church and State), by a legal code formalized or not, by a 'culture' and significant ensembles endowed with powers, by which the *far order* projects itself at this 'higher' level and imposes itself (...). The city is a *mediation* among mediations. Containing the near order, it supports it; it maintains relations of production and property; it is the place of their reproduction. Contained in the *far order*, it supports it; it incarnates it; it projects it over a terrain (the site) and on a plan, that of immediate life; It inscribes it, prescribes it, *writes* it. (...)".

If not legitimate, an interpretation of the above in terms of structure-agency interplay is possible, and has a broad potential: furthermore, the dynamic between the

[10] See the dedication following the date, at the end of the book.

[11] On the 'missed' relation between Lefebvre and Gramsci, see Tosel (2017). Regarding the lack of quotations, it has been noted that L. did not use to quote, and this would depend on his way of writing (dictating to others).

'far order' and the 'near order' implies also *scale*: suggesting to deepen also in that direction the idea of city as *mediation.*

A similar perspective can be found in other passages, such as (e.g.) at p. 126, when discussing the 'critical point', which is, in L. words:

> The distinction between the three levels (global process of industrialization and urbanization – urban society, the specific scale of the city - ways of living and conditions of daily life in the urban) tends to become blurred (…). And yet, this difference between the three levels is more than ever crucial to avoid confusion and misunderstandings, to combat strategies which find in this conjuncture an opportunity to disintegrate the urban into industrial and or residential planning.

There, in my view, the city appears as the place where 'the global' and its power meet daily life, and where societal *structures* 'meet' people's *agency*, and antagonism (or agonism) emerges, producing, or at least opening possibilities for, change.

So, the first facet I would like to focus attention on, is this 'intermediate domain' and the multifarious meaning of city as 'mediation'.

18.1.2 Structure and Form

Going further on along this line of thought, there comes the relationship with *form.*

In the chapter dedicated to 'On urban form', L. clarifies that "The ambiguity, or more exactly, the polysemy or plurality of meanings, of this term, 'form', has already been remarked upon" (p. 133), but that "The only way to clarify the meaning of the term is to begin from its most abstract acceptance. Only scientific abstraction without contents, distinguished from verbal abstraction and opposed to speculative abstraction, enables transparent definitions. Therefore, to define form, one must begin from formal logic and logico-mathematical structures. Not so as to isolate or fetishize them, but, on the contrary, to catch their relation to the 'real'" (ibidem). Thus 'form', which can be abstracted but is not independent from content, is not 'morphology', and the first example that L. proposes to understand this point is the 'form' of the contract; more precisely, he introduces the contract as an example of what he means by 'form'. What does it entail?

> There is no form without content. No content without form. What offers itself to analysis is always a unity of form and content. Analysis breaks this unity. It allows the purity of form to appear, and form refers back to content. Yet, this indissoluble unity broken by analysis, is conflictual (dialectical). (…) For dialectical reason, contents overflow form, and form gives access to contents. Thus, form has a double 'existence'. It is and is not. It has reality only in contents and yet detaches itself from them. It has a mental and a social existence.
>
> Mentally the contract, is defined by a form quite close to logic: reciprocity; socially, this form regulates countless situations and activities; *it confers upon them a structure* … (p. 135, the italics is mine).

In L.'s view, form is what 'contains' and 'shapes' kinds of relationships, established also through structured processes: there are praxis, and/or modalities that are reiterated and determine range of norms, and these are *formalized (structured)* through and within juridical tools (institutions) such as the contract. Here emerges the link between structures and form, and possible directions to deepen it.

Interestingly, L. follows this line of thought to explain what a city is. In the chapter dedicated to form he adopted categories that are common to semiology and formal logic giving to the urban form a different meaning from the one architects and planners usually refer to (that of urban morphology). L. schematizes the concept of form into two main types, which he calls 'mental' and 'social' forms as we saw for the contract, and uses this same approach to explain the *urban form*, which is his last example (from the more 'abstract' which is the logical form) and the more 'complex' (aware of having left aside 'repetition', considered by many as the 'supreme form'—see p. 138). L. defines the urban form mentally as the form of simultaneity, while socially as the form of encounter (cf. pp. 137–138), and another kind of *mediation,* between its twofold nature, *oeuvre* and product:

> VIII. *Urban form*
>
> Mentally: simultaneity (of events, perceptions, and elements of a whole in the 'real').
>
> Socially: the encounter and the concentration of what exists around, in the environment (assets and products, acts and activities, wealth) and consequently, urban society as privileged social site, as meaning of productive and consuming activities, as meeting between the *oeuvre* and the product.

18.1.3 The *City* as *oeuvre* and the *Urban Phenomenon* in History

In L.'s urban theory, *city* is interpreted as the 'temporary' form that *the urban* can have in a certain space and time. This means that the fundamental phenomenon is the urban, as 'simultaneity and encounter', within which the city took form under certain conditions, resulting as the meeting point of *oeuvre* and product.

As I mentioned already, L. did not provide one unitary definition of what the city is, but three different in a row, explaining that "it must be conceived during this progress" (p. 109).

Thus (pp. 109–110),

> [1] We (…) propose a first definition of the city *as a projection of society on the ground,*[12] that is, not only on the actual site, but at a specific level, perceived and conceived by thought, which determines the city and the urban. Long-term controversies over this definition have shown its lacunae. Firstly, it requires more accuracy. What is inscribed and projected is not only a far order, a social whole, a mode of production, a general code, it is also a time, or rather, times, rhythms. (…) Secondly, (…) it brings to light certain historical and generic or genetic differences, but leaves aside other real differences: between the types of cities resulting from history, between the effects of the division of labour in the cities, between the persistent 'city-territory' relations.

[12] In this regard, I found interesting resonances with Emilio Sereni's approach to the landscape (see the Introduction to: 1961, *Storia del paesaggio agrario italiano*. Bari: Laterza).

> [2] Hence another definition (…): the city *as the ensemble of differences* between cities. (…) This definition reveals itself to be insufficient, as it places emphasis on particularities rather than on generalities, neglecting the singularities of urban life, the ways of living of the city, more properly understood as to inhabit.
>
> [3] Hence another definition, of plurality, coexistence and simultaneity in the urban of *patterns*, ways of living urban life (…).
>
> These definitions (relative to the levels of social reality), are not in themselves exhaustive and do not exclude other definitions. (…) Moreover, there would be the need to emphasize the historical role of the city: the quickening of processes (exchange and the market, the accumulation of knowledge and capitals, the concentration of these capitals) and site of revolutions. Today, by becoming a centre of decision-making, or rather, by grouping centres of decision making, the modern city intensifies by organizing the *exploitation* of the whole society (not only the working classes but also other non-dominant social classes). This is not the passive place of production or the concentration of capitals but that of the *urban* intervening as such in production (in the means of production).

18.2 The Urban

Indeed, L. dedicated many pages to 'the urban', interpreted as the more general phenomenon within which at a certain time in history, the city emerged between *oeuvre* and product. In some chapters—for instance, in 'Continuities and Discontinuities' but also in 'Industrialization and urbanization', we can find a reconstruction of the historical formation process of the city and also of its crisis; the urban is what *resists* to the changes determined (induced) by industrialization (the *inductor* of the crisis), which contains basic features and qualities, and survive after the end of the 'traditional city'.

The urban is conceptualized at first through a negative theory: what is said about it, is what it lacks if compared with the socio-spatial relations and features of the traditional city.

To explain what got lost in moving from the city to the urban, L. mentions, e.g. two housing models: the big public housing estates (*grand ensemble*) and the low-density neighborhoods of the urban suburbs, which need each other for recognition[13] and whose relationship is instrumental within a political strategy of segregation. They both are not city anymore—they represent the contrary of the fundamental quality of 'simultaneity and encounter'—to define them L. introduces the concepts of *habitat* and of *inhabiting*, and, quite stimulatingly, he uses these concepts differently from

[13] "Suburban disorder harbours an order: a glaring opposition of individually owner-occupied detached houses and housing estates. This opposition tends to constitute a system of significations still urban even into de-urbanization. Each sector defines itself (by and in the consciousness of the inhabitants) in relation to the other, against the other" (p. 80). I experienced this myself, doing research on Corviale and Tor Bella Monaca, two huge public housing estates in Rome, adjacent to smaller scale—low density *borgate* of illegal origin.

how they are much often used nowadays. In fact, many scholars use the concept *inhabiting*[14] as positive, e.g. to mean that the *housing question*[15] cannot be reduced to houses, and implies relations which can be realized rather independently from specific spatial features. Thus, the concept of inhabiting is often used to highlight the fact that material conditions are not sufficient—although necessary. Concerning the concept of *habitat*, it is more commonly used with a neutral connotation, its frame being that of ecology. On the contrary, L. uses these concepts to say what is a *non*-city, with the related forms of socio-spatial segregation.

After having provided this 'negative theory' about 'the urban', explaining everything it is not if compared with the 'city', Lefebvre highlights what 'resists' of the city, notwithstanding its crisis and disappearance. What resists is related to the quality of 'simultaneity and encounter' which is also, as already mentioned, the essential quality of urban life, thus the possibility for the 'urban renewal'.

In Lefebvre's view, the 'urban renewal' cannot be a 'return' to the traditional city, as well as any 'return' to the past (including its ideologies): on the contrary, a new *form* of the urban (and of the urban life) should be invented, to correspond to a different society.

Thus, the question became also a 'typical' planning question: if we took seriously the relation, as co-construction, between space and society, what is the kind of city we aspire to have, as the material representation and *form* of the society we are looking for? Which spaces would this society needs to express and realize itself? Following Lefebvre's proposal, the first feature to contrast is segregation in its manifold forms and manifestations.

This issue leads to our second and last key word: *urbanism* (urban planning).

18.3 About Urbanism and Urban Planning

As explained in the short introduction, *urbanism* is the second and last key concept I want to focus on in this contribution: my intention is to look also for 'operational' suggestions concerning planning and the role of planners for realizing the most-awaited 'urban renewal'.

I will just briefly mention here how L. *describes* urbanism and urban planning, and other related inspiring topics, conscious that each one of them would need a much in-depth exploration. For instance, L. talks about 'urban thought' and 'urbanism

[14] In Italy, in particular, many scholars prefer to translate *housing* with *inhabiting* [abitare]. See Pizzo (2020).

[15] According to Lefebvre, the housing question, has "disguised the problems of the city and urban society" (p. 177), somehow following Engels, which considered it as a 'minor' problem that could not be solved 'independently' but together with the more general problems of capitalism. But for Lefebvre, the 'solutions' adopted for answering to the housing needs negate the city (the housing models we mentioned before produce *habitats*, and not city; and determine to *inhabit*, not to fully live the life of the city).

without thought' (p. 81)—to say of a time in which there was no need of an 'urban theory'[16]: this is one of the issues worth to be discussed further.

He refers specifically to *zoning* as a way of doing planning which contributes to segregation, and embeds a 'systemic' perspective on cities, which is strongly criticized since objectives are introduced as they were the result of a consistent analytical process. This point, which reminds of the long-lasting disciplinary debate on 'rational comprehensive' planning and its crisis or, better, contributed at generating that debate, also fits perfectly in the more recent ones about technical knowledge and de-politicization.

From L.'s viewpoint, planning could provide a positive contribution if it assumes a more 'limited' scope, e.g. when oriented to avoid forms of segregation and, conversely, to favour encounter in the urban space, and, more importantly, when its ends and goals are the result of 'strategic decisions' justified and sustained through an ideology: because the 'synthesis' L. talks about to have the most-awaited 'urban renewal' belongs to the political.

This point is also very important, signaling L.'s perspective on the relationship between planning and politics.[17] Planning does not emerge just as a *technical instrument* made available by experts for politicians, but it does not even overlap with *politics*: the difference is subtle but worth to be analysed and discussed further. However, the required synthesis cannot come from planners (or architects, or any other technical-professional knowledge).

L. talks about planning in many different ways, some even funny I would say, e.g. when he speaks about "The planning of men of good will (architects and writers). Their thinking and projects imply a certain philosophy. Generally, they associate themselves to an old classical and liberal humanism. This not without a good dose of nostalgia. One wishes to build to the 'human scale', for 'people'…" (p. 83). The risk is formalism and aestheticism: they try to reproduce forms which no longer have to do with the social and economic structures of contemporary society.

Another kind of planning is that of "administrators linked to the public (State) sector. It sees itself as scientific. It relies sometimes on a science, sometimes on studies which call themselves synthetic (pluri or multidisciplinary). This scientism, which accompanies the deliberate forms of operational rationalism, tends to neglect the so-called 'human factor'". In this regard, L. claims that "These (…) aspects are confounded in the conception of centres of decision-making, a global vision, planning already unitary in its own way, linked to a philosophy, to a conception of society,

[16] "The masters of old had no need for an urban theory to embellish their cities. What sufficed was the pressure exercised by the people on their masters and the presence of a civilization and style which enabled the wealth derived from the labour of the people to be invested into 'Oeuvres'. The bourgeois period puts an end to this age-old tradition. At the same time this period brings a new rationality, different from the rationality elaborated by philosophers since ancient Greece (…)" (ibid.).

[17] The centrality of this dimension, and the need to deeply consider historical transformations of politics, would require a much deeper inquiry. Just as an example, we could consider the idea of urban spaces related to segregation in contemporary populisms and the re-articulation of conflicts (including urban ones) in populist perspectives (on this, see Mouffe 2005, 2013).

a political strategy, that is, a global and total system" (pp. 83–84). And we already saw that the synthesis is not any expert's duty. L. talks also about "The planning of developers. They conceive and realize without hiding it, for the market, with profit in mind". (p. 84). Related to this, L. mentions land use and land regimes issues, as well as urban rent issues, which—I am increasingly convinced—should be put back at the centre of the planning debate.

Planners, and architects in particular, often are trained with the idea that they are the ones able to produce the mentioned *synthesis*,[18] able to overcome forms of socio-spatial segregation, and to create the 'new' city—based on the forms of encounter and exchange of the present society, while L. wrote explicitly that "The architect, the planner, the sociologist, the economist, the philosopher or the politician cannot out of nothingness create new forms and relations. More precisely, the architect is no more a miracle-worker than the sociologist. Neither can create social relations, although under certain favorable conditions they help trends to be formulated (to take shape). Only social life (praxis) in its global capacity possesses such powers—or does not possess them. The people mentioned above can individually or in teams clear the way; they can also propose, carry out and prepare forms.[19] And also (and especially), through a maieutic nurtured by science, assess acquired experience, provide a lesson from failure and give birth to the possible" (pp. 150–151).

From this picture, planning emerges as more 'bounded', its actual role more limited, but also its possible paths more defined, e.g.:

(a) The existence of structural factors and agency factors, and the related idea that the city 'belongs to' an *intermediate domain* between them, that it is a *mediation*, implies also that *urbanism,* to be effective should be positioned at that same 'domain' or 'level'. This means, e.g. that the city and the urban are *necessarily* by-products: the results of many different intentionalities, but also the not-fully-intentional result of a combination of choices.[20] This means also that we (planners) cannot 'plan' everything with the idea of defining the *results*—which will be quite surely different from what we planned (Ferraro

[18] Having "an unquestionable knowledge of the real problems of the modern city, a knowledge which gives rise to a planning practice and an ideology, a functionalism which reduces urban society to the achievement of a few predictable and prescribed functions laid out on the ground by the architecture. Such an architect sees himself as a 'man of synthesis', thinker and practitioner. He believes in and wants to create human relations by defining them, by creating their environment and decor. Within this well-worn perspective, the architect perceives and imagines himself as architect of the world, human image of God the Creator" (p. 98).

[19] For L., participation does not help in this process: it is an 'obsession' (of the experts, technicians and administrators), one of the more explicitly criticized concepts: "In practice. the ideology of participation enables us to have the acquiescence of interested and concerned people at a small price. After a more or less elaborate pretence at information and social activity, they return to their tranquil passivity and retirement. Is it not clear that real and active participation already has a name? It is called self-management. Which poses other problems (p. 145).

[20] I proposed this concept to define landscape—in Pizzo, 2008: it had been an object of a profitable discussion with Luigi Mazza, and we eventually agreed that the same concept should be applied to the city—and now, after many years, I can still appreciate that very stimulating conversation.

1998, pp. 262–263)—but e.g. with the idea of shaping the *conditions* for change to happen, and indicating a *direction*, instead.

(b) Moreover, L. acknowledges and, I would say, emphasizes, a contribution of planning at the analytical level, in terms of descriptions and explanations,[21] something that planners practice more rarely and that it is not usually considered as a planners' task. In fact, planners, as well as all experts with a technical knowledge, are more used to concentrate on the 'how', thinking that the 'why' is more a matter of urban sociologists, or social scientists.

Explanations grounded in the broadest possible knowledge of planning considered as a distinctive approach to socio-spatial reality play, therefore, a pivotal role. Empirically valid explanations are a remarkable contribution that urban planning can and should make; their meta-normative meaning cannot be reduced to the 'analysis and interpretation' before the plan. It is rather a matter of evaluation of planning itself, and entails a specific technical and also historical knowledge of planning: the analysis and appreciation of the 'why' requires its own technical competence as well and needs to be assessed before other steps can be fruitfully implemented; indeed, as every child knows, the 'why' always comes first.

References[22]

Brenner N (2000) The urban question: reflections on Henri Lefebvre, urban theory and the politics of scale. Int J Urban Reg Res 24(2):361–378

Brenner N (2014) Implosions/explosions. Towards a study of planetary urbanization. Jovis, Berlin

Buckley M, Strauss K (2016) With, against and beyond Lefebvre: planetary urbanization and epistemic plurality. Environ Plann d Soc Space 34(4):617–636

Ferraro G (1998) Rieducazione alla speranza: patrick geddes, planner in India, 1914–1924. Milano: Jaca Book

Gribat N, Höhne S, Michel B, Schuster N, Pizzo B (2019) An historical and critical reconstruction of disciplines and interdisciplinarity in urban studies (part. 1). In: Cognetti F, Fava F, Grassi P, Pizzo B (eds) La città interdisciplinare/The interdisciplinary city. Tracce Urbane 6, pp 70–91

Insolera I (2018) Modern Rome. From Napoleon to the XXI century. In: Bozzola L, Einaudi R, Zumaglini M (eds) Cambridge Scholar Publishing, Newcastle Upon Tyne. Italian edition: (2011 [1962[1]]) Roma moderna: da Napoleone 1. al 21. secolo. Einaudi, Torino

Madden D, Marcuse P (2016) In defense of housing. The politics of crisis. Verso, London

Mouffe C (2005) The return of the political. Verso, London

Mouffe C (2013) Agonistics: thinking the world politically. Verso, London

Pizzo B (2008) La costruzione del paesaggio. Officina, Roma

Pizzo B (2013) Why talking about 'The right to the city' in a time of neoliberalization. In: Perrone C (ed) Living landscapes/landscapes for living. Policies, practices, images. Planum Publisher

Pizzo B (2018) The many paths of self-organization. In: Cognetti F, Fava F, Grassi P, Pizzo B (eds) La città è interdisciplinare/the interdisciplinary city. Tracce Urbane 6, pp 48–67

[21] This is why, using Lefebvre to describe phenomenon without explanations is to me a betrayal of Lefebvre's own thought.

[22] All Lefebvre quotations are taken from: Lefebvre, H., Kofman, E., & Lebas, E. (1996). *Writings on cities* (Vol. 63). Oxford: Blackwell.

Pizzo B (2020) Casa: Egemonia della logica di mercato e naturalizzazione della crisi. In: Madden D, Marcuse P (eds) In difesa della casa. Politica della crisi abitativa. Edizione italiana curata e tradotta da B. Pizzo. Edit Press/Territori, Firenze
Pizzo B, Rolnik R (2019) On interdisciplinarity and non-disciplinary research: a conversation. In: Cognetti F, Fava F, Grassi P, Pizzo B (eds) La città interdisciplinare. Per itinerari non tracciati tra saperi urbani/The interdisciplinary city. Untracked itineraries of/for urban knowledges. Tracce Urbane 6, pp 23–34
Savini F (2019) Responsibility, polity, value: The (un)changing norms of planning practices. Plan Theory 18(1):58–81
Schmid C (2018) Journeys through planetary urbanization: decentering perspectives on the urban. Environ Plann d: Soc Space 36(3):591–610
Sereni E (1961) Storia del paesaggio agrario italiano. Laterza, Bari
Tosel A (2017) Henri Lefebvre lettore di Antonio Gramsci? Materialismo Storico 1:297–322

Barbara Pizzo, PhD, Research Professor in Urban Planning, Sapienza Università di Roma. She works in the field of Urban Studies with a critical orientation. The entry point of her inquiry is the urban space, its features and dynamics. This she relates to socio-economic changes, concentrating on the many values which are condensed in the built environment, territory and the heritage, on their creation and (uneven) appropriation. Among her more recent works, she edited and translated D. Madden and P. Marcuse's "In defence of Housing The Politics of Crisis", and co-edited an Urban Studies Reader "Mappe e sentieri. Introduzione agli Studi Urbani Critici", both for EditPress/Territori.

Chapter 19
The Trouble with Henri: Urban Research and the Theory of the Production of Space

Christian Schmid

Abstract Henri Lefebvre's theory of the production of space is today one of the most widely quoted accounts on urbanization and the urban. However, this theory provides some important challenges to fully grasp its implications and to use it in concrete empirical research. This book chapter introduces this theory, analyses some of the difficulties arriving in applying it to urban research and identifies possible paths towards creative applications. It begins with the current state of reception, explains the fundamental structure of this theory, lays out Lefebvre's epistemological strategy and analyses his understanding of the relationship between theory and empirical research. These points are then illustrated with examples of research on the urbanization of Switzerland, on planetary urbanization and on the urban development of Havana.

Keywords Henri Lefebvre · Production of space · Complete urbanization · Planetary urbanization · Havana · Switzerland · Transduction

Henri Lefebvre's theory of the production of space, which he developed over a short period from the mid-1960s to the mid-1970s, is today widely quoted and the subject of intense debates. Just two decades ago, this theory was regarded as almost inaccessible, extremely difficult to apply and was consequently rarely used for empirical research. Since then, it has experienced widespread reception and application, both in the field of urban studies as well as in architecture and urban design.

This process of appropriation and application of Lefebvre's theory has not been untroubled. Time and again, all sorts of confusions and problems have arisen. While these initially concerned mainly questions of interpretation and theoretical construction, more recent discussions have primarily focused on the question of how this theory can be successfully introduced into empirical analysis. Engaging with Lefebvre's theory is indeed a remarkable experience: as in one of Hitchcock's

Translated by Christopher Findlay

This article was first published in: Stanek Ł, Schmid C, Moravánszky Á (eds) (2014) Urban revolution now: Henri Lefebvre in social research and architecture. Ashgate, Farnham, pp. 27–48.

C. Schmid (✉)
Department of Architecture, ETH Zürich, Zürich, Switzerland
e-mail: schmid@arch.ethz.ch

C. Perrone (ed.), *Critical Planning and Design*, The Urban Book Series,
https://doi.org/10.1007/978-3-030-93107-0_19

funniest movies, where the corpse of poor Harry suddenly disappears and then reappears in the most unexpected places, this theory creates all kinds of unforeseen trouble.[1] As soon as it seems that one problem is solved, another appears.

19.1 On the Third Wave of Lefebvre Interpretation

In the past years, the debate on the theory of the production of space has entered a new phase. While, in the 1970s and 1980s, mainly reductionist and one-sided interpretations were dominant in the English- and French-speaking debates, a new and more open appropriation of this theory has asserted itself in recent years.[2]

In the first phase, Lefebvre's texts on urbanization and on space had an ambivalent reception: in France during the early 1970s, Lefebvre was pushed to the margins of intellectual life, mainly by the strong influence of Manuel Castells's vehement critique of his broad and far-reaching conception of the urban revolution.[3] This critique also influenced Lefebvre's reception in the English-speaking world, where his work was almost absent from the debates of new urban sociology, radical geography or urban political economy—an important exception being David Harvey, who was especially inspired by Lefebvre's rigorous analysis of the relationship between urbanization and the dynamics of capital accumulation.[4] However, both theorists pursued different projects: while Harvey developed a relatively rigid and narrow political economy of space, Lefebvre had the much broader vision of a comprehensive theory of the production of space. But this project did not arouse much interest for quite some time.

This situation changed fundamentally in the 1990s, when, under the influence of the spatial and the cultural turn in social theory, Lefebvre's theory was 'rediscovered' and finally received with great interest and even enthusiasm.[5] Accordingly, this second phase of Lefebvre reception was strongly influenced by postmodern and poststructuralist thought, which also manifested a reductionist tendency in presenting Lefebvre as a precursor of postmodernism and trying to narrow down his complex and dynamic theory of the *production* of space to a kind of a spatial ontology (see especially Soja 1989). However, it was very difficult to reconcile a postmodernist or poststructuralist approach with Lefebvre's decidedly historical, materialist and dialectical epistemology.[6]

For some years now, a new and different form of access to Lefebvre's work can be observed, which we may consider as the third wave of Lefebvre interpretation.[7] It is marked first of all by the fact that it relates in a fundamentally different way to Lefebvre's heterodox and open-ended materialism than earlier interpretations. It no longer tries to demarcate its position from Lefebvre's epistemology, or conversely to functionalize his work for a specific theoretical approach, but rather to have an independent and open debate on his thinking.

Second, an informed and intense discussion of the epistemological and historical context of his theory was initiated, together with careful analysis of a wide range of aspects of his complex work.[8] This development was much favoured by the translation of some of his most important texts into English,[9] which made many aspects of his theory accessible to a broader audience—the lack of translations had long posed almost insurmountable obstacles to the English-language interpretation of his work. Thus, in recent years, it has been possible to clarify many open questions that had caused considerable confusion, especially concerning the construction of his theory. Furthermore, after the dominant Lefebvre interpretation had long focused mainly on the question of space, topics such as urbanization and the urban, the state and everyday life found a new or renewed interest.

Third, this reception is characterized by curiosity and an open-minded approach to Lefebvre's work. An increasing number of texts take Lefebvre's theory as a point of departure, place it in a contemporary context, and thus make it fruitful for further reflections and analysis. Among these are efforts to combine Lefebvre's with other approaches and thus to open up new perspectives.

Fourth, there has been an increasing number of empirical applications. More and more studies are being published that not only cite Lefebvre's work—such citations having become almost routine in certain fields of research—but integrate his theory into the very heart of the investigation itself.

In this way, the third wave of Lefebvre interpretation has brought about an important expansion of our horizon: it is rooted in an undogmatic reading, uses Lefebvre's work as a point of departure for further reflection, and is at the same time more precise and more open than previous phases of reception.

19.2 A General Historical-Materialist Theory of Society

Discussion of Lefebvre's work has acquired a new quality in recent years. Key elements of his theory that had long been sources of considerable confusion have been clarified, and the main lines of his thinking can be laid open today. The reconstruction of his theory requires a twofold analysis: first of all, a historical reading that reconstructs the sequence of questions and fields arising in his work and that reveals how they have unfolded in the course of historical development. Second, it requires a synchronic reading that is capable of identifying the overall structure of Lefebvre's theory and of positioning the various aspects in a general theoretical framework.

19.2.1 The Historical Production of Social Realities and Theoretical Concepts

A fundamental aspect of understanding Lefebvre's thinking arises from the fact that he always developed and advanced his theoretical concepts through a thorough engagement with the social reality of his times. Thus, a range of new questions came to the fore in the course of his intellectual trajectory. His point of departure was a critical engagement with the philosophical debates of the 1920s and also with the avant-garde and artistic movements of his time—Dada and Surrealism, and later Situationism. He had not only helped to introduce Hegel and the early works of Marx into the French debate but continuously developed his own heterodox materialism through critical engagement with phenomenology, existentialism and structuralism. In this regard, the 'German Dialectic' based on the works of Hegel, Marx and Nietzsche played a crucial role, and would become a formative constant in his work. From those quite diverse sources, he developed his very specific approach of a heterodox, revolutionary, materialist and dialectical thinking.

This epistemological basis had an immediate effect on his theoretical and empirical analyses, in which a key topic gained prominence even before the Second World War: 'everyday life'. This problematic, which was a very unusual subject for scientific research at the time, marked the consequent application of his strong critique of contemporary philosophical thought, his related claim to wrest philosophy away from pure contemplation, to make it practical and thus also to bring it down to earth—hence his attempts to grasp social reality through concrete investigations. Interestingly, Lefebvre began his empirical work in the countryside, in the Pyrenees, and soon extended it to an analysis of social reality in postwar France.[10] The topic of everyday life remained a recurring theme of his work until the end of his life. He returned to it again and again and illuminated it from various angles. It is also reflected in his philosophical work, which in the 1960s dealt with questions of metaphilosophy, modernization and modernity, and also with language theory.[11]

In the course of Lefebvre's explorations of everyday life, a new phenomenon gradually emerged, indicating a fundamental historic transformation: urbanization. Once more based on his investigations in the Pyrenees, where Lefebvre had observed and analysed the construction of a new town, this new phenomenon gained prominence and soon became a dominant subject of his work.[12] The urban question also proved to be a key aspect of the emerging social protest that culminated in the revolts of the late 1960s and particularly in the events of May 1968 in Paris that marked a decisive experience for Lefebvre.[13] His theoretical and practical explorations of the urban phenomenon led him to the identification of two decisive and interrelated aspects: on the one hand, he analysed the process of urbanization and placed it in the context of industrialization and modernization. On the other, he studied the conditions of the emergence of a (possible, virtual) urban society as a concrete utopia. He finally found and developed the central term signifying the epistemological shift from urbanization to urban society: difference.[14]

Soon after these pioneering works, which marked the beginning of critical urban studies in social science, Lefebvre dedicated himself to a further related question that initially won only little attention, but was later considered his most important contribution by many scholars: the question of space and its social production.[15] This work, which is as comprehensive as challenging and difficult to access, anticipated not only the spatial turn but in some respects also the cultural turn in social science, thus providing the decisive point of reference for postmodern and poststructuralist interpretations. It is also in this work that Lefebvre fully developed and employed his three-dimensional dialectic.

Just two years later, Lefebvre's intellectual trajectory once more took an unexpected turn when, at the very beginning of globalization, he provided a comprehensive discussion of the questions of the state and of *mondialisation*, before he returned to the question of everyday life and expanded it with an analysis of rhythms.[16]

In this impressive sequence of topics and research fields, a determining aspect of Lefebvre's overall *œuvre* becomes visible: it follows the process of historic development, explores new aspects of social reality and identifies a whole range of new processes. This implies, for any appropriation and application of Lefebvre's thinking, conceptualizing social reality as well as the production of theory as a historical process, analytically and empirically following the course of social dynamics and, therefore, consistently advancing his concepts.

19.2.2 A General Theory of Society in Space and Time

The apparently loose sequence of topics and questions emerging through the course of Lefebvre's intellectual trajectory is thus not the incidental result of a cruise through the *Zeitgeist*, but the result of a careful study of the unfolding of social reality that can be embedded in an overall historical and theoretical context. A systematic reading reveals that Lefebvre's theoretical construction displays great rigour and coherence on the one hand, but also offers a flexible framework that facilitates the continuous adoption of new developments and allows them to be integrated into the theory. Accordingly, a systematic reading reveals a very different depiction of Lefebvre's theoretical concepts. Here, three core analytical categories may be distinguished: levels, dimensions and configurations.[17]

The category of 'space–time levels' of social reality refers to the social context of the production of space. Here, Lefebvre identifies first of all a general level or distant order; second, a private level or near order; and third, an intermediate and mediating level, the genuine level of the urban.[18] Within this category, the four central topics that dominate Lefebvre's later work can be related to each other and placed in a general scheme: 'everyday life', the 'urban', the 'state' and 'space'. In this framework, 'space' is to be understood as the comprehensive category, while the other categories refer to specific space-time levels of space.

The category of 'space-time dimensions' of social reality refers to the fundamental aspects of any social practice: the perceived, the conceived and the lived (*le perçu, le conçu, le vécu*). Lefebvre superimposes upon these three phenomenological terms three corresponding terms derived from linguistics and semiotics: spatial practice, the representation of space and spaces as representation. The production of space may therefore be grasped analytically as the totality of three dialectically interlinked production processes that mutually imply each other: the production of material goods, the production of knowledge and the production of meanings. On a more general level, one might state that there are various ways of accessing social reality: it can be perceived (i.e. observed using the five senses); it can be conceived and constructed; and it can also be experienced. From this point of view, the three dimensions constitute a contradictory, three-dimensional or triadic unity.[19]

The category of 'space-time configurations' of social reality relates to the historicity and temporality of the production of space. Lefebvre refers to these configurations in various ways: at the level of the urban, he identifies a succession of the rural, the industrial and the urban field. At the general or global level, he sees a historical development from absolute to abstract and finally to differential space. These configurations refer to specific ways of the production of space that are relatively stable over time. One might also say that reference is made here to the trajectories and pathways, but also to the rhythms of social development.

In these diverse theoretical elements, we find nothing less than a general theory of society that integrates space and time as constitutive aspects into its basic structure. Such a theory offers decisive advantages in several respects: first of all, it can be applied at all scales—it allows general analyses of planetary urbanization as well as studies of a concrete urban situation on a street corner. Second, this theory opens up a broad range of applications ranging from the social sciences to architecture and urban design, and it provides productive links to many of the concepts in urban theory and research that have been developed in the past decades. Third, the famous 'turns' in the social sciences of recent decades, from the linguistic and the spatial to the cultural turn, are already included in this theory. For instance, questions of language and semiotics, but also questions of symbolisms and everyday experiences, are integrated as constitutive elements into this social theory. Long before concepts of the actor–network theory were introduced into urban research, Lefebvre had already explicitly integrated the materiality of the urban and of space into his theory and had opened it up to applications in the field of political ecology. Furthermore, his theory also bridges the divide between economic and cultural studies that was largely responsible for the rifts between political economy and postmodern and poststructuralist approaches. Fourth, not the least of its advantages is that it constitutes a general theory of society in its spatial and temporal conditions. As such, it not only offers some heuristic principles and guidelines but also enables the embedding of a vide variety of questions in an encompassing theoretical framework.

However, these advantages have a price: certain specificities of this theory must be accepted that resist any easy and fast appropriation and application.

19.3 On Lefebvre's Epistemology

Even though the basic construction of Lefebvre's theory has in the meantime gained visible contours, considerable difficulties remain in the appropriation of his theory. This is largely due to the fact that the epistemological base of Lefebvre's thinking exhibits certain characteristics that are not widely appreciated in today's strongly empirically oriented and often theoretically shallow debates.

19.3.1 A Three-Dimensional Dialectic

One of the fundamental characteristics is the dialectical construction of Lefebvre's concepts, which entails that they are not easily accessible and comprehensible. The question of dialectics has repeatedly caused a great deal of confusion. Even today, it often remains little understood and continues to create significant problems. Another difficulty is the unusual version of this dialectic, which Lefebvre himself explained only at a late stage. Unlike the dialectics of Hegel and Marx, Lefebvre's version is three-dimensional; that is, it relates three terms to each other. This is a basic principle that appears as early as in his texts and develops gradually in the course of his writings,[20] until it finally unfolds to achieve its full scope in *The Production of Space*, where it becomes the cornerstone of the three-dimensional theory of the production of space.

This dialectic has an immense potential, but it also makes comprehension of the related concepts difficult. For the dialectical movement of the concepts must be comprehended in its entirety and cannot be reduced to singular quotes—a frequent error in the interpretation of Lefebvre's texts. Often, scholars and researchers try to circumvent this dialectic and apply only selected aspects of the concepts. However, if Lefebvre's concepts are used in a non-dialectical way, they get detached from their underlying 'deep' structure and thus lose a significant part of their analytical power and effectiveness.

19.3.2 Concrete Abstraction

Directly related to this dialectic is the principle of 'concrete abstraction', Lefebvre's 'key to the real' (Lefebvre 1977: 63). In his understanding, it is necessary not only to construct concepts but also to free them from their ontological seclusion and to comprehend them in the context of the concrete social reality itself. Therefore, he aims to trace the genesis of the concepts in the historical process and in the dialectic of mental abstraction and concrete reality. By way of exemplary explanation, Lefebvre refers to the ambiguous concept of the 'commodity' as developed by Marx at the beginning of the first volume of *Capital*: the commodity has an exchange value that is, however, always based on its use value. The use value is the material support of the exchange value, which in turn constitutes the social form of wealth in capitalist societies. The use value is absorbed in the social abstraction of exchange, but it can be realized only in use, in consumption. The commodity is therefore abstract, but at the same time always concrete and imbued with certain properties and qualities. Thus, the commodity has a general, abstract aspect (its exchange value) as well as a concrete, material aspect (its use value). It is necessary to advance from the abstract to the concrete in order to apprehend the contradictory nature of the social world.[21]

In Lefebvre's view, therefore, abstraction is the result of a historical process and is generated by social reality. He conceives key terms such as 'everyday life', 'the urban' or 'state' as concrete abstractions.[22] The application of this principle to space brought Lefebvre to his concept of 'abstract space', which is simultaneously homogeneous and fragmented: it is homogeneous and abstract because it is part of the global dominance of the commodity and because its existence is achieved only through the convertibility of all its constituent parts; at the same time, it is concrete and fragmented because, as such, it is a part of the material social reality and therefore always localized (Lefebvre 1991: 341f.).

19.3.3 Strategy of Knowledge

Based on these insights, Lefebvre also pursues a specific strategy of knowledge (Lefebvre 2000 [1972]: 28). He frequently introduced his terms as approximations, or 'strategic hypotheses', whose scope and range of applicability he explores in the course of his investigation. Accordingly, his terminology is also continuously shifting dialectically—it might be said that it is marked by a fluctuating underlying structure.

In Lefebvre's understanding, terms emerge under certain historical conditions, both in social reality and in theoretical thought (Lefebvre 1980: 15f., 61). This is why terms must always be embedded in their historical context of emergence and understood historically. At the same time, these terms express not only a 'truth' but have also an 'effectiveness': they elucidate reality and simultaneously animate the future. Developing a concept therefore means situating it, consolidating it and also detecting its limitations until it can finally be used as an instrument of analysis and critique. Lefebvre's theoretical and empirical procedure (*démarche*) consists of carrying out precisely this process in the course of an analysis, which is also reflected in his books: he usually sets out by offering a tentative definition of a concept in order subsequently to develop it. As he himself explains, the 'truth' of a concept does not make itself known until the end of a book as the result of a process of enquiry that reveals the dynamic nature of the concept. It is of course evident that this process will not stop at the end of a book—thus the concept will continue to evolve further in theoretical debates as well as in practice. However, this being the case, the theory as a whole must also be continuously developed; it must never stand still.

Thus, Lefebvre's theory not only serves as an analytical framework, but might even become a generating force: it invites one to investigate certain questions that otherwise would have remained unconsidered, and it leads to certain logical conclusions, whether one likes them or not—this is the very practical value of such a theory. The point is therefore to reduce this theory not to a nice bouquet of quotations that may be used to decorate one's empirical studies, but as an instrument of analysis and as a tool for practical application.

19.4 On the Relationship Between Theory and Empirical Research

How then can these complex theoretical concepts be used empirically? This brings us to the next source of possible trouble: while his theory has gradually found better understanding in recent years, and many of his concepts have been debated and clarified, the question of empirical application has long remained opaque. Lefebvre himself did not really offer clarification here, as his books remain elusive when it comes to this question, and the examples he gives are often more illustrative in character than presenting detailed results of concrete field research. In his recent book, *Henri Lefebvre on Space*, Stanek (2011) closed this lacuna. He meticulously traces how Lefebvre developed his theories on the basis of concrete empirical studies and how he repeatedly also participated in the elaboration of concrete architectural and urbanistic projects.[23] It is necessary in this context to clearly contest and rectify certain myths that have been spread time and again and are still used as arguments against empirical research on the basis of his theory: Lefebvre's theory has an empirical background and is designed for empirical application.

However, Lefebvre himself did not develop any sophisticated methodology. He and his colleagues and assistants used the existing methods that were available at the time, based mainly on qualitative methodologies. It follows that there are no simple recipes or models that would allow us to apply his concepts. Many of those cannot be transferred directly, or only with a considerable effort. This is a problem in particular because working with Lefebvre's theory is not simply about applying an existing repertoire of concepts, but about the necessity of continuously advancing the theory as part of an engagement with current developments in society. This is largely due to the specific dialectical relationship between theory and empirical research in Lefebvre's understanding.

19.4.1 Transdisciplinarity

Lefebvre's metaphilosophical point of departure is his comprehensive analysis of social processes. Accordingly he attempts to grasp the totality of social reality and he does not respect disciplinary boundaries, but on the contrary attempts to transcend and abolish them. Thus, he frequently fiercely criticizes the prevalent division of labour in academia and repeatedly states explicitly his opposition to the specialization, the parcelling and the resulting disciplining of knowledge. He understands the splitting up of knowledge into individual disciplines as seriously limiting the possibility of generating new insights (Lefebvre 1980: 37).

As Lefebvre stated in *The Urban Revolution*, grasping the problematic of urbanization in its entire breadth would justify the foundation of a university bringing together a wide range of disciplines; but it would also require the convergence of those disciplines (Lefebvre 2003 [1970]: 54, 56). Thus, his approach consists of rigorous transdisciplinarity: the point is not to split up knowledge and to reduce research questions to isolated aspects; instead, one should strive to comprehend the totality of a phenomenon. However, such an aspiration is a frequent cause of conflicts with the narrow institutional boundaries of the various disciplines.

19.4.2 Transduction

A second important aspect of Lefebvre's research is transduction. His own metaphilosophical epistemology prevented him from attempting to systematize terms on the basis of a (new or old) speculative principle. Instead, his goal was to explore the meaning, the content, the boundaries and ultimately the conditions of the social validity of his concepts, which meant applying a double-edged procedure: a critique of the concepts through practice; and a critique of practice through the concepts. This means that (abstract) concepts must always be related to a concrete reality, and thus exposed to a real confrontation. Lefebvre's ambition was to use his concepts as guidelines in order to understand society, by provoking a confrontation of the philosophical concepts with the non-philosophical world.[24]

Thus, Lefebvre develops his concepts in mutual interaction with empirical research. His empirical approach is neither induction nor deduction, but dialectical transduction.[25] He regards this process as a dialectical movement: the first (dialectical) moment is the comprehension of concrete social practice; the second moment aims to approach lived experience without destroying it; the third and final moment consists of developing the concept that will simultaneously permit comprehension of the (historical) social process, a critique of the present and a prospective illumination of the horizon of the future (Lefebvre 1980: 37). Lefebvre's theoretical concepts can therefore not simply be 'applied', that is, used as a theoretical framework for analysing and explaining various empirical case studies. Rather, the purpose is to confront these concepts with reality, or to immerse them in reality and thus make them productive.

19.4.3 Critique and Project

In order to understand this confrontation it is imperative to have a clear understanding of what Lefebvre's project really was: in a universal sense, no less than a project for the emancipation of humankind (Lefebvre 2016 [1965]). Lefebvre's goal was thus not, as is so often the case in academia, to solve puzzles, but to achieve a change in society and to apply the analysis itself as an instrument of transformation. Such an analysis is always both a critique and a project.[26] In this sense, the critique constitutes a theoretical act and thus an intervention into a social field.

This is what Lefebvre referred to as 'metaphilosophy': the philosophy becomes 'realized', part of the 'world'. This means, however, that it has to give up its status as philosophy and become a project that realizes itself in the world and, through this very realization, negates itself in the process of (dialectical) sublation (Lefebvre 2016 [1965]: 62). Thus, the starting point has to be a radical critique of the present situation, which finally opens up the possibility of practical (and radical) change.

One cannot therefore adopt Lefebvre's theory 'as it is', but must advance it conceptually in constant interaction with social reality. This is why the theory must also constantly be further developed: what is needed is an open and creative way of handling it. This will be explained in the following sections by several concrete examples

19.5 The Three-Dimensional Analysis of the Production of Space

Nowhere are both the problems and the potentials of Lefebvre's theory more evident than in his famous three-dimensional conception of the production of space, the key epistemological element of this theory. As outlined above, what this conception refers to is not a set of three independent dimensions, but an ensemble of three contradictions and thus interdependencies between three poles. Therefore, the goal cannot be to use the three dimensions like drawers to be filled with corresponding empirical examples or as a scaffold that serves to order the abundance of social reality. Rather, it should be understood as an instrument that can be used to actively advance the analysis. Therefore, there are the dialectical relationships between the three poles that are of main interest. The aim must be to regard such relationships as active elements and not to study them independently, but rather to analyse the dialectical interplay of the dimensions. As Lefebvre emphasized again and again, the three dimensions or moments should be neither conflated nor separated.[27]

Thus, there are many analyses that work with a three-dimensional outline, but only few that fully apply it and make it fruitful. This is due not least to the complexity of this theory: empirical studies that fully operate with this three-dimensionality require a huge effort, as they must first analyse spatial practice, that is, the material processes related to the production of space; second, examine the representations of space, that is, discourses, concepts, plans and so on; and finally also integrate the spaces of representation and thus lived experience into the analysis. That means applying a whole range of often demanding and laborious, mainly qualitative, methods. Additionally, the dialectical interdependencies between the three dimensions must be analysed, which is again not an easy task. Nevertheless, it is possible to carry out such analyses, and they can be very fruitful.[28]

19.5.1 Networks, Borders, Differences

In order to allow a broader application, it is often useful to narrow down the research question or to develop a more accessible analytical framework. This is what we did in the project *Switzerland—An Urban Portrait* (Diener et al. 2006). This project, which involved architects Roger Diener, Jacques Herzog, Marcel Meili and Pierre de Meuron, about 150 students of architecture and 20 research assistants, aimed at developing a novel analysis of the urbanization of Switzerland. In order to allow the involvement of students in the empirical research, we developed a simplified theoretical concept or 'field kit' based on Lefebvre's triad that is easy to comprehend and can be applied in the field, but at the same time is solidly anchored in theory. As a result we found a series of three terms that could guide and inspire our investigation: networks, borders and differences.

These three terms relate directly to the three basic concepts of Lefebvre's triad (cf. Schmid 2006): first of all, spatial practice can be operationalized using the concept of *networks*. Urban space is a space of exchange, meeting and interaction. It is permeated by all sorts of networks that can be analysed in order to identify various urban constellations on different scales. Second, urban spaces are always permeated by manifold political, social and cultural *borders*: one may understand these as indicators of the urban condition of a territory and therefore use them as the basis of a critical investigation of various definitions of urban space. The methodological assignment here was always to frame the map in such a way that the borders are in the centre: in this way, the existing representations of space were decentred, while the borders between various areas became visible. It is often the case that maps are left with white spaces beyond the borders, as nothing is shown there, and thus they might conceal key elements of spatial development. An analysis of the 'white realms' of a map can therefore be particularly illuminating. Third, we used a term that often creates difficulties, but that is crucial for any analysis of urban spaces following Lefebvre's approach: *difference*, which denotes a specific aspect of lived space. Our goal here was on the one hand to discover various forms of difference, and, on the other, to investigate differences as constitutive elements of urban territories. These three terms enabled a dynamic definition and analysis of urban areas and the development of a typology of different urbanization processes.

19.6 On the Analysis of Urban Territories

In order to apply this set of tools empirically, we followed Lefebvre's classic procedure by taking his famous thesis of complete urbanization as the starting point of our analysis. This thesis had already proven to be very fruitful on many occasions. The first time that my colleagues and I, as students, applied this thesis to Switzerland,[29] it was a real provocation: we met with vigorous opposition from all sides—the thesis simply flew in the face of everyday observation and of the prevalent rural image of

Switzerland, notwithstanding the fact that Switzerland had already become largely industrialized and urbanized in the nineteenth century. Even years later, this thesis continued to evoke objections. Finally, I came to end my lectures on the topic with a joke: 'No doubt there are still cows on Swiss meadows—but they are not rural cows, they are urban cows!' Indeed, in Switzerland (as in other countries), agriculture has become a modern industry using digital technology and is heavily subsidized by the government. At the same time, the so-called 'rural' areas are far removed from idyllic ideas of village communities: they have long become integrated into manifold urban networks; their territory is covered with all kinds of urban symbols and signals; standards of living and lifestyles have largely become similar to those of urban regions; and everyday life is oriented towards the needs of an urban population.

The contradiction between 'city' and 'countryside' has therefore been fundamentally transformed and can no longer serve as a fruitful starting point of an analysis. This means that the analysis should no longer focus on the differences between 'urban' and 'rural' areas, but examine the differences developing within the urban. It would be completely misleading to start from the assumption that the thesis of complete urbanization implies a homogenization of all territories. On the contrary, urbanization is a process that is constantly bringing forth new urban situations and thus also creating new differences. This also means not reducing urbanization to the production of the built environment, but understanding it as a comprehensive social process and developing a much broader definition of the urban, while also including other aspects in the analysis: the transformation of everyday life; the development of (different) urban lifestyles; the changing patterns of communications and mobility; the manifold interdependencies and linkages of social and economic life (see also Schmid 2013).

19.6.1 Complete Urbanization in Switzerland

In the project *Switzerland—An Urban Portrait* we finally took the thesis of complete urbanization as the basis for a systematic analysis of the urban conditions in Switzerland. The challenge was to proceed with a comprehensive analysis of urbanization for the entire territory, including so-called 'rural' areas, and to find a methodological design that allowed us to detect various forms of urbanization. We developed a methodological procedure that combined methods of social sciences and of architecture and also attributed great epistemological significance to the design process. Maps became the main instruments of this analysis; the texts were ultimately only commentaries on the maps—is was a cartographical analysis in the proper sense of the term. For this empirical enquiry, we applied a multi-step approach. First, we came up with the three theoretical categories—networks, borders, and differences—that would guide our investigation. Second, we developed a set of empirical methods that we called 'drills': a specific combination of classic field research methods such as interviews, participant observation and document research with which we explored the urban conditions in various places in Switzerland. Third, based on this empirical material, we used a specific form of mapping to construct and design the urban spaces of Switzerland, essentially following two methodological guidelines: on the one hand, we superimposed a wide range of indicators in order to grasp urbanization as a multilayered phenomenon, while, on the other, we avoided drawing sharp delineations of the various urban spaces in order to emphasize the blurriness and the processual nature of urbanization. Fourth, based on this cartographic analysis, we developed a typology of urbanized territories. The fifth and final phase saw the validation of this analysis through interviews and conversations with experts. This is particularly important in such a case, since the point is to test and consolidate these representations of space, benefiting from the wealth of implicit knowledge that local experts have, which is often much more substantial than statistical analyses.

The result of this work was a new image of Switzerland that avoided the term 'rural' altogether. We identified five different types of urbanized territories: *metropolitan regions*, *networks of cities*, *quiet zones*, *Alpine resort*, and *Alpine fallow lands*. We thus succeeded in shedding light on urbanization processes that are not usually within the scope of an urban analysis, especially various forms of peripheral urbanization.

19.6.2 On Planetary Urbanization

If we take this analysis one step further, it has manifold implications far beyond Switzerland alone. Urbanization today is a process that involves the entire territory of the planet and not only certain areas. There is no longer an 'outside' of the urban condition. A further-reaching concept is therefore required that is capable of comprehending, depicting and analysing the wide variety of urbanization processes that take place outside of the areas hitherto regarded as 'urban'. This is why Neil Brenner and I developed the concept of 'planetary urbanization' (Brenner and Schmid 2011, 2015) that is also based on Lefebvre's thesis of complete urbanization.

Lefebvre captured this observation in the powerful metaphor of 'implosion–explosion': the tremendous concentration of people, activities, wealth, goods, objects, instruments, means and thought, and, at the same time, the immense explosion, the projection of numerous, disjunct fragments into space (Lefebvre 2003 [1970]: 14). Based on these considerations, we developed a novel conception that understands urbanization as a dialectical relationship of 'concentrated' and 'extended' urbanization. It is designed not only to focus on processes of agglomeration alone, but also to take into account the concomitant wide expansion of urban networks, infrastructural arrangements and sociometabolic processes (Brenner and Schmid 2015).

19.7 Representations of Space and 'Other' Spaces

In Lefebvrean terms, we can understand such new images and definitions of urban areas, as discussed above, as 'representations of space'. As Nietzsche has stated, we cannot see something without giving it a name (Lefebvre 1980: 39). Accordingly, we are not able to act within a space without having developed an idea of what that space looks like. In order to communicate this idea, we require terms, concepts, images, maps that delimit and denote this space. Thus representations of space are defined in a twofold manner—as (conceived) ideas and as (communicated) concepts.[30] They signify something, they prescribe something, they guide our actions and give them a direction. As Lefebvre tells us, such representations are never innocent: they are always associated with power.

What effects do such representations of space have? They privilege certain aspects and suppress others. This raises the question of what is present and what is absent, what is illuminated and what is hidden—what is not shown being often more important than what is shown. It was for good reason that Lefebvre called these phenomena 'representations of space', although, as he himself notes, they are close to ideologies. However, in his later work, Lefebvre preferred to develop a theory of representation instead of further employing the term 'ideology' (Lefebvre 1980: 27).[31] For representations of space have an operational quality: their purpose is to denote something, to illuminate something, to change something. Thus proposing a different representation of space constitutes an intervention in the debate. Lefebvre's theory is self-reflexive at this point and shows how the researcher himself becomes an actor in the process of the production of space. This directly poses the question of the responsibility of the researcher: as Lefebvre notes, the author has to take a personal responsibility in such operations and incurs risks, including the risk of error (Lefebvre 2000 [1972]: 28).

19.7.1 The Project: A New Image of Switzerland

This nexus can be illustrated particularly well in *Switzerland - An Urban Portrait*, which was conceived from the beginning not only as an analysis, but also as a project: its declared purpose was to create a new image of Switzerland, and to substitute it for the old image. This meant advancing the analysis to the point where it effectively turned into a project. Initially, frequent criticism was voiced that this portrait contained no concrete proposals—the notion of architects 'only' analysing without proposing a project was unacceptable to many. They overlooked the fact that this analysis simultaneously constituted a project. Primarily, it comprised a detailed critique of the existing representations of Switzerland, especially of the concept of 'decentralized concentration' that had been developed in the 1930s and been reissued and refined repeatedly ever since. This concept, which portrayed Switzerland as a compromise between 'urban' and 'rural' areas, called for a decentralized settlement of the country, which, however, ultimately constituted one of the key conditions for the complete urbanization of Switzerland. The image of Switzerland in the *Urban Portrait*, on the other hand, was diametrically opposed to such idyllic and outdated conceptions and revealed what many did not like to recognize: the presence of large-scale sociospatial inequalities, different speeds of development and varying qualities of urbanization. Our maps thus revealed what earlier studies—in particular those operating with statistical analyses—had carefully concealed: the unequal development of the territory. Thus, for instance, the category of 'rural areas' as shown in many statistics comprises two completely different types of areas: on the one hand, areas experiencing strong pressure of urbanization (we called them 'quiet zones'), and on the other hand, areas marked by a continuous loss of social energy and activities (which we called 'Alpine fallow lands'). This second type in particular gave rise to intense public debate and strong criticism. The long-time cultivated image of Switzerland as a decentralized, but well-organized, carefully balanced and regionally equilibrated country had developed cracks and fissures, which were clearly addressed and exposed in this new representation of space.

In this context, it was interesting to note other reactions, for example: 'What you're showing is nothing new—we've known that for years.' In reality, however, we have found no publication that depicted Switzerland in a similar way before the *Urban Portrait* was published. Thus our representation was seen as so plausible that some regarded it as a statement of well-known facts. This observation, in turn, takes us back to the triadic analysis: it is impossible to perceive a space without previously having conceived it, and space cannot be conceived without having been experienced. The depiction must therefore be based on experience; it must re-present something that is rooted in a concrete reality.

Interestingly, the new image of Switzerland we proposed has largely established itself in the meantime. It marked the beginning of a paradigm shift in Swiss planning history and has left a significant mark on it, up to the point where it was used as the starting point for a new official planning concept for the spatial development of Switzerland.[32]

19.7.2 Havana Profunda: 'Forgotten' Space and Alternative Project

The approach of the *Urban Portrait* was also adopted in another project, in order to elaborate an analysis for the future urban development of Havana (Peña Díaz and Schmid 2008). This analysis had obviously to cope with completely different conditions and to apply other means than those used in the *Urban Portrait*, mainly because of a blatant scarcity of statistics, data and information. We therefore organized a series of workshops with experts from academia and urban planning to elaborate a new conception of Havana's urban structure.[33] Here, too, the identification and analysis of urban configurations revealed an urban structure that was very different form the hitherto dominant representation of this urban space, which is basically defined in strictly geographical terms: the prevailing planning concept, which has been applied for many years, subdivided the entire urban area of Havana into three zones—a central, a peripheral, and an intermediary zone. In contrast to this representation, two major urban configurations emerged from our analysis: the *Blue Strip* and *Deep Havana*.

The *Blue Strip* is our term for a heterogeneous strip of varying breadth along the coast in which the important institutions, facilities, restaurants and hotels are concentrated. It is the area that continues to determine the image of Havana today—where the image of a tropical, urban Eldorado, reproduced for decades in cinema, literature and advertisements, finds its world-famous iconography. Accordingly, the *Blue Strip* is also the part of Havana that draws the greatest international attention, as well as the crowds of tourists and visitors and the influx of investments. International architectural and research projects also focus almost exclusively on this area. By contrast, the south of the city features a large area that is also quite heterogeneous and comprises very diverse neighbourhoods. While it is located right in the geographical centre of Havana, it forms in many respects a remote, forgotten, neglected and disregarded part of the Caribbean metropolis, which is overshadowed by the famous neighbourhoods along the coast and is far away from the flows of visitors and capital. It also became clear that there are hardly any academic studies on this area. We called this area *Deep Havana*.

In this case, too, the elaboration of an 'other' representation of space was the key to addressing the issue of social and sociospatial inequalities that have developed in Havana. Although the full analysis has yet to be published, the two terms we coined have already entered the local debate on planning and urban development.

As these examples show, the full reach of the three-dimensional analysis unfolds here:

What representations of space reveal, illuminate, make explicit, are just as significant as what they leave in the dark and conceal. This allows us to identify 'forgotten' spaces, underappreciated spaces, disregarded spaces.

The mere exposure already has an effect in this context: it is important to disclose another reality and to draw a different map. The analysis reveals something and brings it to the surface; the point is to make something visible, to show a reality that was always present, but is not depicted, expressed and exposed in dominant representations and is therefore absent from discourse and not subject to debate. Introducing this reality and thus making it accessible to experience is one of the decisive moments of analysis and marks the beginning of a possible alternative project.

19.8 Beyond Lefebvre[34]

How should the theory of the production of space be handled? These examples show only a few of the possibilities for mobilizing and applying this theory in a productive way, and of course there are many other possible ways for engaging with this theory. However, based on the reflections presented above, we can nevertheless sketch a few elements of a possible research strategy.

First of all, Lefebvre's theory must be taken seriously if it is to deploy its full potential. It should not be regarded as a *quantité négligeable*; nor should it be used as a quarry of ideas and concepts. Rather, its principles of construction should be illuminated and the full potential of its effectiveness should be exhausted.

Second, it is important to remember that the most important texts on the theory of the production of space were written more than four decades ago, which means that their further development is inevitable. We live in a completely different world today: new developments have arisen, and we require new concepts to be able to understand this world. It is therefore important that Lefebvre's work not be canonized, but continuously expanded in engagement with reality. Additionally, theoretical advances that have since been achieved should also be acknowledged and taken into account.

First and foremost, employing Lefebvre's concepts means applying them, which implies following the core of Lefebvre's procedure: always confronting the theory with concrete experiences, experimenting, continuously engaging with concrete practice in order to develop the theory further, immersing the theory in reality and making it fruitful—and thus ultimately also going beyond Lefebvre.

Notes

1. Alfred Hitchcock, The Trouble with Harry, Paramount 1955.
2. See Kipfer et al. (2008, 2012).
3. Castells (1977); cf. also Coornaert and Garnier (1994). Castells later revised his view on Lefebvre's theory considerably; cf. Castells (1983).
4. See Harvey (1973) and (1982). One of the few other scholars who positively received Lefebvre's theory was Gottdiener (1985).
5. See, e.g., Soja (1989, 1996), Gregory (1994), Dear (2000).
6. Cf. Kipfer et al. (2008) and Schmid (2008).
7. Cf. Schmid (2005, 2022), Kipfer et al. (2008).
8. The list of important publications on Lefebvre's work has become very long in the meantime. To quote just a few: Hess (1988), Kofman and Lebas (1996), Shields (1999), Elden (2004), Schmid (2005, 2022), Merrifield (2006), Goonewardena et al. (2008), Brenner and Elden (2009), Stanek (2011), Ajzenberg et al. (2011), Stanek et al. (2014).
9. Since the English translation of La Production de l'espace in 1991, a whole series of further translations of books and articles on the city and urbanization, on everyday life and modernity and also on the state has been published.
10. Lefebvre (2008 [1947, 1958], 2000 [1972]), cf. Stanek (2011).
11. Lefebvre (2016 [1965], 1995 [1962], 1966).
12. Lefebvre (1996 [1968], 2003 [1972]).
13. Lefebvre (1969 [1968]).
14. Lefebvre (1970). Cf. Schmid (2005, 2022), Kipfer (2008).

15. Lefebvre (1991 [1974]).
16. Lefebvre (1976a, b, 1977, 1978, 2006 [1981], 2004 [1992]), Brenner and Elden (2009).
17. The theory of the production of space is reconstructed in detail in Schmid (2005).
18. Cf. also Goonewardena (2005).
19. Cf. Schmid (2008).
20. Cf. e.g. Lefebvre (2002 [1962], 1966, 1986).
21. Cf. also Stanek (2008).
22. Lefebvre (2003 [1972]: 87, 1991: 341).
23. Cf. also Bittner and Weber (2009).
24. Lefebvre (2016 [1965], 2000 [1972]: 28).
25. Cf. Lefebvre (2002 [1962]: 118; 2003 [1972]: 5).
26. Cf. Stanek (2011).
27. Cf. e.g. Lefebvre (1991 [1974]: 12, 413).
28. Cf. e.g. Stanek (2014), Bertuzzo (2009), Schmid (2012).
29. Hartmann et al. (1986).
30. Lefebvre (1980: 8) refers here to the German terms Vorstellung and Darstellung.
31. Cf. also Goonewardena (2005), Wachsmuth (2013).
32. Bundesamt für Raumentwicklung (2012).
33. This was a participative process involving around 50 experts in varying compositions; see also Peña Díaz and Schmid (2008).
34. This title refers to the conference 'Urban Research and Architecture: Beyond Henri Lefebvre' (ETH Zurich, 24–26 November 2009).

References

Ajzenberg A, Lethierry H, Bazinek L (eds) (2011) Maintenant Henri Lefebvre: Renaissance de la pensée critique. L'Harmattan, Paris

Bertuzzo E (2009) Fragmented Dhaka: analysing everyday life with Henri Lefebvre's theory of production of space. Franz Steiner Verlag, Stuttgart

Bitter S, Weber H (2009) Autogestion, or Henri Lefebvre in New Belgrade. Sternberg Press, Berlin

Brenner N, Elden S (eds) (2009) Henri Lefebvre—space, state world: selected essays. University of Minnesota Press, Minneapolis, MN

Brenner N, Schmid C (2011) Planetary urbanization. In: Gandy M (ed) Urban Constellations. Jovis Verlag, Berlin, pp 10–13

Brenner N, Schmid C (2015) Towards a new epistemology of the urban? City 19(2–3):151–182

Bundesamt für Raumentwicklung (2012) Raumkonzept Schweiz. are, Bern

Castells M (1977) The urban question: a Marxist approach. Edward Arnold, London

Castells M (1983) The city and the Grassroots: a coss-cultural theory of urban social movements. University of California Press, Berkeley and Los Angeles, CA

Coornaert M, Garnier J-P (1994) Présentation: Actualités de Henri Lefebvre. Espaces Et Sociétés 76:5–11

Dear M (2000) The postmodern urban condition. Blackwell, Oxford

Diener R, Herzog J, Meili M, de Meuron P, Schmid C (2006) Switzerland—an urban portrait. Birkhäuser, Basel
Elden S (2004) Understanding Henri Lefebvre. Continuum, New York
Goonewardena K (2005) The urban sensorium: space, ideology and the aestheticization of politics. Antipode 37(1):46–71
Goonewardena K, Kipfer S, Milgrom R, Schmid C (eds) (2008) Space, difference, everyday life: reading Henri Lefebvre. Routledge, London
Gottdiener M (1985) The social production of urban space. University Texas Press, Austin, TX
Gregory D (1994) Geographical imaginations. Blackwell, Oxford
Hartmann R, Hitz H, Schmid S, Wolff R (1986) Die urbane revolution—Thesen zur Stadtentwicklung Zürichs. In: Ginsburg T, Hitz H, Schmid C, Wolff R (eds) Zürich ohne Grenzen. Pendo, Zürich, pp 150–65
Harvey D (1973) Social justice and the city. Edward Arnold, London
Harvey D (1982) The limits to capital. Blackwell, Oxford
Hess R (1988) Henri Lefebvre et l'aventure du siècle. Éditions A.M. Métailié, Paris
Kipfer S (2008) Hegemony, everyday life, and difference: how Lefebvre urbanized Gramsci. In: Goonewardena K, Kipfer S, Milgrom R, Schmid C (eds) Space, difference, everyday life: reading Henri Lefebvre. Routledge, New York, pp 193–211
Kipfer S, Goonewardena K, Schmid C, Milgrom R (2008) On the production of Henri Lefebvre. In: Goonewardena K, Kipfer S, Milgrom R, Schmid C (eds) Space, difference, everyday life: reading Henri Lefebvre. Routledge, New York, pp 1–23
Kipfer S, Saberi P, Wieditz T (2012) Henri Lefebvre: debates and controversies. Prog Hum Geogr 37(1):115–134
Kofman E, Lebas E (1996) Lost in transposition: time, space and the city. In: Kofman E, Lebas E (eds) Writings on cities: Henri Lefebvre. Blackwell, Oxford, pp 3–60
Lefebvre H (1966) Le langage et la société. Gallimard, Paris
Lefebvre H (1969 [1968]) The explosion: Marxism and the French Upheaval. Monthly Review Press, New York and London
Lefebvre H (1970) Le manifeste différentialiste. Gallimard, Paris
Lefebvre H (1976a) De l'État. Tome 1: L'État dans le monde moderne. Union Générale d'Éditions, Paris
Lefebvre H (1976b) De l'État. Tome 2: Théorie Marxiste de l'État de Hegel à Mao. Union Générale d'Éditions, Paris
Lefebvre H (1977) De l'État. Tome 3: Le mode de production étatique. Union Générale d'Éditions, Paris
Lefebvre H (1978) De l'État. Tome 4: Les contradictions de l'État moderne. Union Générale d'Éditions, Paris
Lefebvre H (1980) La présence et l'absence. Contribution à la théorie des représentations. Casterman, Paris
Lefebvre H (1986) Le retour de la dialectique. 12 mots clefs pour le monde moderne. Messidor/Éditions sociales, Paris
Lefebvre H (1991 [1974]) The production of space. Blackwell, Oxford
Lefebvre H (1995 [1962]) Introduction to modernity: twelve preludes. Verso, New York
Lefebvre H (1996 [1968]) The right to the city. In: Kofman E, Lebas E (eds) Writings on cities: Henri Lefebvre. Blackwell, Oxford, pp 63–182
Lefebvre H (2016 [1965]) Metaphilosophy. Verso, London
Lefebvre H (2000 [1972]) Everyday life in the modern world. Continuum, New York
Lefebvre H (2002 [1962]) Critique of everyday life volume 2. Foundations for a sociology of the everyday. Verso, New York
Lefebvre H (2003 [1972]) The urban revolution. University of Minnesota Press, Minneapolis, MN
Lefebvre H (2004 [1992]) Rhythmanalysis: space, time and everyday life. Continuum, New York
Lefebvre H (2006 [1981]) Critique of everyday life volume 3. From modernity to modernism. Verso, New York

Lefebvre H (2008 [1947, 1958]) Critique of everyday life volume 1: introduction. Verso, New York
Merrifield A (2006) Henri Lefebvre: a critical introduction. Routledge, New York
Peña Díaz J, Schmid C (2008) Deep Havana. In: Gugger H, Spoerl H (eds) Havana lessons. Lapa, ENAC, EPF Lausanne, Lausanne, pp 156–167
Schmid C (2005) Stadt, Raum und Gesellschaft: Henri Lefebvre und die Theorie der Produktion des Raumes. Franz Steiner Verlag, Stuttgart
Schmid C (2006) Theory. In: Diener R, Herzog J, Meili M, de Meuron P, Schmid C (eds) Switzerland—an urban portrait. Birkhäuser, Basel, pp 163–223
Schmid C (2008) Henri Lefebvre's theory of the production of space: towards a three-dimensional dialectic. In: Goonewardena K, Kipfer S, Milgrom R, Schmid C (eds) Space, difference, everyday life: reading Henri Lefebvre. Routledge, New York, pp 27–45
Schmid C (2012) Travelling Warrior and complete urbanization in Switzerland: landscape as lived space. In: Girot C, Truniger F (eds) Landscript 1: landscape, vision, motion. Jovis, Berlin, pp 138–155
Schmid C (2013) Patterns and pathways of global urbanization: towards comparative analysis. In: Acebillo J, Lévy J, Schmid C (eds) Globalization of urbanity. Actar, Barcelona, pp 51–77
Schmid C (2022) Henri Lefebvre and the Theory of the Production of Space. Verso, London
Shields R (1999) Lefebvre, love and struggle. Spatial dialectics. Routledge, London
Soja EW (1989) Postmodern geographies: the reassertion of space in critical social theory. Verso, London
Soja EW (1996) Thirdspace. journeys to Los Angeles and other real-and-imagined places. Blackwell, Oxford
Stanek Ł (2008) Space as concrete abstraction: Hegel, Marx and modern urbanism in Henri Lefebvre. In: Goonewardena K, Kipfer S, Milgrom R, Schmid C (eds) Space, difference, everyday life: reading Henri Lefebvre. Routledge, New York, pp 62–79
Stanek Ł (2011) Henri Lefebvre on space: architecture, urban research and the production of space. Minnesota University Press, Minneapolis, MN
Stanek Ł, Schmid C, Moravánszky Á (eds) (2014) Urban revolution now: Henri Lefebvre in social research and architecture. Ashgate, Farnham.
Wachsmuth D (2013) City as ideology. Society and Space, Environment and Planning D forthcoming

Christian Schmid is a geographer and urban researcher. He is a Professor of Sociology at the Department of Architecture, ETH Zürich. His scientific work is on planetary urbanization, comparative urban analysis and theories of urbanization and space. He is a founding member of the International Network for Urban Research and Action (INURA). His book on Henri Lefebvre's theory of the production of space will be published in English by Verso in 2022. He is currently working on research projects on the comparison of urbanization processes in large metropolitan regions and territories of extended urbanization.

Chapter 20
Gilles Deleuze & Félix Guattari, *Mille Plateaux*, 1980: "The Good Use of Philosophy"

Gabriele Pasqui

Abstract *Mille Plateaux*, even more than *L'Anti-Oedipe*, has aroused much interest in the world of planning theory in recent decades, becoming one of the cornerstones of an approach to the discipline that some have defined as post-structuralist (or sometimes post-modern). Concepts such as that of rhizome, or that of deterritorialization and reterritorialization, have exercised a fascination that derives perhaps from the transversal and transdisciplinary nature of the philosophical reflection of Deleuze and Guattari. In this chapter, however, the reading of *Mille Plateaux* tries to show the conditions of a rigorous use of philosophical texts, theories and concepts in the fields of planning theory and practice. *Mille Plateux* is a book deeply rooted in philosophical tradition and problems, with very strong theoretical ambitions, and each of its "applications" must be carefully controlled. For this reason, the reading aims to place the volume in the political, cultural and theoretical atmosphere in which it was conceived, to reconstruct the form and style of the text, to reflect on its possible uses. The goal is above all methodological: to highlight cautions, limits and the possibility of an active use of texts of a highly theoretical nature in the context of planning theory and practices.

Keywords Planning theory · Deterritorializzation · Concepts · Atmosphere · Consistency plan

20.1 Why (and How) Planners Should Read Philosophical Books?

In these pages I will propose a reading of a philosophical book. The book is *Mille Plateaux: Capitalisme et shizophrénie, vol.2*, published by Deleuze and Guattari in 1980.

Planners and architects, in the past, have often read, quoted and used philosophical books. Only to give some examples, I remind here John Forester and Jurgen

G. Pasqui (✉)
Department of Architecture and Urban Studies, Politecnico di Milano, Milan, Italy
e-mail: gabriele.pasqui@polimi.it

C. Perrone (ed.), *Critical Planning and Design*, The Urban Book Series,
https://doi.org/10.1007/978-3-030-93107-0_20

Habermas, with reference to the communicative turn in the field of planning theory (Forester 1989), or Bernardo Secchi and Michel Foucault, about discourses and narratives in urban planning (Secchi 1984). We could also remember the importance of authors as Simmel, Heidegger, Husserl and the phenomenological school, the structuralist turn in the 1960s, Jacques Derrida and the so-called deconstructionist approach in architecture. The dialogue between architects and philosophers about the city, in some phases of twentieth century cultural debate, was very important and fertile.

The question is: why, and how, planners (and architects) should read philosophical books?

In my reading of *Mille Plateaux* I will not answer directly to this question. Rather, I will show which—from my point of view—should be the conditions of a rigorous use of philosophical texts, theories and concepts in the fields of planning theory and practice.

I propose this exercise for two reasons. First, I would suggest that the encounter with texts, authors and disciplines that are far from planning is always insidious, asking caution and "exactness". Exercising the practice that Albert O. Hirschman called *trespassing*, means learning to respect texts, discourses and authors to dialogue with. This respect means developing a deep understanding not only of the contents of these texts, but also of the context producing texts and their "atmosphere".

On the other side, I argue for the practice of the "pleasure of the text" (Barthes 1980), when in reading philosophical books. In other words, I would like to evoke, for you and with you, an encounter.

Having in mind these two objectives, the choice of *Mille Plateaux* is not a casual one. This is a very difficult, long book (more than 600 pages in the English translation). It is also a book that few readers have really read and studied from the first to the last page. This book asks for a reader with time and patience.

Moreover, in the last years *Mille Plateaux* has become important in the planning debate, being a term of reference for the attempts to reformulate new and promising foundations of planning theory (Healey and Hillier 2008).

In the last years Deleuze e Guattari have been at the centre of an increasing attention in the planning theory field, above all thanks to the work of Jean Hillier and Gareth Abrahams (Hillier 2008a, b; Hillier and Abrahams 2013; Abrahams 2016). In many texts Hillier has proposed a Deleuzoguattarian-inspired planning theory, using *Mille Plateaux* as one of the main references. Following Hillier, other planning scholars have proposed theoretical reflections and empirical applications of this Deleuzogattarian approach, intended as a radical alternative in exploring foundations for planning theory (Hillier and Scott-Cato 2010; Nyset et al. 2010; Purcell 2013; Ansaloni and Tedeschi 2015; Purcell and Born 2016).

In my reading of *Mille Plateaux* I would like to show conditions and risks of these contributions, proposing four different steps to approach the book by Deleuze e Guattari: encounter, atmosphere, text, uses.

I will read *Mille Plateaux* in the English translation by Brian Massumi for University of Minnesota Press (1987). The original edition dates back to 1980: the book was published by Les Editions de Minuit (Paris).

In the next pages I would also like to invite you to read the book, if you haven't done it yet. For this reason I suggest a biographical introduction to Gilles Deleuze and Félix Guattari, written by Dosse (2010). As I will try to show, an in-depth understanding of texts implies recognizing also "lives below the texts". We should imagine Gilles and Félix working together, side by side, sitting in Deleuze's home in Paris: discussing, inventing, correcting. We should also have in mind Paris in the second half of the 1970s, and we should listening to Deleuze's lessons at Paris Vincennes, understanding the political urgency of philosophical discussions and ideological controversies.

Knowing the author's biography, is therefore a necessary condition to encounter a text. Through biographies we can at least glimpse the invisible power that lays back to authors: discourses, powers, institutions, atmospheres (Sini 2016). Moreover, introductory books to Deleuze (and Guattari) can be useful, because *Mille Plateaux* is not an isolated book. The book is also the product of a complex chain of texts, a "station" in a trajectory that started twenty years before. I suggest three introductory books, in three different languages: Clairebrook (2001), Zourabichvili (2003) and Ronchi (2015).

These books do not represent a necessary step to read *Mille Plateaux*, but may be helpful in taking seriously the complexity of philosophical thought, and in understanding that the voice of thought should be listened and respected before using and "applying" philosophy to planning (or to whatever).

20.2 Encounter

A book is always encountered in a specific context of practices. Not only research practices, but also existential experiences, political trajectories, affections and emotions.

For me, re-reading *Mille Plateaux* after many years was the consequence of another reading: the conversation between Jean Hillier and Gareth Abrahams, in a series about "Exploring foundations for planning theory" published by Aesop editions (Hillier and Abrahams 2013).

Hillier's text is very ambitious. She makes a proposal for a spatial planning theory based on Deleuze and Guattari's ontology, and especially on the application of three families of concepts, taken from *Mille Plateaux* and other texts written by the Authors: assemblage, smooth and striated space, machines.

As Hillier writes: «The work of Deleuze and Guattari is relatively little known (and less understood) by planning scholars and practitioners. The aim of this short book is to highlight some of the concepts which Deleuze and Guattari develop in their work and which might be useful for those involved in spatial planning research and practice to consider» (Hillier and Abrahams 2013: 10).

What Hillier proposes is clear: try to apply some concepts in order to support theoretically a non-linear, experimental approach to planning. In this sense, Deleuze and Guattari's concepts can become parts of a new planning "methodology". «Their work, especially their constructivist pragmatics or cartography, offers a potential methodology» (Hillier and Abrahams 2013: 64).

I am not sure that the interesting effort by Jean Hillier points in the right direction. Can we really think to *Mille Plateux* as a tool for building up a new "methodology"? It is interesting that "method" is a word that has never been used by Deleuze and Guattari. *Mille Plateaux* does not propose any methodology: its aim, as we will see, is forging concepts.

Moreover, Jean Hillier proposes a discussion of Deleuze and Guattari's concepts in order to understand what spatial planning practices do (and could do). In his direction, she writes for example that a Deleuzian inspired spatial planning «would appreciate that the construction of the region/city/neighbourhood/house is never complete but always/already in change» (Hillier and Abrahams 2013: 54). And more: «Inspired by Deleuze and Guattari, I have argued for broad trajectories and visions of strategic spatial planning to act as frames of references (planes of immanence) for more specific local short-term plans and projects (plans of organisation)» (56). Are we sure that the concept of "plane of immanence", that is an ontological concept based on a non-dualistic conception of being, could be translated and transferred in the context of a description of planning processes?

What Deleuze and Guattari call "plane" have nothing to do with plans. «There is an altogether different conception of a plane. Here, there are no longer forms or development of forms: nor are these subjects or the formation of subjects. There is no structure, any more that there is a genesis. There are only relations of movement and rest, speed and slowness between unformed elements [...]. There are only *haecceitas*, affects, subjectless individuations that constitute collective assemblage» (Deleuze and Guattari 1980, eng. tr.: 266). If we want really to understand what an immanence/contingence plane is, we should go back to Duns Scoto's theory of being uniqueness and Spinoza's theory of substance.

These examples (but I could show other exercises of translation of *Mille Plateaux*'s concepts in the field of planning theory) were for me a strong incentive to propose a different approach to the text. A more cautious approach, aimed at not confusing the use of concepts with their application.

My new encounter with *Mille Plateaux* is then moved by the strong necessity to distinguish clearly the different possible ways of reading philosophical books by planners. Encountering a philosophical text is not a matter of "application" (of words, concepts, theories), it is a matter of use. As Deleuze and Guattari say clearly, concepts should be used to create something new, to concatenate concepts in our problems in order to "make something" with them.

20.3 Atmosphere

A text is conceived in a particular time and place, in an "atmosphere" and becomes part of an archive. The archive is a chain of readers, interpreters, individual and collective practices, more or less institutionalized, in which a text is taken and captured.

What is the "atmosphere" backing *Mille Plateaux*? How it has become part of an archive?

First, it is important to remind that *Mille Plateaux* is the second part of a fortunate book that Deleuze and Guattari published in 1972, *Anti-Oedipe*, that was considered a sort of "manifesto" of '68 social and political movements (Deleuze and Guattari 1972).

The books have the same subtitle: *Capitalism and schizophrenia*, and the second has explicitly been denominated "Volume 2". Therefore, a general comprehension of *Mille Plateaux* needs at least a reading and a general understanding of its "Volume 1". Moreover, the political climate in which *Mille Plateaux* was written was completely different from that of *Anti-Oedipe*, as the capitalistic restructuring of the Seventies was able to contrast social and political movements, and in some sense capitalism was winning. So many hopes that had guided Deleuze and Guattari were running out.

For all these reasons, it is very important to understand that both *Anti-Oedipe* and *Mille Plateaux*, in the years following their publications, became part of complex assemblages of social, political and psychiatric practices.

They were militant books, read not only by professional philosophers and discuss not only in the intellectual circles. Their effects were not limited to academics, but were spread in a complex chain of social actions and practices. Searching in Youtube the videos of Deleuze's lessons in Vincennes (Université Paris 8) in the 1970s, we can partially capture the atmosphere in which *Mille Plateaux* was conceived.

Vincennes: a big classroom with people sitting everywhere. No distinctions between the professor desk and the audience. The attendant are not only students: there are activists, psychologists and psychoanalysts, artists. Old and young. Male and female. Many of them are smoking, taking notes, looking at Deleuze who is mesmerizing and fascinating them with its voice. The atmosphere is informal, but it is clear that a master is speaking: everyone is silent and concentrated.

We have to understand that this was not only a lesson of a professor of philosophy. It was a context of politics, ideology, powers. In the biography of Deleuze and Guattari Francois Dosse reminds that sometimes a maoist group based in the Philosophical Faculty of Université Paris 8 (called by Deleuze "the Maoist gang"), guided by Alain Badieu (an important philosopher, that will become Deleuze's friend), aggressively interrupted the lessons to discuss urgent ideological and political issues.

For us, today it is difficult to revive that atmosphere. Difficult, but necessary: *Mille Plateux*, was written slowly in many years, raised in that atmosphere, inside the violent discussions and controversies about *Anti-Oedipe*, criticized as a manifesto of irrationalism, an hymn to the '68 rebellion, but also a betrayal of orthodox marxism.

In this conflictual context *Mille Plateux*, as all the great philosophical books, has many enemies: Freud and psychoanalysis; Lacan and structuralism, orthodox and Freudo-marxism, Chomsky's linguistics, but also the French Communist Party and many others.

Against these enemies, the book is implacable, ferocious. At the same time, it proposes a path of liberation, against all those who propose to inhibit and punish desire. Reading *Mille Plateaux* means re-animate this erotic and political dimension of the text.

On the other side, we cannot never forget that *Mille plateaux* is a philosophical book. If Felix Guattari has a training as psychiatrist and psychoanalyst, Deleuze is a "classical" philosopher. He studied for many years the history of philosophy and its main interests are genuinely "metaphysics". In contrast with the hermeneutical line, dominant in the continental philosophy of twentieth century, Deleuze is a philosopher who wants to tell how the world is, more than analysing and deconstructing texts. In this sense, Deleuze and Derrida, that have many traits in common, have completely different theoretical styles.

Deleuze has written amazing books on Spinoza, Nietzsche, Bergson, and has produced philosophical treatises as *Differénce et répétition* (1968) and *Logique du sense* (1969). He has worked on metaphysical concepts as the "continuous variations", inspired by the medieval theologian Duns Scoto, or the "event", taken by the antique stoicism.

The last great book written with Felix Guattari is a critical appraisal of their relationship with philosophy, whose title is explicit: *Qu'est-ce que la philosophie?* (Deleuze and Guattari 1991).

On the other hand, we cannot forget that Deleuze has written inspiring books on cinema and painting, theatre (he was a good friend of the Italian dramatist and actor Carmelo Bene) and literature; on Bacon, Sacher Masoch, Proust and Kafka (with Felix Guattari). From this point of view *Mille Plateaux*, is a book of philosophy, working on philosophical concepts, but is also a text that cannot be reduced to any disciplinary closure.

20.4 Text

Mille Plateaux is a complex book. It requires time and concentration. It is organised in three main parts: a famous Introduction ("Rhizome"); fourteen Chapters, that have not necessarily to be read in sequence; a Conclusion that is a dictionary of the main concepts introduced and discussed in the book. Each chapter has a title preceded by a date. For example, the first chapter is entitled: "1914: One or Several Wolves?", and the date refers to the year in which Sigmund Freud published his clinical case "The Wolfman". The title of the second chapter is "10,000 B.C.: The Geology of Morals (Who Does the Earth Think It Is?)", and the date refers to a long phase of anthropological transition before the rise of the Neolitics. Sometimes the dates are clear, and refer directly to the contents of the chapter. Sometimes the connection is less evident. In general, each chapter has more or less explicit connections with the others, but can also be read as a "singularity".

I propose three keys to approach the text: *style*, *concepts* and *fields* (*champs*).

Every great philosophy book exhibits its own style. It is not a question of "*belle écriture*". Style is unavoidably entrenched with thought, and this cannot be ignored. Moreover, we cannot forget that *Mille Plateaux* it is the result of the meeting between two authors, but it is also, for their explicit admission, a "collective" work.

Let's read the first words of the introduction: «The two of us wrote Anti-Oedipus together. Since each of us was several, there was already quite a crowd. Here we have made use of everything that came within range, what was closest as well as farthest away. Why have we kept our own names? Out of habit, purely out of habit» (Deleuze and Guattari 1980, eng. tr.; 3).

Mille Plateau's style is not a detail, a "dress" for abstract and conceptual thought. Anyone using this book, has to deal with this style. The most striking feature of the style of *Mille Plateaux* is that it is a book without rules, easily moving between theories, examples, disciplines. A book guided by its greatest urgency: forging concepts.

Reading a great philosophical book, it's important listen to its style. I propose two examples. The first one is taken by Chap. 2: «God is a Lobster, or a double pincer, a double bind. Not only do strata come at least in pairs, but in a different way each stratum is double (in itself has several layers). Each stratum exhibits phenomena constitutive of a *double articulation.* Articulate twice, B-A, BA» (Deleuze, Guattari, 1980, eng. tr.: 40). The second example is taken by Chap. 6 ("November 28, 1947: How Do You Make Yourself a Body Without Organs?"): «The BwO: it is already under way the moment has had enough of organs and wants to slough them off, or loser them. A long procession. The *hypochondriac body*: the organs are destroyed, the damage has already been done, nothing happens anymore […]. The *paranoid body*: the organs are continually under attack by external forces, but are also restored by outside energy […]. The *schizo body*, waging its own active internal struggle against the organs, at the price of catatonia […]. The *masochist body*: it is poorly understood in terms of pain; it is fundamentally a question of BwO. It has its sadist or whore sew it up: the eyes, anus, urethra, breasts and nose are sewn shut. It has itself strung up to stop the organs from working flayed, as if the organs clung to the skin: sodomized, smothered, to make sure everything is seal tight» (Deleuze and Guattari 1980, eng. tr.:150).

We should not be shocked facing this writing, as well as we should not exalt this style. We have to understand. In these paragraphs the Authors forge concepts (double articulation, Body without Organs) that are firstly ontological. These concepts answer to classical metaphysical questions (the uniqueness or duality of being; the possibility of multiplicity, the connection between being and event). The Body without Organs is a plan of absolute immanence: it is Life before articulation, *zoé* before *bìos.*

Only taking seriously these conceptual articulations, strictly involved in the style of theory presentation, we can think to use *Mille Plateaux*'s concepts.

As Deleuze and Guattari say, philosophy is a machine forging concepts. Some of these concepts already exist (they have their own history: Deleuze and Guattari 1991), but are taken in new assemblages. Others are completely new, generated *ex-novo* inside the book "workshop".

What is a concept? This question opens *What Is Philosophy?* (Deleuze and Guattari 1991). Answering their title's question, Deleuze and Guattari seek to place philosophy in relation to science and art, all three being modes of thought, showing no subordination among them. Thought, in all its modes, struggles with chaos against opinion. Philosophy is the creation or construction of concepts, and a concept is an intensive multiplicity, inscribed on a plane of immanence, and peopled by "conceptual personae" which operate the conceptual machinery. A "conceptual persona" is not a classical subject, for thinking is not subjective. Concepts are complex, and they intersect each other. For all these reasons, they work only in connection with their own problems.

Concepts are assembled each other, and create new conceptual relations and connections. For example, the concept of deterritorialization/reterritorialization has a strong connection with smooth and striated spaces and with the lines of flight. The concept of Body without Organs is related to consistency plan. The concept of rhizome is assembled with multiplicity. Refrain is a concept that refers to the couple difference/repetition.

It is not possible introducing here all the concepts forged in *Mille Plateaux*. If we have understood what a concept is, we can be sure that it cannot be "applied". Rather, concepts can be used in our specific assemblages, making them work with our concepts and questions.

The third aspect I would like to underline, after style and concepts, is that *Mille Plateaux* is at the same time a philosophical book and a book that breaks all the disciplinary rules. In which field of knowledge, in which *champ*, should we put a book that quotes musicians and historians, anthropologists and psychoanalysts, literates and biologists, semiotics and painters, musicians and linguistics?

In each Chapter Deleuze and Guattari dialogue and work with different sciences and disciplines, that become part of a new and original way of knowledge-writing. To give only an example, in Chap. 2 geology, biology, chemistry, anthropology, linguistics and semiotics are evocated.

The consequence is that, even if the concepts are philosophical, they are intrinsically plural. For example, the theory of assemblage is basically territorial (Deleuze and Guattari 1980, eng. tr.: 502) but it is at the same time an ontology and an anthropology. Only assuming this complexity it is possible to work with *Mille Plateaux*'s concepts.

Reading *Mille Plateaux* with the eyes of spatial planners, it is important to underline that planning items are never directly called into question. On the other side, the text machine produces maps, living on the consistency plan. These maps are not zenithal representations (Deleuze hates representation!): they are flight lines that inhabit the same consistency plan of the events.

The concept of deterritorializazion/reterritorialization (D/R) is forged to show how "someone" (not a classical subjet, but an anonymous individual, a group, a population) is always deterritorializing and reterritorializing, inside specific social, political and institutional conditions. These processes of D/R are linked to the "production of humanity", in the long period between Paleolithic and Neolithic.

Discussing the problems of "stratification" Deleuze and Guattari write: «The hand must not be thought as an organ but instead as a coding, a dynamic structuration, a dynamic formation. The hand as a general form of content is extended in tools, which are themselves active forms implying substances, or formed matters; finally, products are formed matters or substances, which in turn serve as tools. Whereas manual formal traits constitute the unity of composition of the stratum, the forms and tools and products are organised into parastrata and epistrata that themselves function as veritable strata and mark discontinuities, breakages, communications and diffusions, nomadism and sedentarities, multiple thresholds and speeds of relative deterritorialization in human population» (Deleuze and Guattari 1980, eng. tr.: 60–61).

All these problems, and many others, have no direct connections with planning theory. On the other hand, these conceptual maps could be used to understand the anthropological roots of planning and governing practices, and the connections between anthropological, biological and geological conditions in the processes of territorial articulation and stratification.

20.5 Use

What can we do with *Mille Plateaux*? How can we build up our own "war machine", making new assemblages with the concepts forged in the book?

First, we can also decide not to do anything with it. The encounter with a text, for each of us, is marked in the first instance by desire and pleasure: the desire for encounter and the pleasure of reading. From this point of view, *Mille Plateaux* can fascinate his/ her reader, but also annoy him/ her. Someone may think that this is a confusing book, not respecting the rigorous standards of scientific writing. Perhaps we should not reach any conclusion too quickly: before saying that a philosophical text is incomprehensible, we have to suspect that we did not understand it, that we did not read it with the necessary attention, patience and seriousness.

How to take *Mille Plateaux* seriously? How to assemble its concepts with ours? I propose here three rules that could be used not only for this book. I suggest that these rules should be useful working with all philosophical texts.

The first rule is that we should not think of bending the texts to our intentions. *Mille Plateaux* is not a normative text, it cannot tells us what we should or should not do. For this reason, any interpretation in a normative key is improper, when used as a declination in the planning field of the supposed political indications of the book.

The discussion between Hillier and Purcell, about the possibility or impossibility of a form of planning that should be coherent with a Deleuzeoguattarian inspiration inside the State apparatus (Hillier and Abrahams 2013) is, from this point of view, misleading. The problems of *Mille Plateaux* are completely different, and the genealogy of the State apparatus produced in the last Chapters of the book (and especially in Chap. 13: "7000 B.C.; Apparatus of Capture") has no direct connections with the problems of territorial government and governance.

The second rule is that the encounter with a text should not start from its theses or doctrines.

For example, considering *Mille Plateaux*, all the binary oppositions should be suspected and suspended. In the book it is not possible to identify dualistic contrapositions between smooth and striated spaces, or between plain of immanence and plain of organization. The binary tables proposed by Jean Hillier (Hillier and Abrahams 2013: 29, 44) betray the spirit and the letter of *Mille Plateaux*. In fact, for Deleuze and Guattari there is not pure deterritorialization, pure rhizome. There are continuous movements of reterritorialization, as well as continuous making of folds and strata. The multiplicity on the plane of immanence is the effect of these movements, irreducible to every dualism and to every dialectic, to every *Aufhebung*.

The third rule is related to the assumption of the point of view of the Authors when constructing something with their concepts. For example, in *Mille Plateaux* the issue of government in the perspective of the State never arises. Nevertheless, the theme of the Apparatus of capture can be considered a way to read the State changes in the transition between nomadic and sedentary populations. Rather than applying the concepts of *Mille Plateaux* to some normative theory of planning, I argue that we should use them to understand the historical and transcendental conditions of possibility for the human action to control and to govern territories.

In conclusion, I would like to show briefly two possible concatenations with *Mille Plateaux*'s concepts.

The first line is the understanding of the connections between processes of deterritorialization and reterritorialization, and the disarticulation of the link between sovereignty and territoriality. I already developed this issue in the past with reference to urban populations (Pasqui 2008, 2016), using the concept of deterritorialization. In Chap. 13, describing the Apparatus of capture, Deleuze and Guattari describe the cities as phenomena of *transconsistency*, as *networks*. Cites are «threshold of deterritorialization, because whatever the material involved, it must be deterritorialized enough to enter the network, to submit to the polarization, to follow the circuit of urban and road recoding» (Deleuze and Guattari 1980, eng. tr.: 432). The opposition between cities and the State, proposed in these pages by Deleuze and Guattari, is rooted in the open nature of the cities, and it is linked with the complex problems of codification of the cities by the State.

There is no need to underline the importance of this approach for reading the processes of pluralization in the contemporary cities, as well as the complex and bidirectional links between State and urban dynamics.

Moreover, *Mille Plateaux*'s can be used to build up an interpretation showing how the link between city, territory, sovereignty and citizenship has always been highly problematic: the radical pluralism typical of planetary urbanization processes brings into play new forms and new figures of deterritorialization and reterritorialization.

In contemporary cities individuals, groups, populations deterritorialize and reterritorialize along peculiar plots and trajectories, defining new maps. As Deleuze and Guattari say, there is no "absolute" deterritorialization. Planning and designing should consider carefully these plural forms of D/R, that can leave multiple possibilities. As Deleuze and Guattari say: «the Earth girded, encompassed, overcoded, conjugated as the object of a mortuary and suicidal organization surrounding it on all sides, *or* the Earth consolidated, connected with the Cosmos, brought into the Cosmos following lines of creation that cut across it as so many becomings» (Deleuze and Guattari 1980, eng. tr.: 510).

We cannot really govern these planetary processes, through spatial planning. Aware of how difficult this could be, we might produce new places for multiplicity and pluralism, leaving space to a new sense for the vertical connection between Earth and Cosmos. At the very beginning of planning practices, thousands of years ago, this connection was visible in the "folds" man produced on earth, in the specific forms of reterritorialization that forged cities and cultivates territories. Now we are called into question to reinvent this possibility, recognizing multiplicity in a promise of justice.

The second line tries to think planning practices in their making, through a process of active suspension of each one's own action, allowing to observe the supports, the scriptures and the diagrams within which our practice is taken (Pasqui 2018).

Along this line, the concept of consistency plan becomes precious. Each practice is an assemblage of other practices, organizing and becoming in a unitary consistency/immanence plane. The deviation from our own practices, is not transcendent (Deleuze hated transcendence!). It is rather a question of making a chain with other practices, thus exhibiting new assemblages.

This exercise is obviously a practice, through which it is possible, in a plane of consistency, to observe our own practices with their peculiar tools, and with the invisible power lying behind its back.

Mille Plateaux offers us an extraordinary example to carry out this operation, since the whole book is both a production of the concepts, and the display of the tools that allowed to build them.

So, let's get to work and make assemblages with *Mille Plateaux*!

Acknowledgements I would like to thank Paola Briata, who read a first version of this paper and helped me to improve the text. Any responsibility for errors or inaccuracies is obviously mine.

References

Abrahams G (2016) Making use of Deleuze in planning: proposals for a speculative and immanent assessment method. Routledge, London

Ansaloni F, Tedeschi M (2015) Ethics and spatial justice. Unfolding non-linear possibilities for planning action. Plann Theory 15(3):316–332

Barthes R (1980) The pleasure of the text. Will and Wung Publisher, New York

Clairebrook C (2001) Gilles Deleuze. Essential guide for literary studies. Rotledge, London

Deleuze G (1968) Différénce et ripètition. PUF, Paris

Deleuze G (1969) Logique du sense. Minuit, Paris

Deleuze G, Guattari F (1972) L'Anti-Oedipe. Capitalisme et schizophrénie, vol. 1. Minuit, Paris

Deleuze G, Guattari F (1980) Mille Plateaux. Capitalisme et schizophrénie, vol. 2. Minuit, Paris, (trans: Massumi B), A thousand plateaus. Capitalism and schizophrenia, University of Mibnesota Press, Minneapolis, London, 1987
Deleuze G, Guattari F (1991) Qu'est-ce que la philosophie? Minuit, Paris
Dosse F (2010) Intersecting lives. Columbia University Press, New York
Forester J (1989) Planning in the face of power. California University Press, Berkley
Healey P, Hillir J (eds) (2008) Foundations of the planning enterprise: critical essays on planning theory. Routledge, London
Hillier J (2008a) Plan(e) speaking: a multiplanar theory of spatial planning. Plan Theory 7(1):24–50
Hillier J (2008b) Stretching beyond the horizon. Ashgate, Aldershot
Hillier J, Abrahams G (2013) Deleuze and Guattari. Jean Hillier in conversation with Gareth Abrahams. In: Planning, Aesop
Hillier J, Scott-Cato M (2010) How could we study climate-related social innovation? Applying Deleuzian philosophy to transition towns. Environ Politics 19(6):869–887
Nyset T, Ploger J, Holm T (2010) Planning beyond the horizon: the Tromso experiment. Plan Theory 9(3):223–247
Pasqui G (2008) Città, popolazioni, politiche. Jaca Book, Milano
Pasqui G (2016) Populations and rhythms in contemporary cities. In: Pucci P, Colleoni M (eds) Understanding mobilities for designing contemporary cities. Springer, Heidelberg, pp 49–65
Pasqui G (2018) Città, saperi, pratiche. Donzelli, Roma, in progress
Purcell M (2013) A new land: Deleuze and Guattari and planning. Plan Theory Pract 14(1):20–38
Purcell M, Born B (2016) Planning in the spirit of Deleuze and Guattari? Considering community-based food projects in the United States and Mexico. J Urban Geogr 38(4):521–536
Ronchi R (2015) Gilles Deleuze. Feltrinelli, Milano
Secchi B (1984) Il racconto urbanistico. Einaudi, Torino
Sini C (2016) Inizio. Jaca Book, Milano
Zourabichvili F (2003) Le vocabulaire de Deleuze. Ellipses, Paris

Gabriele Pasqui is a full Professor of Urban Policies and Planning. Coordinator of the Research Project "Excellence Department–Territorial Fragilities" funded by the Italian Ministry of University and Research. Former Director of the Department of Architecture and Urban Studies (DAStU), Politecnico di Milano (2013–2019). His key scientific interests are interpretations of contemporary cities dynamics, urban populations and policies, strategic planning, urban governance and urban regeneration tools. His books: Le città, I saperi, le pratiche (Donzelli, 2018); Coping with the Pandemic in Fragile Cities (Springer, 2022).

Chapter 21
Georges Didi-Huberman, *La survivance des lucioles (2009). The Thickness of Time: Going Beyond the Surface of the Present to Understand Contemporary Territories*

Lidia Decandia

Abstract This article proposes a rereading of Didi-Huberman's work *La survivance des lucioles*, a book with a philosophical and political approach that does not appear to have anything to do with the territory, the city or planning. The writer of this article is nevertheless convinced that, thanks to the questions posed and the epistemological apertures it suggests, it is actually a precious instrument for revisiting the emergence of the new urban and territorial phenomena in a new light. By helping us cast off the idea that what is contemporary can be compacted onto the surface of the present and reduced to pure visibility of forms, the article frees us from the paradigms to which cartographic vision has accustomed us. It invites us to pay attention to the temporal statute of survival and the value and meaning of what exists but is yet devoid of form. As Didi-Huberman seems to suggest, attention to these aspects, often neglected by our disciplinary fields, could provide us with new lenses to transcend the apocalyptic gaze of much of contemporary literature and help us grasp, within the depths of contemporary territories, tiny things hidden from view, which have survived though forgotten: hidden energies, vital sparks that could nurture our present times and open up to other possible futures.

Keywords Survival · Fireflies · Temporality · Transcend cartographic vision · Contemporary · Wandering gaze · Community

21.1 Genealogies

The book we are going to talk about might appear to have nothing to do with the territory, the city or planning. It has a philosophical and political approach, remote from the themes we deal with in our discipline. Yet I believe that if we start with some of the questions it poses and some of the epistemological apertures it suggests, this book could become a precious instrument to reread in a new light the emergence

L. Decandia (✉)
Department of Architecture, Design and Urbanism, Sassary University (Alghero Campus), Sassary, Italy
e-mail: decandia@uniss.it

C. Perrone (ed.), *Critical Planning and Design*, The Urban Book Series,
https://doi.org/10.1007/978-3-030-93107-0_21

of new urban and territorial phenomena. In particular, it helps us cast off the idea that contemporary society can be compacted onto the surface of the present and reduced to pure visibility of forms.

For it poses at least two fundamental questions: that of the temporal statute of survival, and that of the value and meaning of what exists but is devoid of a definite shape and consequently assumes an erratic character, unable to be grasped, luminescent and iridescent. Two questions that for some time our discipline, modelled from a zenithal, cartographic viewpoint, which still in many ways deeply enmeshes us, thought it should disregard and to a certain extent eliminate from the forms of representation of the territory and the city.

It was not perhaps a coincidence that Georges Didi-Huberman, the author of the book we are dealing with, obtained his Ph.D. at the École des Hautes Études en Sciences Sociales (EHESS) in Paris in 1981, under the guidance of Louis Marin. This philosopher, together with other authors, gave an essential contribution through his works to the deconstruction of the implicit assumptions upon which both the perspective and the cartographic vision were founded.[1]

Marin (1973, 1994), Didi-Huberman helped to highlight the fact that both these forms of representation, understood as authentic paradigmatic visions of a way of considering time and space, have led us to reduce our knowledge of the world to an exceedingly visual matter. What the perspective or cartographic eye actually wants to represent is basically the unmoving pattern, the evidence of the look and shape of an object already given, already coded, untouched by time, delimited, stable, inanimate and motionless, viewed in a simultaneous 'now' that immortalises the instantaneous vision of an individual gazing at a building that has persisted, escaping the 'becoming' of the world.

This kind of vision thus provides a satisfactory instrument only to represent figures clearly silhouetted against the background. By doing this, it ignores what is "un-steady" (Zambrano 1997/2004, p. 15); what appears and disappears in vibrant transitivity; what "arises suddenly, lights up and disappears or ceases. But it does not simply pass without leaving a trace" (ibidem); what sparkles, trembles or flashes imperceptibly, fluctuates and steals away, like the shapeless and the indistinct: the teeming, murmuring, the background noise, the silent cross-referencing of things. That state of the visible in a nascent condition, in which things seem not to have taken shape yet, but where vibrations of meaning not yet accessible are nevertheless moving. But that is not all. It limits reality just to the material form life has generated. And in this way it causes contact to be lost with what produced it and made it live. It ignores what goes beyond the appearance of the present: what has already happened; the past that "never stops passing" (Deleuze 1966/2001, p. 49) and which coexists with the present; but also what does not yet exist; what is incessantly becoming and drags away the form; what could, can and will take shape and will enable the thrust of life towards the future.

[1] On the deconstruction of the implicit assumptions on which the perspective and cartographic visions are based I take the liberty of referring to Decandia (2008) and the copious bibliography on the subject given therein.

Of course the book does not hint at all at the question of perspective or cartographic representation, but everything unfolds on a different plane. Yet if we keep these genealogical matrices in the background, which certainly nurtured our author, it will help us, I am convinced, to free ourselves of this way of seeing and open up promising tracks forward for us to understand and interpret the consistency of our contemporary reality.

The two powerful images—fireflies and survival—that already appear in the title, move directly along this path.

The image of fireflies hints at the glow produced by miniscule beetles and given off during their dance of love: fragile, intermittent lights, released from a chemical substance, luciferin, by which they manage to give an expression of strength and the power of desire.

The image of survival refers to something that was presumed dead, obsolete or come to an end, but comes back to the surface of the world again, in other places or times in history. An existence of the past that continues to survive (Didi-Huberman 2002/2006, p. 34).

Two words therefore that lead us to focus again on just the issues we have mentioned. Fireflies evoke a kind of visibility that is unsteady and hazy, something that appears and disappears and can certainly not be grasped by the mechanical eye of a perspective and cartographic device; survival calls up things already passed, hence an idea of temporality that problematises a notion of a linear, chronological past, making it more complex and impure, weaving it with sediments of long duration, "of ageless latencies and brusque revivals" (Didi-Huberman 2002/2006, p. 84).

The matters brought to mind by these images provide the cornerstone of the book and supply the instruments to make our understanding and interpretation of the present more complex.

21.2 Two Excellent Witnesses

To dwell on these matters Didi-Huberman quotes two excellent witnesses: Pier Paolo Pasolini and Giorgio Agamben. These two authors are chosen by Didi-Huberman to develop a discussion grounded on unquestionable intellectual respect.

The author, as Centanni observes, summons them, almost like "the accused parties in a hermeneutical trial he is staging" (Centanni 2010), precisely because he considers that in both these thinkers there is—as he himself affirms—"great impatience about the present; but this impatience is always linked with infinite patience about the past" (Didi-Huberman 2009/2010, p. 65). This is why, as he says himself, "we need them: because they look at the contemporary world with a violence always sustained by vast research on the consistency of time. For this, too, they cause a scandal: because they bring up the unexpected...because they often bring us face to face with the return of things removed in the past" (ibidem).[2]

21.3 The Disappearance of Fireflies: Towards a Criticism of the Apocalyptic Vision

A cue is offered for this issue, for the title and images, and for us to begin to compare these two authors by Pasolini's famous article on the disappearance of fireflies of 1975 (Pasolini 1975/2000).

Pasolini had already talked about fireflies in 1941 in a previous work (Pasolini 1986) "taking them as a symbol to express the joy of love, the resources of desire, the erratic glow of innocence, simultaneously erotic, joyful and creative; an alternative to the too dark or too bright times of reigning Fascism" (Centanni 2010). In this article of 75, published by the Corriere della Sera (Pasolini 1975/2000) 30 years after his first writing, Pasolini theorised, on the contrary, on the disappearance of fireflies. He used this powerful, suggestive image to fiercely criticise the success of the consumer society.

For, according to Pasolini, the triumphant consumer society that he saw as even more devastating than the previous Fascist regime, had managed to finally wipe out, through the darkness and the "piercing, cruel and fierce light of goods" (Didi-Huberman 2009/2010, p. 24), the symbols of innocence that fireflies represented, producing a decisive anthropological mutation with a millenarianist cadence. This society, dominated by the power of capital and its "ability to mould minds and lives, working alongside them through the sphere of needs" (Cherchi 2005, p. 22), would manage, according to Pasolini, to do what not even the Fascist regime had managed to do: pursue a general process of homologation, an authentic cultural genocide, finally wiping out the differences in languages and cultures produced by the thousand histories developed in the various national contexts.

If we wanted to find a relation between Pasolini's thought and our ideas on the city, we could without doubt state that this same powerful consumer society is the one that has led, in our case, to the homologation not only of cultures but also of cities, territories and landscapes. By contributing in a decisive manner to changing mentalities and behaviour systems above all, what Debord had theorised as the "society of the spectacle" (Debord 1967/2000) had had just as devastating effects in this direction.

In particular, by triggering, through the imposition of highly decontextualized models, a loss of relations and contacts between the different societies and their life

[2] As Didi-Huberman maintains, Pasolini did know how to create from a poetic point of view in his works "a continuous connection between the images of the present and the energy drawn from survival, from the sensitive archaeology of gestures, songs, dialects, the ruins of the architecture of Matera and the suburbs of Rome" (Didi-Huberman 2010, p. 65). In the same way, Agamben, an authentic philosopher of paradigms, was able in his works to go backwards in history and carry out an authentic philosophical archaeology, to produce ideas able to cope with contemporary life "in the considerable, complex consistency of its intertwined temporalities" (idem, p. 43).

environments, it caused the loss of those kinds of knowledge belonging to tradition, which conveyed through oral memory unitary models, techniques and languages of construction and production of the landscape and cities.

But that is not all. The consumer society itself, especially with the invention of tourism and its huge accumulation of spectacles, has powerfully transformed the diversities of places. Those places that were the outcome of remarkable histories where every sign, every artefact, every thread of the territory and landscape referred to meanings deriving from specific economic, cultural and social horizons were turned, we might say, into an image, a sign, a backstage. Just as in the game of perspective representation, these vestiges of ancient cities and landscapes, deprived of the life that created them, have been used simply as materials, as resources to play with in that economy of the holiday that financial capitalism was creating in this restoration phase.[3] Also in the case of territories we might say, following in Pasolini's footsteps, that destruction has finally come, just as fireflies have disappeared forever.

Hence, if Pasolini's way of thinking seems to take on a tone of total pessimism and denouncement of the disappearance of any type of survival, Didi-Huberman finds again "an undeniable relation between the diagnosis without appeal of present times" (Didi-Huberman 2010, p. 45) as in Pasolini in some of the features of Agamben's thought.

Agamben, too, used the word "destruction" to explain present times in his first, book Infancy and history (Agamben 1978), written a few years after Pasolini's letter on the disappearance of fireflies. In taking up the famous essay by Benjamin on the narrator (Benjamin 1936/1995), Agamben actually states: "All discussions on experience must today begin with recognising that it is no longer something given to man, contemporary man [...] has been dispossessed of experience: as a matter of fact, the inability to have and convey experiences is perhaps one of the few certain facts he possesses about himself" (Agamben 1978, p. 5).

In pointing out a situation of latent apocalypse, in which the devastation of bodies and the spirit now seems to have been achieved, Agamben gives the example, not by chance, of life in a big city, in particular that of a contemporary visit to a museum or a tourist pilgrimage destination. As the author states, "faced with the greatest marvels of the earth (let us say the Palacio de los Leones in the Alhambra), the vast majority of human beings refuse nowadays to enjoy the experience: they prefer the camera to enjoy the experience. It is not a question here, of course, of condemning this reality, but of acknowledging it" (ibidem, p. 7). Also in this case, the effects of the society of the spectacle, we might say, seem to have reduced places forever—depriving them of any reference and consistency—to a backstage for passing tourists to photograph, subjected and subservient to the powerful lights of the terrible reign of glory, the crowds, cheering people, entirely modelling their behaviour, as Agamben himself showed in his other subsequent book (Agamben 2007).

---

[3] For further study on the causes and effects of homologation induced on the territory by the consumer society, I take the liberty of referring to Decandia (2017) and the bibliography given therein.

Like Pasolini, Agamben, too, therefore, offers us an apocalyptic vision, in fact as Didi-Huberman suggests "an apocalyptic way of seeing the times at work, especially the present time (Didi-Huberman 2009/2010, p. 48).

"When Pasolini declares that there are no more human beings or Agamben, in turn, states that contemporary man has been "dispossessed of experience", we find ourselves exposed each time to the blinding light of an apocalyptic time and place" (ibidem), to the blinding light in which the end of the world triumphs so that another may take place. In effect, the Apocalypse represents, as Didi-Huberman recalls, a feature of the Judeo-Christian tradition in which "the magnificent panorama of radical destruction takes place so that a higher truth, equally radical, may reveal itself" (ibidem). It is not a coincidence that in the Apocalypse heavenly Jerusalem, one of the archetypes on which the actual tradition of the utopia and ideal cities is founded, appears on the horizon, just at the moment when Babylon, the seat of the contradictory reality of human action, devoured by corruption, conflicts and injustices is definitively destroyed. Heavenly Jerusalem descends from the sky, "situated as it is, in a different place, vague in the time and space of the imagination" (Baczko 1981, p. 865) and reveals itself indeed as a great dazzling light, "resplendent in the glory of God. Its splendour is, *in fact*, similar to that of a very precious gem, like crystal jasper stone" (Apocalpyse, 1972, 21, 9–11). It was a great light that absorbed all the other small lights in its devouring glow, killing all the other little surviving elements that give us experience (Didi-Huberman 2010, p. 48). A transcendent light that kills all other glimpses of possible truths, fatally temporary, empirical, intermittent, fragile, discordant and fleeting like fireflies.

And it was not a coincidence that in another book, The kingdom and the glory (Agamben 2007), after consolidating the idea of radical destruction, Agamben then builds up only the possibility of transcendence. As Didi-Huberman observes "his thought oscillates between the extremes of destruction and a sort of redemption" (Didi-Huberman 2010, p. 47). Redemption nevertheless impossible to realise. "An Apocalypse without Apocalypse", as Derrida would say (quoted in Didi-Huberman 2010, p. 50). For he "does not see any alternative to the terrible glory of the spectacle" (idem, p. 62). Not even the possibility of a countersubject, or a counterpower that might in a certain sense take up again the "tradition of the oppressed that fight relentlessly against Fascism" (idem, p. 55). This would be the only way to be saved, which he hoped for, to get out of the dark times of the Apocalypse underway. The only chance to "overturn the content of a tradition underway" (ibidem). But like Agamben he maintains that the people no longer exist (Agamben 2007). The spectacular dominance of the media over every aspect of social life has in fact turned us into completely absorbed peoples.

Hence, Centanni observes "that the retreat into nostalgia of the latest Pasolini in his firefly article (1975) and the collision on the apocalyptic horizon in The Kingdom and the Glory (2007) of the latest Agamben are in harmony in signalling the hopelessness of the present" (Centanni 2010), adopting, as often also happens in much discussion on the city "the dark, steel grey tone, of an unhappy conscience, condemned to its own horizon, its own closure" (Didi-Huberman 2009/2010, p. 63).

21.4 Fireflies Have Not Disappeared: Let Go of the Zenithal Gaze and Adopt a Wandering Gaze

> If we widen our vision of the horizon which, immense and motionless, spreads beyond us, or on the contrary, focus our gaze on the image that, minuscule and unstable, passes us by, we perceive very different things. The image is the firefly of passing flashes, the horizon floods final states with light, the unchanging times of totalitarianism or the final times of the Judgement. To see the horizon, the afterlife, means to not see the images that brush against us. Little fireflies give shape and light to our immanence, the "fierce reflectors of the great light devour every form and every glow – every difference – in the transcendence of ultimate aims. To grant our sole attention to the horizon means to not be able to look at the most common of images (Didi-Huberman 2009/2010, p. 69).

Faced with Pasolini and Agamben's apocalyptic despair, Didi Huberman nevertheless attempts to offer a different vision which greatly interests us and seems extremely fruitful also for the theme of the city and the territory to which we immediately want to transfer it. For he asks himself, criticizing Pasolini and Agamben's pessimism: is it really true that fireflies have disappeared? Or is it rather that we are no longer able to see them?

He invites us, then, first and foremost, to not be content with this sentence of disappearance precisely so as to not take sides with those who would like to make us believe that this process of homologating destruction really is decisive.

It is one thing, he says "to blame the totalitarian machine, but another to so quickly grant it a decisive victory without reserve [...] to suggest such a thing means precisely to give credit to what the machine wants us to believe: it means to see only total darkness or the blinding light of the reflectors. It means to see only the whole. To not see, therefore, the interstitial space, intermittent nomad, *what the actual perspective device prevented us from glimpsing and representing*,[4] placed in an unlikely manner, apertures, possible elements, 'in spite of everything" glows (Didi-Huberman 2009/2010, p. 28).

Perhaps—Didi Huberman goes on to suggest—fireflies disappear because our reading instruments do not even manage to intercept them: "they disappear only in so far as their spectator gives up following them. They disappear from his view because he stays still where he is, which is not the right place to be able to glimpse them" (idem, p. 30). Of course under the powerful light of reflectors everything appears without shadows or depth, but if we sharpen our gaze and learn "not to reduce all contemporary depth to the current situation of flatness and trivialities necessary for the society of the spectacle" (idem, p. 40), it is possible to notice little lights aiming at us that continually move away from us, vital survivors which, though weakly, fleetingly, put us in touch with a much more extensive temporal reality than that we are used to dealing with: certainly not compacted on the surface of the present.

To manage to glimpse these little surviving lights that are flashing, we cannot however be content with looking at territories from above, as the cartographic gaze has got us into the habit of doing, or paying attention to what large numbers reveal.

---

[4] My italics.

We must use the moving, itinerant gaze of the traveller, able to immerse himself in the folds of the land, in the cracks, where the spotlight does not reach.

The gaze that can only observe close up the unique elements, fragments, passing flashes, the details that hide from the high point of our reading instruments. Those tiny things hidden from view which could reveal, on the other hand, were we able to recognise and interpret them, important drifts towards understanding the needs of our time in greater depth. We must, therefore, pay attention to marginal data, to revealing clues, using methods of knowledge closer to those of hunters and detectives, trained, as the historian Ginzburg maintains, " to penetrate secret, hidden things on the grounds of elements that are hardly appreciated or unnoticed, in the refuse or trash of our observation" (Ginzburg 2000, p. 162).

Nor must we take on classifier models to try to understand them. "It would be criminal and stupid, in effect, to put fireflies under a reflector thinking we will be able to observe them better. And it is no use studying them without first killing them, piercing them with a pin and attaching them to an entomologist's board, observing them like ancient things imprisoned in the amber of millions of years" (Didi-Huberman 2009/2010, p. 33). To know fireflies we need to see them in the bosom of their survival: to see them dancing alive in the deep of the night, even if this night will be blown away by some reflector. We need to listen, collect their stories and emotions and pass them on.

Only this careful, delicate exploration will help us understand that sometimes on these over-exposed territories there still exist cases of forgotten survival, stars that arrive from afar and have more memory and future than those who gaze at them. Precious, rare pearls, fragments emerging from the pile of ruins of the past that can help us enter into a much more complex relationship with temporality than that we are accustomed to dealing with. An "impure" temporality in which the past is not separate from the present and which, indeed for this reason, neither implies destruction accomplished nor redemption begun" (Didi-Huberman 2009/2010, p. 77). But it seems somewhat to be composed of deposits, latencies, continuous stirring, assembling and dismantling, fractures, and anachronisms, caesuras, regressions, and returns often unexpected. A temporality in which "time—as Didi-Huberman again suggests in another important book of his—rather than flow, works" (Didi-Huberman 2002, p. 294), with a continuous movement that causes sliding, falls and rebirths, burials and resurrections, decomposition and recomposition, tensions and latencies, and hits and backlashes.

21.5 Communities of Fireflies Dance in the Dark and Open up Towards the Future

If we could go beyond the depressed, apocalyptic contemplation characterising many intellectuals of our time we might, furthermore, realise that vital resources and sparks exist in our present, mixed with surviving energies, that drag form away and open up towards the future. Small craving communities, like fireflies indeed, that in spite of everything maybe do not just resist but, without promising a resurrection, emit flashes of light, dance in the dark and produce little embryos of original forms of urbanity. An urbanity "in which memory and hope reciprocally exchange their signals" (Didi-Huberman 2009/2010, p. 48) to illuminate pathways of possible futures. More and more remote from the atmospheres and simulacra of the ancient cities, they begin to emerge, undoubtedly in microbial forms, outstanding, plural signals of dispersed creativity, teeming and widespread, vital places of extraordinary strength and communication, rich, intriguing stories and emotions.

In the repeatability of days, in the minute, capillary operations of everyday life, new operations of acclimatisation are implemented that do not always leave footprints and traces visible on the territory, but perhaps reinterpret the spaces of peripheral districts of the old city, reappropriating them in minute forms; but also the silent voids and abandoned craters of our urban scores, singling out new centralities, but also marking the territories with brand new expressive qualities (Attili et al. 2007).

In everyday places, often in the most anonymous, difficult ones of the outskirts, people create islands of resistance to the rapid times of the homologated city. With molecular, widespread action they use the spaces, change them, create new forms of relationship with places. In rethinking new ways of being together and developing common goods, inventing new relational environments, they re-establish via unusual forms of occupation, new relations between public and private, create different forms of participation and self-management of social spaces; they produce hidden writings, creative and potential, that transgress the well-ordered text of the old city.

It is in this way that workshops are produced and experimented with, in forms far from those proposed by the official culture, where the classic dichotomies between city and country, real and virtual, inside and outside are challenged. Different scales and times are interwoven. Unrest is created. New spaces and words are invented. Cuts and wounds are mended, sewing the scars with golden thread. Pearls and coral able to feed the future are dug from seas of the past that each territory brings with it, linking up memory and dream once more. New ways of building relational goods are conceived, of thinking and experiencing sociality. New interactions with nature are experimented with. New collective knowledges are produced, but also hypermedia languages, ludic, metaphorical and sensitive, that displace the usual technical codes and those of disciplines.

These dancing communities are not war machines, but craving machines, as we said. For this reason they are fragile, but bold, obstinate and powerful. Like weaving spiders, they build and rebuild their webs, or like wild plants that, with their strength, crack the cement of the single, homologating thoughts. The tiniest and most fragile, coloured ones, that know, in spite of everything, how and where to wedge themselves in, holding on, multiplying and reproducing not to win wars but to take life ahead. "Firefly peoples [...] seeking however they can their freedom of movement, evading the reflectors of the "kingdom", do all they can to reach what they are craving, give off their flashes of light and project them towards others" (Didi-Huberman 2009/2010, p. 93). Of course we are speaking of flaws, of flashing images.

"These little fireflies that give shape and a glow to our fragile immanence do not allude indeed to any transcendence" (Centanni, p. 2010). They do not promise any resurrection "but slit the opaque veil of the 'too real' and expose the background of immanent fullness (not far and transcendent) that underpins reality and nowhere else: the base thanks to which our time has a unique, precious value" (ibidem).

But it is precisely for this that we need to pay attention to them. Precisely because darkness is thick and there is every reason to be pessimistic and we are not meant to think of a messianic time when a new city with its great light may descend from heaven; we must keep our eyes open at night, keep moving to root out the fireflies that dance in the dark, with the hope that their intermittent light may offer some slash of light to our time.

As Didi-Huberman teaches us, in taking up Benjamin again "the aesthetic and political urgency in times of catastrophe does not actually consist […] of drawing the logical consequences of the decline up to its horizon of death, but in finding the unexpected resources of this decline in the depth of those images that move like fireflies or isolated stars" (Didi-Huberman 2009/2010, p. 74).

The word 'decline' does not in fact simply mean 'end', but rather declination, inflection, movement. It evokes not a disappearance but what is "on its way to disappearing" (ibidem), but has not yet disappeared and as such is susceptible to unpredictable drifting. "Atoms—as the cosmologist Lucretius reminded us—decline uninterruptedly but their fall, in an endless clinamen, permits exceptions with incredible consequences. It is enough that an atom make a slight deviation from its parallel trajectory for it to collide with the others, giving origin to a world" (ibidem). And this is the great teaching we receive from many periods of decadence and crisis, as Walter Benjamin and Alois Riegl have taught us, taken up again by Didi-Huberman (ibidem, p. 74). Often it is precisely at these moments that a particular vitality arise from the break-up of old patterns which is capable of opening up towards other possible worlds.

It is in these periods that indeed the dynamic strength emanating from the collapse might bring to light the sudden germinating power of some seed locked for hundreds of years in the rooms of time, but also liberate remains, residues, survivals, firefly-peoples, fields of strength, sheltered at the edges, in the folds, in the interstices, in those cracks not illuminated by bands of light. It is from here that productive energy could emerge and a nascent point of new origin be generated, able to make the past encounter the present and start up a new constellation (Benjamin 1997), new figures of thought, but also new spaces and new forms.

In this sense the little communities of glows, dancing in the night, with their indestructible character of desire and the tenacity of a project might be, in fact, not just witnesses of survival, but "prophecies of a future history (Didi-Huberman 2010, p. 83). This is why we must, with strength, "say yes to the night flecked with glows, and not be content with describing the no of the light that blinds us" (idem, p. 92).

References

Agamben G (1978) Infanzia e storia. Distruzione dell'esperienza e origine della storia. Einaudi,Tourin

Agamben G (2007) Il Regno e la Gloria. Per una genealogia teologica dell'economia e del governo. Homo sacer 2.2. Neri Pozza, Venice

Apocalisse (1972) Angelini C. (a cura di). Einaudi, Tourin

Attili G, Decandia L, Scandurra E (2007) Storie di città. Verso un urbanistica del quotidiano. Edizioni interculturali, Rome

Baczko B (1981) Utopia. Enciclopedia, vol 14. Einaudi, Tourin, pp 876–920

Benjamin W (1936) Der Erzäler Betrachtungen zum Werk Nikolai Lesskovs, in "Gesammelte Scriften: Sieben Bände", Suhrkamp Verlag, Frankfurt; trad.it. 1995, Il narratore. Considerazioni sull'opera di Nikolai Leskov, in "Angelus Novus. Saggi e frammenti", Einaudi, Tourin, pp 247–274

Benjamin W (1997) Sul concetto di storia, Bonola G. e Ranchetti M. (a cura di). Einaudi, Tourin

Centanni M (2010) Luce rara. Una lettura politica di Come le lucciole di Georges Didi-Huberman, in "Engramma" <www.engramma.it> n 84, ottobre 2010

Cherchi P (2005) Crais. Su alcune pieghe profonde dell'identità. Zonza, Sestu

Debord G (1967) La société du spectacle, Buchet—Chastel, Paris; trad. it. 2002, La società dello spettacolo. Massari Editore, Bolsena

Decandia L (2008) Polifonie urbane. Oltre i confini della visione prospettica. Meltemi, Rome

Decandia L (2017) I territori esterni alla città nella quarta Rivoluzione urbana. In: Decandia L, Cannaos C, Lutzoni L (eds) I territori marginali e la quarta rivoluzione urbana. Il caso della Gallura. Guerini Associati, Tourin

Deleuze G (1966) Le bergsonisme, Presses universitaires de France, Paris; trad. it. 2001, Il bergsonismo e altri saggi, Einaudi, Tourin

Didi Huberman G (2002) L'Image survivante. Histoire del'art et temps des fantômes selon Aby Warburg, Éditions de Minuit, Paris; trad. it 2006, L'immagine insepolta. Aby Warburg, la memoria dei fantasmi e la storia dell'arte. Bollati Boringhieri, Tourin

Didi Huberman G (2009) Survivance des lucioles, Les Editions de Minuit, Paris; trad. it. 2010, Come le lucciole. Una politica delle sopravvivenze. Bollati Boringhieri, Tourin

Ginzburg C (2000) Spie. Radici di un paradigma indiziario. In: "Miti, emblemi, spie. Morfologia e storia". Einaudi, Tourin, pp 158–193

Marin L (1973) Utopiques, jeux d'espaces. Minuit, Paris

Marin L (1994) De la représentation, Gallimard Seuil, Paris; trad.it., 2001, Della rappresentazione, L. Corrain (a cura di), Meltemi, Rome

Pasolini PP (1975) Il vuoto del potere in Italia. In: "Corriere della sera", 1 febbraio 1975; ripubblicato con il titolo L'articolo delle lucciole, in Scritti corsari, Milan, Garzanti, 2000, pp 128—131

Pasolini PP (1986) Lettere 1940–1954, Nico Naldini (a cura di), Tourin, Einaudi, p. 36 e sgg

Zambrano M (1977) Claros del bosque, Fundación María Zambrano, trad. it. 2004, Chiari del bosco. Bruno Mondadori, Milano

Lidia Decandia, PhD, is an Associate Professor of Urban Planning at the Sassari University (Alghero campus), where she teaches Design and Context and Urban and Regional History. At the same faculty, she founded and is in charge of "Matrica: a laboratory of Urban Fermentation". She is a member of the Urban Planning PhD Committee at the, "La Sapienza" University of Rome.